Python 语言运维开发

基于 Django 和 Vue 的自动化发布系统实战

陈 刚 高立伟 编著

内容简介

前后端分离开发模式已是大中型软件系统开发的主流技术。在运维研发领域,鲜有讲解这一开发模式的实战类书籍。本书两位作者结合各自的前后端技术优势,以实战形式逐步带领读者建立一个基本可用的自动化部署系统。

本书的前端采用 Ant design Vue 作为 UI,将后端提供的数据流作专业的网页呈现。本书的后端框架是以 Django 最主流的 Python 语言的 Web 框架为基础,使用 Django rest framework 第三方库作为后端开发实现,为前端提供其所需要的数据流。希望读者在理解本书代码后,可以独立开发一套适合自己公司的自动化部署系统。由于 Web 开发涉及较多的基础知识,希望读者具备 Python、Javascript、CSSS 和 HTML 的基础知识,或通过其他学习途径获得。

图书在版编目(CIP)数据

Python 语言运维开发:基于 Django 和 Vue 的自动化发布系统实战 / 陈刚,高立伟编著 . —上海:上海交通大学出版社,2022.9

ISBN 978-7-313-26907-2

Ⅰ.①P… Ⅱ.①陈… ②高… Ⅲ.①软件工具-程序设计 Ⅳ.①TP311.561

中国版本图书馆 CIP 数据核字(2022)第 105029 号

Python 语言运维开发:基于 Django 和 Vue 的自动化发布系统实战
Python YUYAN YUNWEI KAIFA:JIYU Django HE Vue DE ZIDONGHUA FABU XITONG SHIZHAN

编　著:陈　刚　高立伟

出版发行:上海交通大学出版社　　　　　地　址:上海市番禺路 951 号
邮政编码:200030　　　　　　　　　　　电　话:021-64071208
印　刷:三河市祥达印刷包装有限公司　　经　销:全国新华书店
开　本:710 mm×1000 mm　1/16　　　　印　张:36.5
字　数:631 千字
版　次:2022 年 9 月第 1 版　　　　　　 印　次:2022 年 9 月第 1 次印刷
书　号:ISBN 978-7-313-26907-2
定　价:129.00 元

版权所有　侵权必究
告读者:如发现本书有印装质量问题请与印刷厂质量科联系
联系电话:021-33854186

前　言

背景缘起

这本书的出现，是一种神奇的际遇。

作者曾在 2019 年写过一本书《Python 3 自动化软件发布：Django 2 实战》，内容如其名，主要讲解的是使用 Django 框架实践运维开发领域的技能。书中使用的是 Django 模板结合 Jquery 等技术来实现前端的 Web 功能，前后端没有实现分离。此书发售后作者并没有关注销售情况，但不时有读者通过电子邮箱询问此书的 GitHub 项目地址（那本书有小小的错误，书籍对应的 GitHub 地址有误），让作者知道至少对一些朋友帮助很大。

2021 年，作者将之前使用 Go 语言和 Vue 框架的开发经验又汇聚成了另一本书《Go 语言运维开发：kubernetes 项目实战》，在这本书中使用了前后端分离的开发模式，向读者讲解了如何快速地通过 Web 界面将 docker 镜像部署到 K8s 集群中。

在这之后，剧艳婕编辑找到作者，问有没有可能将《Python 3 自动化软件发布：Django 2 实战》这本书升级到第二版，因为有读者不断地反馈意见，希望能将运维开发的能力延升到前端专业领域。有了前两本书的经验，作者心动了，毕竟有资源、有市场、有能力，这种机会也不是人人都能遇上。

但若只是纯粹在第一本书的稿件之上植入前后端分离的技术，会让作者感觉这是偷懒行为，过不了作者自己这关。因此这次，作者再一次挑战自己，将难度升级！作者重新写了一个 Demo 项目（毕方 BiFang），系统地使用了 Django Rest Framework 来实现所有的后端 API 功能。前端方面，作者和本书另一作者高立伟合作，使用 Ant-design-vue 框架开发，力求将更实用的运维研发技术为读者做全面细致的呈现。

全部代码重写，全部书稿重写，通过近 10 个月业余时间的努力，终于搞定！

如果读者有缘读到此处，希望今后可以共同努力，向着运维全栈工程师的方向出发。

全栈工程师可以理解为：掌握多种技能，胜任前端与后端，能利用多种技能独立完成产品开发的人，也就是常说的"一条龙""打通关"。而在本书中，作者想特指的是运维全栈工程师。他不但要具有常规开发系统平台的技能，在运维领域还要具有一些网络配置、服务器运维和开源项目二次开发的能力。人生苦短，作者们不可能在所有领域内都是顶尖的，但须有全局能力和眼界，并在某一方面有所擅长。

你说，这样的能力，重不重要？

而本书就希望在这两方面做第二次尝试，至于是否尝试成功，还需要读者和市场的检验。

本书内容

第1章将读者带入毕方这个Demo级的自动化部署项目的场景当中，分析了毕方项目主要的功能模块、用户的操作流程及部署拓扑图和开发环境的介绍。希望读者在读完本章之后，对本书后面的章节有一个提纲式的导航图，本书后面所有的实践都在第1章里有对应的知识点。

第2章演示了如果没有自动化部署平台，作者们日常是如何手工操作的。这章将带领读者熟悉Bookinfo虚拟微服务构架的部署过程。这个Bookinfo项目涉及了Java、Ruby、Python和Node.js主流开发技术。然后，为保证覆盖全面，作者还手写了一个简单的Go Web程序，经过编译后，将之部署于测试环境中。本章最后将这些源代码放在测试环境的GitLab中。

第3章的开头先实现了一个简单的文件服务器（File-Server），将其用做一个极简的制品仓库，用来存放编译之后的制品。本章后面主要内容为通过GitLab本身提供的CI/CD功能，实现自动化CI的手工配置及触发。这章以实践的方式带领读者配置Docker版本的gitlab-runner，启用Pipeline Trigger触发CI过程，并用SonarQube做静态代码质量扫描。Jenkins一直以来都是CI/CD领域的主力军，Jenkins的使用技能已在作者的第一本书中进行了讲解。为了熟悉新的领域，这里先用了GitLab的CI/CD来实现同样的过程。

第4章在第3章的基础之上，主要讲解了如何使用SaltStack来实现软件包的远程部署功能。本章不但实践了SaltStack的安装配置功能，还讲解了如何使用saltpie这个Python的第三方库来调用Salt-API的接口，实现软件包的远程部署。最后，结合本书项目和本章知识，实现一个deploy.sh的部署脚本。

有了前4章的知识了解和能力积累，从第5章开始，进入毕方项目的后端开发能力实践。本章为读者讲解了如何搭建一个DRF（Django REST framework）的开发环境、requirements.txt文件的作用、PyCharm IDE的使用，并开始撰写一个毕方后端项目（bifangback）的初始代码。

第6章主要讲解毕方后端项目的数据库设计，在讲数据库设计之前也简述了Django ORM的基本知识。然后，就Git、SaltStack、Environments、Project、App、Server和Releaase等所有的数据表做了ORM建表语句的讲解和主要字段的作用。同时，使用ORM生成了一批初始的测试数据。

第7章主要讲解了毕方后端项目的用户模块功能，包括用户和用户组的API、用户注册和JWT认证的实现。本章及以后的章节主要基于DRF来实现所有的API，而不是用原生的Django来实现。因为在前后端分离的项目里，使用DRF来实现后端API，明显比Django更规范、快速和专业。

第8章讲解了毕方后端的项目和应用模块的实现。作为首个规范的API实现，本章详细地讲解了第一个步骤的代码，如序列化文件、过滤器文件、分页文件、各种功能类视图、路由文件及Postman的测试过程。有了这些知识打底，之后很多模块的实现就比较相似了。所以本章选择了详细讲解，而之后章节实现类似的功能就选择了略讲，这点请读者注意。本章后面针对毕方后端项目还设计了一个针对组件级的权限管理功能。

第9章主要讲解了发布单及环境流转功能的实现。包括发布单列表、新增和触发GitLab CI/CD的Pipeline Trigger实现。在实现的过程中使用了Python的python-gitlab三方库，对这个库的使用，本章也有相应的代码可供读者使用。接下来，环境流转功能实现比较简单，为了保持实践的连续性，也做了对应的代码说明。

第10章算是毕方后端实现的核心功能章节，从代码上详解了自动化部署过程。四个主要函数：Deploy()、task_run()、cmd_run()和salt_cmd()是实现自动化部署的主要流程。这个流程相当于从代码上模拟了第3、4章的手工化实现过程。如果读者觉得有不完善之处，在自己今后的实践过程中可以加入更多的场景化功能。

第11章作为后端实现的最后一章，本章做了三个方面的数据展示功能实践。一是发布单历史数据，二是服务器部署历史数据，三是简单的dashboard数据。这三个方面的数据限于本书篇幅，故而选择了3个典型功能来实现。在真正的企业化实践中，应该有更多、更完善的数据需要收集和展示。

从第 12 章开始进入毕方前端项目（bifangfront）项目的实践开发。本章主要讲解了前端项目的造型和环境搭建。在对比了 React、Angular、Vue 后，作者们选择使用 Ant-design-vue 这样一个快速框架来实现毕方的前端项目。为了知识的连续和概念导入，本章同时也讲解了 css、less、scss/sass 这些细节概念及 Vue CLI 的基本使用。

第 13 章为对前端不是很熟悉的读者准备了"餐前甜点"。本章使用了 Angular、React、Vue 来实现一个前端项目的 ToDoList 这样一个 Demo。这个 ToDoList 相当于编程语言界的 "Hello world"。通过这个 ToDoList 的学习，相信读者对接下来的毕方前端项目开发代码不会再有陌生之感。

第 14 章主要讲解了前端开始模式和一些公共服务的配置。如前端向后端请求的契约格式、认证方式、Mock 数据模拟服务、数据请求的封装及 Vue Router 的路由管理，这些都是作者们在学习前端开发过程中的核心技能。希望读者能通过本章的学习和实践，从而敲开前端开发的大门。

第 15 章主要讲解了登录页面的设计和搭建。在开始之前需要对 Vue 当中的 Flex 布局，对 template、style、script 这样的页面结构有一个系统的了解。这些内容在本章都有相应的解答。接下来，VueX 针对 Vue 中的全局数据管理也做了介绍和毕方前端项目的实现。最后，在设计好的前端登录页面上也讲解了如何做路由和 Mock。

第 16 章算是毕方前端项目开发的核心章节。本章讲解了毕方前端项目的主界面设计，这里涉及了 Layout 布局、SideMenu 侧边导航组件的实现、AadminHeader 顶部导航组件的实现、PageView 内容组件的实现以及管理模块的讲解与实现。有了这些功能，结合前面章节的前端内容，演示了用户页面、用户组页面等的搭建过程。

第 17 章主要讲解了项目、应用和服务模块的设计过程，它们都涉及相同的功能，如列表查询、编辑数据、删除数据和查看详情等，这些页面的搭建在本章都有实践级的讲解。同时，针对后面应用级的权限管理，本章也有前端的相应实现。

第 18 章主要是配合后端的发布单管理和环境流转，设计了相应的前端实现。如发布单列表、新增发布单、构建发布单和环境流转等，这些页面的搭建都在本章进行了详细的讲解。

第 19 章主要实现了发布单部署和 Dashboard 的前端模块功能。发布单的部署历史及服务器的部署历史，这些非标准的前端页面功能的实现对于读者之后的前端开发生涯，有很强的学习借鉴价值，希望读者能好好掌握。

第 20 章作为本书的最后章节，讲解了如何进行服务器的联调。在联调之前，前后端的项目可以分别独立开发，不用相应依赖进度。这得益于后端的 Postman 测试及前端的 Mock 技术。但项目最终的上线还是要经过测试联调的。本章主要讲解这方面的实践，最后也讲解了如何将前端项目进行打包部署。

本书的主要内容介绍至此。相较于前后端开发的所有内容，本书并没有一一涉及，如单元测试的方法、接口测试用例的编写、CI/CD 过程中的质量门禁和制品晋级等。这些技能需要读者在工作实践中逐步学习积累。

阅读顺序

本书一共 20 章，建议的阅读顺序是先读前 4 章，了解一下开发毕方项目的前置知识。如果感觉这些知识的掌握不够系统，可以再多读几遍，跟着书中示例实践，并且找到其他资料相互参照学习。

如果读者的 Python 和 Django 基础较好，接着可以阅读第 5~11 章。这些章节详解了如何使用 Django Rest Framework 开发后端的所有功能。待读者掌握这些技能之后，阅读本书后面的章节会有比较好的实践连续性。

如果读者的 JS 和 Vue 基础较好，接着可以阅读第 12~19 章。这些章节详解了如何使用 Ant-design-vue 开发前端的所有功能。读者了解这些章节的内容后，可以再返回阅读第 5~11 章，把这块知识补起来。建议最后阅读第 20 章，学习一些开发实践中的联调知识。

读者对象及必备知识

本书不是一本细致讲解 Python 及 JS 的入门书，或是普及 GitLab 和 Saltstack 的教程，而是一本以实战为特点的书籍。所以希望读者在实践本书的项目前，最好已系统地学习过 Python 及 JS 的知识点，操练过 vue.js 官网的文档。那么，相信读者可以在学习本书后取得更大的收获。

如果误打误撞，读者先有机会看到本书，看完本书及所涉及的项目代码，兴趣大发，再去系统地学习 Python、vue.js、GitLab 等知识，那真是善莫大焉！那一定与本人及本书无关，而是读者您自身的能力和际遇。

世界万千法门，指月之手，得意忘形，大而化之。

本书适合运维开发领域从业者或感兴趣的读者阅读。对于已有一定前后端开发技能的人，本书也贡献了在运维开发行业的详细实现经验，相信对于他们也有一定的借鉴价值。

致谢

首先，感谢剧艳婕编辑，从网上彼此陌生到合作愉快。是她让作者相信，努力之后能够开花结果，冥冥中一切自有天注定。感谢她在博客园发现了作者，然后成就了作者。恩本不言谢，知遇在心头。

掐指一算，这已是作者在最近四年内出的第三本书了。这小小的成就，来自家人的陪伴和鼓励，朋友的督促和信任，同事及领导的理解与支持。

在今年，作者换了工作，也转了行业，希望能在新的领域，有新的成就。

——陈刚

感谢刚哥在写书过程中给予的指导，有幸跟大牛合作，学到了很多，是现实版的从写书小白到入门。犹记得偶然跟刚哥畅聊技术沉淀，两人一拍即合决定创作，恍如昨日。感谢他在这段六个月的旅途中持续的督导和信任，铭记在心。

这是作者人生中写的第一本书，也算在芸芸众生中留下一丝印记，感谢家人的陪伴和支持，尤其是两个开心果汐汐和七七，感谢领导和同事的支持，作者会继续努力续写自己新的篇章。

——高立伟

纠错

由于作者能力有限，如果读者发现本书中的错误，欢迎发送到邮箱：aguncn@163.com，请以"xxx 章节 xx 页的内容或代码有错误"为标题。

在写作本书时，作者参考了网上的大量同行文档，在此一并鞠躬感谢。另外，毕方项目的代码也参考和使用了互联网上很多的开源库框架和公开代码，在此也要一并感谢这些朋友，是 Ta 们，让这个 IT 世界更美好、更强大。

本书代码

本书所有代码均收录于如下 GitHub 地址：

前 言

https://github.com/aguncn/bifang
https://github.com/aguncn/bifang-book
https://github.com/aguncn/bifang-go-demo
https://github.com/aguncn/bookinfo-ratings
https://github.com/aguncn/bookinfo-reviews
https://github.com/aguncn/bookinfo-details
https://github.com/aguncn/bookinfo-productpage

陈 刚 高立伟

目 录

第 1 章 毕方项目简介 1
- 1.1 自动化部署需求 2
- 1.2 自动化部署系统主要模块 4
- 1.3 自动化部署系统操作流程 15
- 1.4 自动化部署系统部署拓扑图 21
- 1.5 小结 24

第 2 章 Demo 项目应用的手工部署 25
- 2.1 Bookinfo 之 productpage 组件 26
- 2.2 Bookinfo 之 details 组件 28
- 2.3 Bookinfo 之 reviews 组件 30
- 2.4 Bookinfo 之 ratings 组件 32
- 2.5 Go 语言的 demo 组件应用 33
- 2.6 使用 GitLab 管理所有源代码 35
- 2.7 小结 40

第 3 章 实现 GitLab 的 CI/CD 功能 41
- 3.1 file-server 文件服务器的实现 43
- 3.2 Docker 版 gitlab-runner 的安装配置 48
- 3.3 启用 GitLab 的 Pipeline triggers 功能 51
- 3.4 GitLab Pipeline triggers 的 Python API 示例 55
- 3.5 将 SonarQube 集成进 GitLab 的 CI/CD 流程 59
- 3.6 小结 62

第 4 章　使用 SaltStack 实现远程部署功能 63

4.1　SaltStack 简介 63
4.2　SaltStack 的安装配置 66
4.3　启用 Salt-API 功能 70
4.4　结合部署脚本、软件包和 saltypie 来实现远程部署 73
4.5　小结 84

第 5 章　Python、Django 与 DRF 的开发环境 85

5.1　Python 环境安装 85
5.2　Django 及 DRF 库安装 88
5.3　安装开发毕方（BiFang）所有的第三方库 92
5.4　PyCharm 安装配置 95
5.5　新建 bifangback 项目的 app 100
5.6　本章 GitHub 代码拉取运行 109
5.7　小结 110

第 6 章　毕方（BiFang）数据库设计 112

6.1　Django model 与 ORM 112
6.2　git model 115
6.3　SaltStack model 123
6.4　environment model 126
6.5　project model 128
6.6　app model 129
6.7　server model 133
6.8　release model 138
6.9　permission model 141
6.10　history model 144

6.11 小结 148

第 7 章 后端用户模块 149

7.1 bifangback 默认首页 149

7.2 bifangback 用户与用户组的 API 152

7.3 bifangback 用户注册 156

7.4 bifangback 用户 JWT 认证 163

7.5 小结 171

第 8 章 后端项目及应用模块 172

8.1 实现 bifangback 项目列表 API 172

8.2 实现 bifangback 新增项目的 API 179

8.3 实现 bifangback 查看具体项目的 API 182

8.4 实现 bifangback 修改具体项目的 API 184

8.5 实现 bifangback 删除具体项目的 API 186

8.6 实现 bifangback 应用增删查改的 API 187

8.7 实现 bifangback 基于应用的权限管理 193

8.8 小结 201

第 9 章 发布单及环境流转 202

9.1 实现 bifangback 发布单列表 API 202

9.2 实现 bifangback 新增发布单的 API 207

9.3 实现 bifangback 软件构建的 API 211

9.4 实现 bifangback 发布单环境流转的 API 219

9.5 小结 222

第 10 章 自动化部署 224

10.1 实现 bifangback 发布单部署列表 API 224
10.2 实现 bifangback 部署服务器列表 API 225
10.3 实现 bifangback 自动化部署 API 225
10.4 小结 237

第 11 章 后端数据展示 238

11.1 实现 bifangback 发布单历史 API 238
11.2 实现 bifangback 服务器部署历史 API 241
11.3 实现 bifangback 简单的 dashboard 244
11.4 小结 248

第 12 章 前端项目选型与搭建 249

12.1 项目开发语言选型 249
12.2 ant-design-vue 251
12.3 css、less 与 scss/sass 252
12.4 使用 Vue CLI 搭建项目 254
12.5 项目引入 Antd 组件库 259
12.6 小结 260

第 13 章 前端框架 ToDoList 实现 261

13.1 ToDoList 介绍 261
13.2 Angular 实现 ToDoList 263
13.3 React 实现 ToDoList 273
13.4 Vue 实现 ToDoList 283
13.5 小结 291

第 14 章　前端开发模式以及公共服务配置 292

14.1　前后端分离开发架构简介 292
14.2　请求认证方式 294
14.3　mock 数据模拟服务 298
14.4　环境变量配置 301
14.5　Service 数据请求服务 302
14.6　路由管理器 Vue router 309
14.7　小结 314

第 15 章　登录页面设计与搭建 316

15.1　前端登录界面设计 316
15.2　Flex 布局 318
15.3　通用布局组件简介 321
15.4　前端登录页面搭建 326
15.5　全局数据仓库 Vuex 333
15.6　路由以及 mock 服务配置 336
15.7　小结 340

第 16 章　主界面及管理员模块设计与搭建 341

16.1　主界面框架设计 341
16.2　主界面布局组件 342
16.3　侧边导航组件 SideMenu 348
16.4　顶部导航组件 AdminHeader 354
16.5　内容组件 PageView 358
16.6　管理员模块 361
16.7　用户组页面搭建 365
16.8　用户页面搭建 373

16.9 管理员基础服务配置 ... 381

16.10 小结 ... 389

第 17 章 项目应用、服务器模块设计与搭建 ... 390

17.1 项目与应用 ... 390

17.2 项目管理页面搭建 ... 399

17.3 应用管理页面搭建 ... 408

17.4 项目与应用基础服务配置 ... 427

17.5 服务器管理模块 ... 438

17.6 服务器管理页面搭建 ... 439

17.7 服务器基础服务配置 ... 450

17.8 小结 ... 457

第 18 章 发布单生成、流转模块设计与搭建 ... 458

18.1 发布单生成模块 ... 458

18.2 发布单列表页面搭建 ... 463

18.3 发布单部署历史页面搭建 ... 478

18.4 发布单生成基础服务配置 ... 482

18.5 环境流转模块 ... 490

18.6 环境流转页面的搭建 ... 493

18.7 环境流转基础服务配置 ... 501

18.8 小结 ... 504

第 19 章 发布单部署、Dashboard 模块设计与搭建 ... 505

19.1 发布单部署 ... 505

19.2 待部署列表页面搭建 ... 510

19.3 部署发布单 ... 516

19.4 部署模块基础服务配置 526

19.5 Dashboard 数据面板模块 530

19.6 Dashboard 页面搭建 535

19.7 Dashboard 基础服务配置 546

19.8 小结 550

第 20 章 前后端服务联调 551

20.1 前端接口服务转发 551

20.2 后端服务本地启动 554

20.3 前后端联调 558

20.4 前端项目打包部署 560

20.5 小结 564

第 1 章 毕方项目简介

> 模仿是人类一切学习的开始,然后才是创新,最后是你自己做主。好的艺术家模仿皮毛,伟大的艺术家窃取灵魂。
>
> ——毕加索

开宗明义,本书以一个实战项目——毕方(BiFang)自动化部署系统的开发过程贯穿始终,带领读者将一些编程语言(Python、Javascript、Go)的语法和技能应用到日常的编程开发工作中。

带着明确的目标和方向,才不会让我们掌握的编程语言只流于语法的细节和知识的累积——列表有序列、字典无序、元组不可更改、循环里可有 continue 和 break 关键字、类有 init 初始化方法等,而是将编程语言当作工具,如书写中的笔墨纸砚,它们只是工具,不会自动形成有意义的文字,是人的大脑主控,让笔墨纸砚完成了沟通的工作。在工作中也是如此,若能以使用工具的心态对待各种编程语言和软件系统,就能组合运用各种技巧,完成既定的开发任务。

项目取名毕方,是因为本书作者喜爱祖国传统文化中的一些神鸟名称。毕方名字出于《山海经·西次三经》,其中记载:"有鸟焉,其状如鹤,一足,赤文青质而白喙,名曰毕方,其鸣自叫也,见则其邑有讹火。"毕方神鸟之形如图 1-1 所示。

图 1-1 毕方神鸟(图片来源于网络)

1.1　自动化部署需求

关于软件自动化部署的需求，可以从两个场景出发来确定。第一个场景出自作者的亲身经历。本人长期工作于运维一线，在没有成套的工具之前，通常是在开发同事完成软件开发和构建、测试部门同事完成软件之后，将软件包提交到运维同事；运维同事接到提交过来的软件后，会将其上传到服务器；在允许的维护窗口时间内，将正在运行的旧版软件服务停止，再将新版软件服务启动起来，完成一次服务变更。这一变更过程可以完全手工完成，也可以辅以一些自动化脚本完成。

这一过程看似正规合理，但待运维同事熟悉了流程之后，无异于建筑工地的搬砖，毫无技术含量且容易疲惫出错。如果产品经理不断地提出产品功能，又或软件开发质量不高、bug 频出，这一服务变更就会更加频繁。

作者本人多年前就遇到过一些极端的场景。按计划周五下午 3 点前要完成的交付，由于产品紧急功能增加、重要 bug 修改，原定周五下午 3 点交付给运维同事测试好的软件包，不断地拖到下午 5 点、7 点、9 点、11 点乃至周六凌晨 1 点、3 点……而运维同事由于处于执行链的末端，在这漫长而又不得不等待的过程中无所事事、百无聊赖。好几次都拖到凌晨四五点才完成线上变更，再完成一些收尾验证工作，可能就看到第二天的太阳了。

那时我就在想，如果有一个系统可以实现自动部署，研发和测试想几点发版就自己在 Web 上操作几点发版，多方便！就算要运维同事负责操作，也可以远程在家里通过手机或电脑操作一下，不就解放了这么漫长的等待时间了吗？工作不误，家庭社交时间也不误。就是在那时，我用自己的 Ruby(rails) 和 HTML 知识写了一版最简单的 Web 在线发布系统。

第二个场景，让我们将眼光放长远些。在最近的软件开发变革运动中，敏捷开发、CI/CD（持续集成/部署交付）和 DevOps 都是非常热门的概念。网上总结过一张图，用来表示这三个概念的关系，如图 1-2 所示。

本书无意于在概念和理论上做过多辨析，而在意 CI/CD 工具链的实现。当一个企业实现了基本的 CI/CD 工具链后，就可以说有一些敏捷开发和 DevOps 的雏形了，向下可以整合公司的 cmdb 资源和 ITSM 流程，向上可以探索微服务及云原生技术的更多可行性。

第 1 章 毕方项目简介

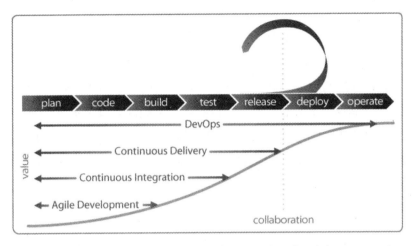

图 1-2 敏捷开发、CI/CD 及 DevOps 范围（图片来源于网络）

CI/CD 流程的打通解决了运维同事的搬砖工作，使运维同事可以朝着运维研发、SRE 的方向升级。同时，标准化的执行流程可以减少手工运维的失误。所有的操作都有记录和统计数据，便于追溯和不断改进。

在确定了自动化部署的场景后，其需求如下：

- 对于开发人员：可以管理自己的项目和应用（可以对应于一个微服务及其下属多个组件），对其进行增删查改。开发人员可以将机器加入对应的应用中（手工增加或数据导入同步）。他可以对其应用进行授权，定义哪些用户有生成软件包的权限，哪些用户有环境放行流转的权限，哪些用户有部署的权限。在日常操作中，开发人员先生成一个应用的发布单，通过构建（静态语言需要编译打包压缩，动态语言纯打包压缩）生成软件包，在这个过程中可以进行单元测试，集成软件代码质量管理。
- 对于测试人员：测试人员按测试进度将发布单放行流转到指定的环境中。
- 对于运维人员：将发布单部署到指定环境的指定机器上，完成一次服务变更。
- 对于系统管理人员：拥有所有开发人员、测试人员和运维人员的权限，维护整个系统的平衡运行和故障解决。
- 如果一家公司有 SRE 岗位和角色，则完全可以抛开开发人员、测试人员和运维人员的职能定位，在协同沟通的基础上，单人快速完成上述流程。

毕方自动化部署系统结合开源工具，后端以 Django 为框架，前端以 Vue 为框架，实现 CI/CD 的上述需求。

对于大的公司，可以有比较多的技术人力来支援这一过程。对于中小型公司，不一定有专门的人手来做这些开发工作。这时，就可以选用一些开源的工具来支撑这一流程。而本书其中一个目的就是希望读者看完本书，习得这些技能之后，不管是基于毕方自动化部署系统进行修改，还是自己完全重新设计一个更贴合自己公司的 CI/CD 部署系统，以及修改第三方的开源自动化部署代码，都能够应付自如。

1.2 自动化部署系统主要模块

毕方自动化部署系统主要分为八大模块，分别为用户模块、项目及应用模块、服务器管理模块、发布单生成模块、环境流转模块、部署模块、管理员模块和 Dashboard 模块。

1. 用户模块

用户模块主要提供与用户相关的功能，比如用户注册、用户登录等。

（1）用户模块功能清单如表 1-1 所列。

表 1-1 用户模块功能表

模块名称	功能名称	所属页面	功能描述
用 户	用户注册	注册页面	①必须参数校验 ②二次密码比对 ③单击登录进入登录页面
	用户登录	登录页面	①登录信息校验 ②通过校验进入 dashboard 首页 ③单击注册进入注册页面

（2）用户模块功能说明。

① 注册时，用户名、二次密码和电子邮箱均为必填项。

② 登录后，前端会获取后端传过来的 json web token（jwt），并且保留到本地，以后每次请求后端时，在 header 里附上此 token。

（3）用户模块截图如图 1-3 所示。

图 1-3 毕方（BiFang）用户注册登录

2. 项目及应用模块

一个项目可以包含多个应用，而一个应用必须属于一个项目。这两者的结合使得层次结构更为清晰，让有相关性的应用归为一类，更容易管理一些大型的部署。

项目及应用模块主要提供项目和应用的增删查改以及权限管理功能。

（1）项目及应用模块功能清单如表 1-2 所列。

表 1-2 项目及应用模块功能表

模块名称	功能名称	所属页面	功能描述
项目及应用	项目管理	项目列表页面	① 新增项目 ② 编辑项目 ③ 删除项目 ④ 搜索项目
	应用管理	应用列表页面	① 新增应用 ② 编辑应用 ③ 删除应用 ④ 查看应用详情 ⑤ 搜索应用
	权限管理	应用页面	① 新建发布单权限的用户新增与删除 ② 环境流转权限的用户新增与删除 ③ 部署权限的用户新增与删除

（2）项目及应用模块功能说明。

① 由于项目的信息较少，故未提供项目详情查看功能。

② 应用的设置项比较多，为了方便自助操作，相关输入框附有填写说明。

③ 权限在应用一级设置，而不是在项目一级设置。

④ 权限分为三类：新建发布单（包括构建软件包）权限、环境流转及部署权限。

⑤ 应用创建者和管理员（属于admin用户组）默认拥有所有权限，其他用户需要授权。

⑥ 所有删除操作都有二次确认动作，防止误删除。

（3）项目及应用模块截图如图1-4~图1-6所示。

图1-4　毕方（BiFang）项目管理

图1-5　毕方（BiFang）应用管理

图 1-6 毕方（BiFang）应用权限管理

3. 服务器管理模块

服务器管理模块主要提供服务器的增删查改等功能。

（1）服务器管理模块功能清单如表 1-3 所列。

表 1-3 服务器管理模块功能表

模块名称	功能名称	所属页面	功能描述
服务器	服务器管理	服务器列表页面	① 新增服务器 ② 编辑服务器 ③ 删除服务器 ④ 搜索服务器

（2）服务器管理模块功能说明。

① 服务器和端口组合，形成一条唯一记录。

② 同一个服务器的不同端口，可以部署不同的应用。

③ 服务器和端口组合必须属于一个环境（测试环境和线上环境等）。

④ 服务器和端口组合必须属于一个应用。

（3）服务器管理模块截图，如图 1-7 所示。

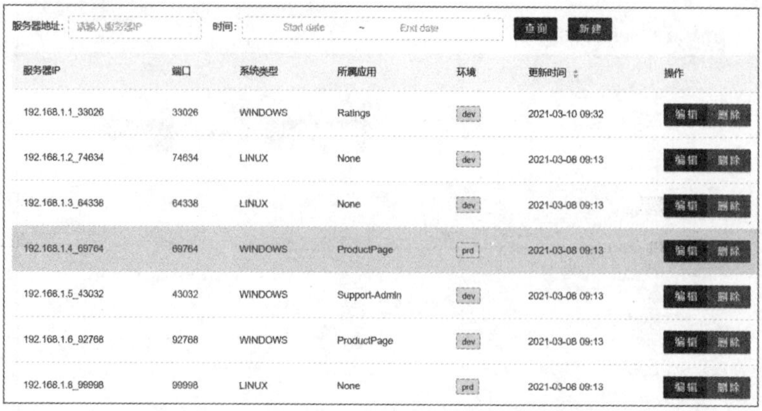

图 1-7 毕方（BiFang）服务器管理

4. 发布单生成模块

发布单生成模块主要有发布单新建、列表、详细及软件包构建功能。

（1）发布单生成模块功能清单如表 1-4 所列。

表 1-4 发布单生成模块功能表

模块名称	功能名称	所属页面	功能描述
发布单	新建发布单	新建发布单页面	① 发布单号自动生成 ② 选择相应项目下的应用 ③ 输入 Git 代码分支 ④ 输入发布单描述
	发布单列表	发布单列表页面	① 所有发布单按时间排序 ② 可按项目应用搜索发布单
	构建发布单	发布单列表页面	① 显示发布单具体信息 ② 构建此发布单软件包 ③ 失败后可重复构建 ④ 构建完成后可快速进入环境流转环节
	发布单详情	发布单详情页面	① 显示发布单所有信息 ② 显示发布单操作历史

（2）发布单生成模块功能说明。

① 发布单名称由时间戳和两位随机字母构成，预防冲突。

② 输入的 Git 分支会传到 GitLab 的 CI/CD 流水线中。

③ 构建时，会通过 trigger token 触发 GitLab 的 CI/CD 流水线。

④ 在发布单详情中也会展示发布单的历史操作记录。

（3）发布单生成模块截图，如图 1-8~ 图 1-11 所示。

图 1-8　毕方（BiFang）新建发布单

图 1-9　毕方（BiFang）发布单列表

图 1-10　毕方（BiFang）构建发布单

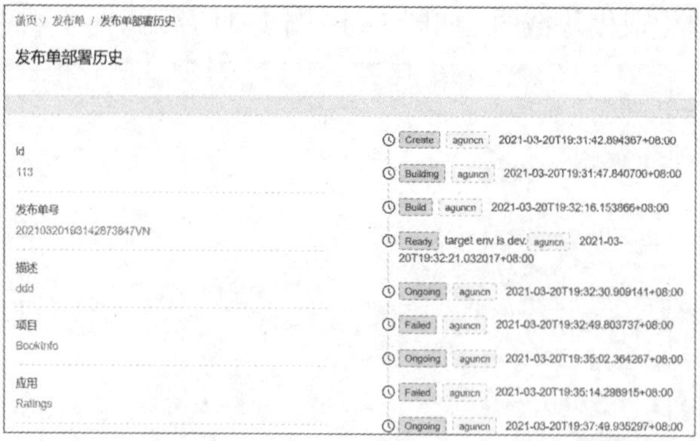

图 1-11 毕方（BiFang）发布单详情

5. 环境流转模块

环境流转模块主要提供环境流转功能，以备发布单在此环境进行接下来的部署操作。

（1）环境流转模块功能清单如表 1-5 所列。

表 1-5 环境流转模块功能表

模块名称	功能名称	所属页面	功能描述
环境流转	环境流转	环境流转页面	①显示发布单的环境列表 ②选择指定发布单的指定环境进行流转 ③流转信息二次确认 ④显示流转成功或失败消息

（2）环境流转模块功能说明。

① 当发布单不在所属环境时，不可以在该环境部署。

② 一般应由测试人员拥有此权限，以免随便部署中断正在进行的测试任务。

（3）环境流转模块截图如图 1-12 所示。

图 1-12 毕方（BiFang）发布单环境流转

6. 部署模块

部署模块主要提供发布单的部署回滚、日志查看等功能。

（1）部署模块功能清单如表 1-6 所列。

表 1-6　部署模块功能表

模块名称	功能名称	所属页面	功能描述
部署	部署列表	部署列表页面	① 显示可部署发布单列表 ② 显示发布单部署状态 ③ 进入发布单部署页面 ④ 搜索具体发布单
	发布单部署	发布单部署页面	① 显示发布单信息 ② 显示部署批次 ③ 显示及选择部署服务器 ④ 显示服务器上已部署的主备发布单 ⑤ 进入查看服务器日志页面 ⑥ 进入部署页面 ⑦ 进入回滚页面
	服务器日志	日志查看页面	① 显示部署批次 ② 显示操作细节 ③ 显示出错细节（如有）
	部署	实时部署进度页面	① 显示部署服务器 ② 显示部署状态 ③ 显示出错细节（如有） ④ 更新部署批次
	回滚	实时回滚进度页面	① 显示回滚服务器 ② 显示回滚状态 ③ 显示出错细节（如有） ④ 更新部署批次

（2）部署模块功能说明。

① 发布单的状态是以其在指定环境里所有的服务器上的部署进度而言的。

② 服务器上的备用发布单以 tips 的方式显示。

③ 只支持回滚到上一次的部署，不支持无限回滚。

④ 每次选好服务器就单击一次部署按钮，算一个部署批次。所以，同一个发布单可能产生多个部署批次。

⑤ 服务器日志会显示当前发布单的所有批次内容，并且给出批次的标示。

⑥ 部署和回滚，在后端实现上大体相同，只是传的参数稍有差别。

（3）部署模块截图，如图 1-13~图 1-15 所示（回滚或部署等截图，在书后配合代码讲解时给出）。

图 1-13　毕方（BiFang）部署列表

图 1-14　毕方（BiFang）发布单部署

图 1-15　毕方（BiFang）服务器部署日志

7. 管理员模块

管理员模块主要提供用户组和用户的增删查改的管理功能。

（1）用户模块功能清单如表 1-7 所列。

表 1-7　管理员模块功能表

模块名称	功能名称	所属页面	功能描述
管理员	用户组管理	用户组管理页面	① 新增用户组 ② 更改组名 ③ 删除用户组 ④ 搜索用户组
	用户管理	用户管理页面	① 新增用户 ② 修改用户所属用户组 ③ 删除用户

（2）管理员模块功能说明。

① 只有在 admin 用户组中的用户，登录之后才能看到这个侧边管理菜单。

② 要所有用户均退出一个用户组后，才能删除这个用户组。

（3）管理员模块截图如图 1-16~ 图 1-18 所示。

图 1-16　毕方（BiFang）用户组管理

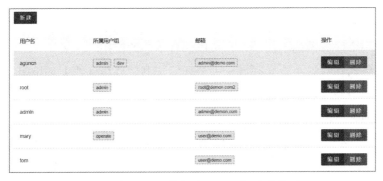

图 1-17　毕方（BiFang）用户管理

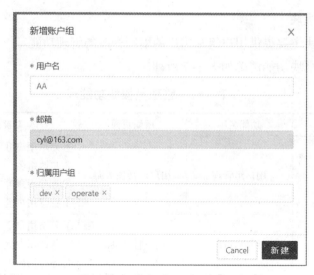

图 1-18　毕方（BiFang）新增用户

8. Dashboard 模块

Dashboard 模块主要提供几个统计数据，如项目、应用、发布单总数；以及发布最频繁的项目、出错率最高的项目和最近的操作等。

（1）Dashboard 模块的功能清单如表 1-8 所列。

表 1-8　Dashboard 模块功能表

模块名称	功能名称	所属页面	功能描述
Dashboard	Dashboard	Dashboard 页面	① 显示项目、应用、发布单总数 ② 显示发布单数量最多的 top5 应用 ③ 显示发布出错数量最多的 top5 应用 ④ 显示用户最近的操作记录

（2）Dashboard 模块功能说明。

① 服务器和端口组合，形成一条唯一记录。

② 同一个服务器的不同端口，可以部署不同的应用。

③ 服务器和端口组合必须属于一个环境（测试环境、线上环境等）。

④ 服务器和端口组合必须属于一个应用。

（3）Dashboard 模块的模块截图，如图 1-19 所示。

第 1 章 毕方项目简介

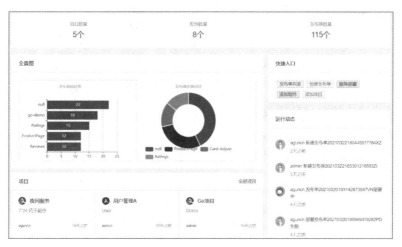

图 1-19 毕方（BiFang）Dashboard

1.3 自动化部署系统操作流程

为了使读者更深入地理解毕方（BiFang）自动化部署系统，本节将讲解此系统的主要操作流程。先以流程图展现，再辅以文字说明，把此系统更动态地展示出来。

1. 用户注册及登录流程图

图 1-20 为用户注册及登录流程图。在注册页面需要填写用户名（不和已有用户名冲突）、二次密码（需要相同）和用户邮箱。用户注册成功或失败会有顶部即时的消息提醒。在用户登录页面输入用户名和密码，单击【登录】。如果用户名和密码正确，则进入系统的 dashabord 页面；如果用户和密码不正确，在登录页面会有信息提示，用户可重新输入。

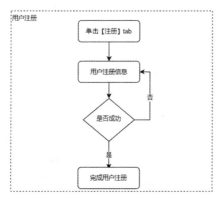 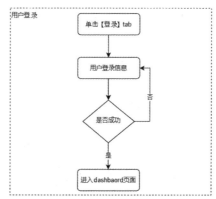

图 1-20 用户注册及登录流程图

- 15 -

2. 项目管理流程图

图 1-21 为项目管理流程图，通过侧边导航栏【项目应用】→【项目】进入。新建项目的页面需要输入项目 ID、项目英文名称与中文名称、项目描述等信息。这里设计了一个项目 ID 的信息，考虑的是有的企业的项目 ID 是统一固定的。对于删除项目这种比较危险的操作，系统会有二次确认，以防误删。

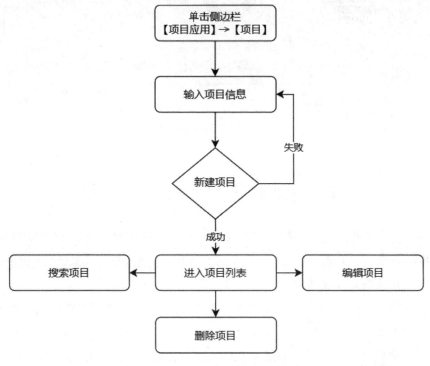

图 1-21　项目管理流程图

3. 应用管理流程图

图 1-22 为应用管理流程图，通过侧边导航栏【项目应用】→【应用】进入。新建应用的页面，需要输入应用 ID、应用英文名称与中文名称、应用描述、所属项目、GIT 服务器、GIT 项目 ID、Git Trigger Token、编辑脚本路径、部署脚本路径、应用压缩包和服务端口等信息。这些信息比较多，每一个输入的作用会在书后详细讲解。

应用的管理除了具有项目管理的常规功能外，BiFang 在规划书的代码规模时，将权限的颗粒度设计在应用这一级，所以每个应用管理里还有权限管理的功能。可以分别为每个应用设置新建发布单、环境流转和部署权限的用户列表。

如果读者对这样的权限不满意，或是不能实际应用于公司的具体场景时，完全可以设计其他维度的权限体系，但前提是学会本书的内容，"授人以鱼，不如授人以渔"。

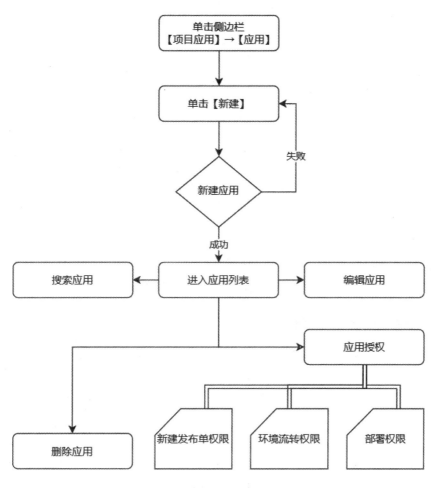

图 1-22　应用管理流程图

4. 服务器管理流程图

图 1-23 为服务器管理流程图。通过侧边导航栏【服务器】→【列表】进入。新建服务器的页面需要输入 IP、端口、所属应用、所属环境、操作系统和描述等信息。为了能在同一个服务器上部署多个应用，BiFang 支持输入同一个 IP 的多个不同的端口，只要端口不冲突即可。

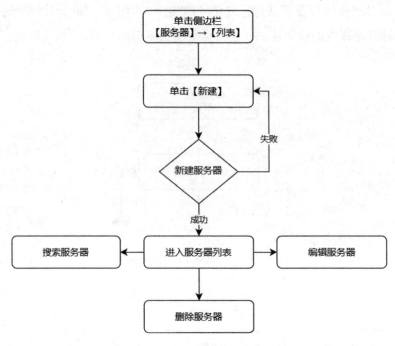

图 1-23 服务器管理流程图

5. 新建发布单与构建流程图

图 1-24 为发布单新建和构建流程图。新建发布单页面需要填写发布单所属项目、应用组件、Git 代码分支及发布描述。而发布单号由系统自动生成，其生成规则：由 Python 的时间戳加上两位随机数组成，避免发布单号冲突。

在发布单生成后，可在发布单列表里选择相应的发布单进行构建。如果在构建软件包的过程中出错，可在调整源代码或相关配置后再重复构建。如果软件包成功生成，则此发布单不可再进行构建操作。

在发布单列表中单击具体发布单号，可进入查看发布单详情页面。在此页面除了可以看到发布单的所有详细信息，还可以看到此发布单的操作历史，比如何时新建、何时构建、何时流转以及何时部署等信息。

发布单的新建与构建都由后端服务进行权限控制，隶属于同一个权限。如果一个用户没有指定应用组件的新建发布单权限，则单击创建与构建按钮，会出现无操作权限的消息提示。

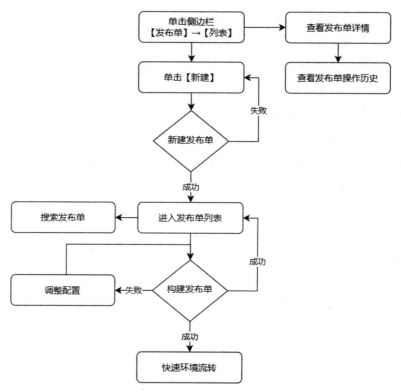

图 1-24　新建发布单与构建流程图

6. 发布单流转流程图

图 1-25 为发布单环境流转流程图。此页面操作比较简单，选择指定发布单的指定环境进行流转。如果失败，一般是代码或权限问题，需要联系管理人员解决。

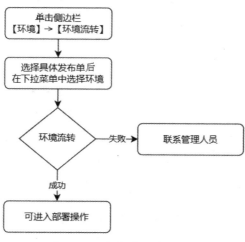

图 1-25　发布单环境流转流程图

发布单的环境流转权限由后端服务进行权限控制。如果一个用户没有指定应用组件的环境流转权限，则单击流转按钮，会出现无操作权限的消息提示。

7. 应用部署流程图

图 1-26 为应用部署流程图。在部署列表中选择一个具体的发布单，进入部署页面。在具体的部署页面，用户可选择一个或多个服务器来部署发布单或进行回滚操作（回滚，在成功部署了 2 个发布单之后才有意义，且在任何部署页面操作回滚按钮效果都一样，因为 BiFang 只支持一次回滚）。

在部署或回滚的操作过程中，前端浏览器会有模态窗口显示相关信息。如果出错，则及时中止后续的部署流程，并显示错误的 bug 信息。如果成功完成，则显示执行的所有步骤。这些操作日志在完成部署或回滚后可以通过单击具体服务器的日志来回看回滚。

发布单的部署或回滚共用一个权限，由后端服务进行权限控制。如果一个用户没有指定应用组件的部署权限，则单击部署或回滚按钮，会出现无操作权限的消息提示。

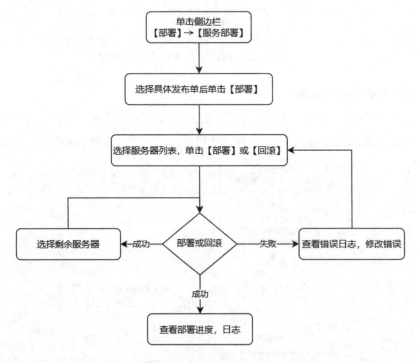

图 1-26　用户注册及登录流程图

8. 管理员操作流程图（略）

相信读者掌握了前面的操作流程后，管理用户组和用户的流程就比较简单了，在此不再赘述。

1.4　自动化部署系统部署拓扑图

本章节以图示的方式讲解毕方（BiFang）自动化部署系统的部署拓扑图，让读者对毕方（BiFang）系统的总体部署有一个梗概性的感性认识。更细节具体的内容，后面会一一展开。

1.4.1　部署拓扑图

毕方（BiFang）自动化部署系统的部署拓扑如图 1-27 所示。

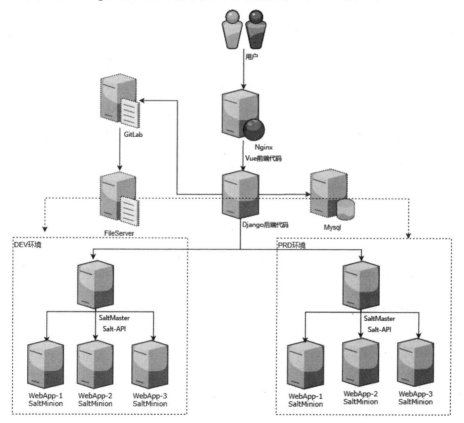

图 1-27　毕方（BiFang）自动化部署系统的部署拓扑图

1.4.2 部署拓扑图说明

在图 1-27 中，可以划分为 4 个部分来理解。

（1）由 Nginx 和 Django 代码组成的自动化部署系统，这也是本书主要讲解的代码实现部分。前端主要以 Vue 作为开发架构。开发完成后，将生成的静态文件置于 Nginx 的配置目录下。后端主要以 Django 作为开发架构，在开发阶段，Django 服务器上以 Sqlite 内置数据开发，即可满足需求。如果用于生产线上环境，可能还需要一个类似 Mysql 的数据库服务器做支撑。

（2）由 GitLab 和 Fileserver 组件的 CI/CD 部分。GitLab 既用于源代码的版本化管理，也用于软件包的构建。并在 FileServer 文件服务器上将构建后的软件包按项目、组件应用、发布单号的目录形式组织好。构建软件包由 BiFang 触发。而各个测试、开发和生产环境的服务器可从 FileServer 上下载所需的软件包及部署脚本。

（3）SaltStack 的远程管理部分由一个或多个 Saltmaster 及其管理的 Saltminon 构成。BiFang 支持每个环境一个 Saltmaster 集群，以达到操作安全隔离的作用，也可以只用一个 Saltmaster 连接所有机器。Saltmaster 向上开发 Salt-API 供 BiFang 远程调用，向下向各个 minion 客户端发送远程命令，从而达到 BiFang 在指定服务器上执行指定命令序列，并返回实时结果的目的。

（4）各个环境的各个机器部分需要部署 Saltminion，以便接受 Saltmaster 的管理。同时，它们也运行各自的服务。而这些服务的软件更新命令也接受 Saltmaster 的指令。

这 4 部分相互作用，就可以完成一个基于 Web 的自动化部署功能。

1.4.3 开发环境介绍

在毕方（BiFang）自动化部署系统的本地开发过程中，本书用一台 Windows 10 笔记本电脑并连接上网，然后在 Windows 10 系统中安装 VirtualBox，采用桥接网络模式虚拟出两个 CentOS 8.3 的系统来完成代码编写和环境测试，如图 1-28 所示。

第 1 章 毕方项目简介

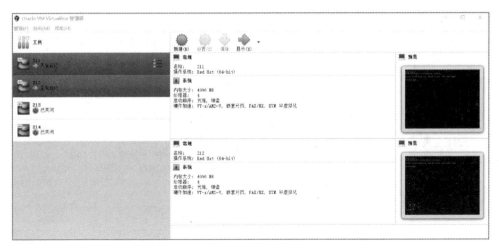

图 1-28 在 Windows 10 上部署 VirtualBox 虚拟机

这 3 台机器的配置如表 1-9 所列。由于使用桥接模式，它们均可连接互联网。

表 1-9 毕方（BiFang）自动化部署系统开发环境列表

主机名称	IP	系统	配置（CPU/内存/硬盘）
10 主机	192.168.1.10	Win 10	12 核 /16G/500G
211 主机	192.168.1.211	CentOS 8.3	4 核 /4G/50G
212 主机	192.168.1.212	CentOS 8.3	4 核 /4G/50G

每台机器安装的主要软件和用途如表 1-10 所列。

表 1-10 测试环境机器软件及用途

主机名称	主要软件	用途
10 主机	Python 3.8.6 Pycharm Node.js v14.15.3 Vscode go1.14.10 Liteide BowPad Git 客户端 MariaDB	● 编写基于 Django 框架的 Python 代码 ● 编写基于 Vue 构架的 JS 代码 ● 修改基于 Go 语言的 FileServer 代码 ● 修改任何代码文件（BowPad） ● 运行 BiFang 的前后端测试代码 ● 将代码纳入 GitHub 和 GitLab 管理 ● 测试 Django 的 MariaDB 数据库连接

续表

主机名称	主要软件	用途
211 主机	Docker Gitlab-ce FileServer Saltmaster Saltminion Salt-api Python3.8.6 Java 1.8.0_281 Ruby 2.5.5p157 Node.js v14.15.3	● 容器化部署 GitLab 服务器 ● 运行 FileServer ● 运行 Saltmaster、Saltminion、Salt-api ● 运行 Bookinfo 示例项目
212 主机	Docker Gitlab-runner Sonarqube Saltminion Python3.8.6 Java 1.8.0_281 Ruby 2.5.5p157 Node.js v14.15.3	● 容器化部署 gitlab-runner ● 容器化部署 Sonarqube ● 运行 saltminion ● 运行 Bookinfo 示例项目

限于本书篇幅，这里不再详细地讲解每一个软件的安装及配置细节，但会在需要的情况下，给出主要软件的大致精简命令。至于一些细节可能需要读者自己从软件官网或是其他途径找更多的文档来解决，敬请谅解。

如果有必要，我们也会根据读者反馈要求，在网上直播、录播、或以写文章等方式来细致讲解这些细节。

1.5 小 结

本章带领读者熟悉了本书实战的主要项目——毕方（BiFang）自动化部署系统的概况，包括自动化部署的需求从何而来，它包含哪些主要的功能模块，操作毕方（BiFang）的主要流程以及主要的部署拓扑。最后还介绍了在测试环境下的开发代码、测试的机器配置以及软件安装情况。

相信读者看完此章，对于要学习的项目已有了清晰的了解。为了更加逼近实战，第 2 章将模拟一个书店的微服务项目，将这个微服务项目的主要组件应用清晰地展示给读者。

围绕这个场景，一起进入第 2 章的学习吧。

第 2 章　Demo 项目应用的手工部署

> 永远要像你不需要金钱那样地工作，永远要像你不曾被伤害过那样地爱，永远要像没有人在注视你那样地跳舞，永远要像在天堂那样地生活。
>
> ——泰戈尔

毕方（BiFang）自动化部署系统的主要用途就是将开发的代码以流水线及可视化的方式，持续地集成和部署（CI/CD），而作为开发这个系统的人员，需要先了解手工操作这一过程的细节，才可以完成这个任务。本章的主要任务就是为读者展示这一过程的细节。由于水平有限，所以这个 Demo 项目是改造自一个网上开源的微服务项目，让它作为本章乃至本书的所有示例部署。

这个微服务项目就是 Istio 官网提供的一个 Bookinfo 的项目，它原本是用来演示在 kubernetes 集群上部署服务网格（Service Mesh）产品 Istio 之后的效果。本书把它稍加改造，变成了可以在一个服务器上部署的四个微服务组件，这样就可以让我们在后面的演示中言之有物、视之有像。

Bookinfo 由 4 个单独的微服务组成，该应用程序用于显示书籍有关的信息，类似于在线书籍商店的单个目录条目。页面上显示的是书的说明、详细信息（ISBN、页数等）和一些书评。该应用程序是由多种语言编写的，即微服务以不同的语言编写。这四个单独的微服务分别如下：

- productpage：productpage 微服务调用 details 和 reviews 微服务来填充页面（python）。
- details：details 微服务包含图书的详细信息（ruby）。
- reviews：reviews 微服务包含书评，它还调用 ratings 微服务（java）。
- ratings：ratings 微服务包含书的排名信息（node.js）。

其中，reviews 微服务提供了 3 个版本：

- 版本 v1 不调用 ratings 服务。
- 版本 v2 调用 ratings 服务，并将每个等级显示为 1~5 个黑星。
- 版本 v3 调用 ratings 服务，并将每个等级显示为 1~5 个灰色星号。

该应用程序的端到端体系结构如图 2-1 所示。

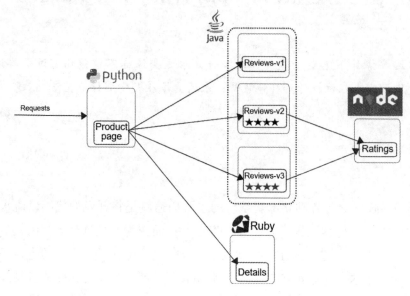

图 2-1　Bookinfo 的端到端结构图

原版 bookinfo 的代码位于 GitHub 下 URL 中：https://github.com/istio/istio/tree/master/samples/bookinfo/src。

改造主要基于两个方面，一是让所有 4 个微服务在访问其他微服务时都通过 127.0.0.1 网络；二是让 reviews 服务只提供一个版本以便简化复杂度，且将 Java 的构建过程从 gradle 更改成 maven，输出 jar 格式软件包。经过改造后的所有组件代码均已放置在 GitHub 仓库上。

下面就以 GitHub 上放置的仓库代码为源，开始手工部署并运行这些微服务组件。根据第 1 章测试环境的整理表格，本章的所有代码均假定在 212 主机上测试（IP 地址为 192.168.1.212，如果读者的开发测试环境与此不同，可做相应转换）。

2.1　Bookinfo 之 productpage 组件

GitHub 网址：https://github.com/aguncn/bookinfo-productpage，这是一个基于 Flask 框架、简单的 Web 请求应答程序。

1. 克隆代码

（1）如果 git 命令未安装，可使用如下命令先行安装：

```
yum install git
```

（2）克隆代码后在 212 主机安装：

```
git clone https://github.com/aguncn/bookinfo-productpage.git
```

（3）进入 bookinfo-productpage 目录，后面的操作在此目录内进行：

```
cd bookinfo-productpage
```

（4）bookinfo-productpage 目录里的文档结构如下：

```
-rw-r--r--  1 root root  1149   Mar 27 10:57   Dockerfile
-rw-r--r--  1 root root   539   Mar 27 21:08   .gitlab-ci.yml
-rw-r--r--  1 root root 12560   Mar 27 10:57   productpage.py
-rw-r--r--  1 root root   308   Mar 27 10:57   README.md
-rw-r--r--  1 root root   521   Mar 27 10:57   requirements.txt
drwxr-xr-x  2 root root    55   Mar 27 10:57   script
-rw-r--r--  1 root root   354   Mar 27 10:57   sonar-project.properties
drwxr-xr-x  3 root root    44   Mar 27 10:57   static
drwxr-xr-x  2 root root    48   Mar 27 10:57   templates
-rw-r--r--  1 root root    21   Mar 27 10:57   test-requirements.txt
drwxr-xr-x  3 root root    18   Mar 27 10:57   tests
```

- Dockerfile 是 Istio，原本是用来生成镜像的，不用理会。
- .gitlab-ci.yml 文件用于 GitLab 的 CI/CD 功能，现在没有讲解到，不用理会。
- script 目录是毕方系统要求的文件目录，现在没有讲解到，不用理会。
- sonar-project.properties 是 GitLab 与 sonar-qube 集成的，不用理会。
- 后面几个组件都有这几个共同的文件。

2. 安装 pip 依赖包

Python 由荷兰数学和计算机科学研究学会的 Guido van Rossum 于 20 世纪 90 年代初设计的，作为一门叫作 ABC 语言的替代品。Python 提供了高效的高级数据结构，还能简单有效地面向对象编程。Python 语法、动态类型，以及解释型语言的本质，使它成为多数平台上写脚本和快速开发应用的编程语言，随着版本的不断更新和语言新功能的添加，逐渐被用于独立的、大型项目的开发。

因为测试环境使用的是 CentOS 8.3，所以自带 Python 3.6。读者可以升级为 3.8，也可以直接在这个版本下测试，暂时省去升级的麻烦。此示例代码在 Python 3.6 下运行也没问题。需要注意的是，在默认安装的系统上，没有 Python 及 pip 命令，而只有 Python3 和 pip3 命令，效果一样。代码如下：

```
pip3 install -r requirements.txt
```

3. 运行程序

```
python3 productpage.py 8001
```

4. 测试效果

在 Win10 上访问：http://192.168.1.212:8001/productpage?u=normal，正常输出如图 2-2 所示。

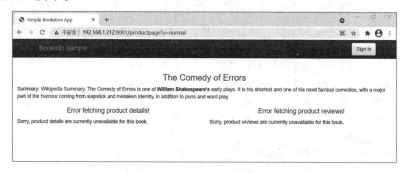

图 2-2　productpage 浏览器测试效果

可以看到，因为只有 productpage 组件启动，所以网页上报获取 details 和 reviews 组件数据失败。莫慌，下面一个个来完善。

2.2　Bookinfo 之 details 组件

GitHub 地址：https://github.com/aguncn/bookinfo-details，这是一个 Ruby 语言写的 Web 程序。

1. 克隆代码

（1）克隆代码后在 212 主机安装：

```
git clone https://github.com/aguncn/bookinfo-details
```

（2）进入 bookinfo-details 目录，后面的操作在此目录内进行：

```
cd bookinfo-details
```

（3）bookinfo-details 目录里的文档结构如下：

```
-rw-r--r-- 1 root root 4493 Mar 27 20:36 details.rb
-rw-r--r-- 1 root root  894 Mar 27 20:36 Dockerfile
-rw-r--r-- 1 root root  539 Mar 27 21:08 .gitlab-ci.yml
-rw-r--r-- 1 root root   61 Mar 27 20:36 README.md
drwxr-xr-x 2 root root   55 Mar 27 20:36 script
```

2. Ruby 环境安装

Ruby 是一种简单快捷的面向对象（面向对象程序设计）脚本语言，在 20 世纪 90 年代由日本人松本行弘（Yukihiro Matsumoto）开发，遵守 GPL 协议和 Ruby License。它的灵感与特性来自 Perl、Smalltalk、Eiffel、Ada 以及 Lisp 语言。

由于 CentOS 中并未预装此语言环境，所以需要自行安装。最简单的 yum 安装命令如下：

```
yum install ruby
```

3. 运行程序

```
ruby details.rb 8002
```

4. 测试效果

在 Win10 上访问：http://192.168.1.212:8001/productpage?u=normal，正常输出如图 2-3 所示（此时，需要保证 productpages 和 details 服务同时启动）。

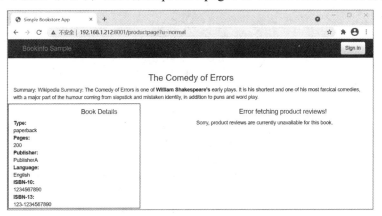

图 2-3　details 浏览器测试效果

可以看到，除了 2.1 节的 productpage 组件内容，Book Details 里也出现了内容。这说明 Ruby 程序已成功启动，且 productpages 网页请求 details 内容成功。我们继续。

2.3　Bookinfo 之 reviews 组件

GitHub 网址：https://github.com/aguncn/bookinfo-reviews，Istio 官方本来是用 ibm 的一个 Java 框架，使用 gradle 构建成 war 包的应用程序。我结合网上其他开源代码，将之改造成一个国内更通用的 Spring Boot 框架通过 maven 构建，生成 jar 包的应用程序。

1. 克隆代码

（1）克隆代码后在 212 主机安装：

```
git clone https://github.com/aguncn/bookinfo-reviews.git
```

（2）进入 bookinfo-reviews 目录，后面的操作在此目录内进行：

```
cd bookinfo-reviews.git
```

（3）bookinfo-reviews 目录里的文档结构如下：

```
-rw-r--r-- 1 root root  157  Mar 27 21:08 Dockerfile
-rw-r--r-- 1 root root  539  Mar 27 21:08 .gitlab-ci.yml
-rw-r--r-- 1 root root  325  Mar 27 21:08 manifest-v2.yml
-rw-r--r-- 1 root root  303  Mar 27 21:08 manifest-v3.yml
-rw-r--r-- 1 root root  330  Mar 27 21:08 manifest.yml
-rw-r--r-- 1 root root 9114  Mar 27 21:08 mvnw
-rw-r--r-- 1 root root 5811  Mar 27 21:08 mvnw.cmd
-rw-r--r-- 1 root root 1374  Mar 27 21:08 pom.xml
-rw-r--r-- 1 root root 3704  Mar 27 21:08 README.md
drwxr-xr-x 2 root root   55  Mar 27 21:08 script
-rw-r--r-- 1 root root  227  Mar 27 21:08 service-v2.yaml
-rw-r--r-- 1 root root  227  Mar 27 21:08 service-v3.yaml
-rw-r--r-- 1 root root  225  Mar 27 21:08 service.yaml
drwxr-xr-x 4 root root   30  Mar 27 21:08 src
drwxr-xr-x 6 root root  151  Mar 27 21:14 target
-rw-r--r-- 1 root root  589  Mar 27 21:08 virtual-service.yaml
```

真正起作用的只有以下几个目录和文件：mvnw、mvnw.cmd、pom.xml、src 和 target。熟悉 Java 语言的读者都了解这些文件和目录的作用。其他几个不了解的文件都是 Istio 为了准备 K8s 部署的文件，与本书主题无关，无须理会。

2. Java 及 maven 的环境安装与构建

Java 是一门面向对象编程语言，不仅吸收了 C++ 语言的各种优点，还摒弃了 C++ 里难以理解的多继承和指针等概念，因此 Java 语言具有功能强大和简单易用两个特征。Java 语言作为静态面向对象编程语言的代表，极好地实现了面向对象理论，允许程序员以优雅的思维方式进行复杂的编程。

Java 具有简单性、面向对象、分布式、健壮性、安全性、平台独立与可移植性、多线程、动态性等特点。Java 可以编写桌面应用程序、Web 应用程序、分布式系统和嵌入式系统应用程序等。现代的 Java Web 程序多半采用 maven 或 gradle 构建应用程序包，此处采用国内最为流行的 maven 程序来进行构建。

（1）maven 安装（同时会安装 Java 语言环境）：

```
yum install maven
```

（2）设置 maven 的国内下载源（略过）。

（3）构建 jar 软件包：

```
mvn clean package -Dmaven.test.skip=true
```

如果构建完成，正常情况下 target 目录下会出现 rob-reviews-1.0.jar 文件。

3. 运行程序

```
java -Dserver.port=8003 -jar target/rob-reviews-1.0.jar
```

4. 测试效果

在 Win10 上访问：http://192.168.1.212:8001/productpage?u=normal，正常输出如图 2-4 所示（此时，需要保证 productpages、details、reviews 服务同时启动）。

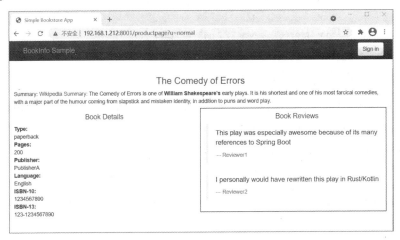

图 2-4　reviews 浏览器测试效果

可以看到，网页上又多了 Book Reviews 的内容，这说明 productpage 组件已可以正常获取 reviews 组件的内容了。

The show must go on！

2.4　Bookinfo 之 ratings 组件

GitHub 网址：https://github.com/aguncn/bookinfo-ratings，这是一个基于 Node.js 写的后端 Web 服务（这种编程，读者需要区别于前端的 Vue、React、Angular 等的工作机制）。

1. 克隆代码

（1）克隆代码后在 212 主机安装：

```
git clone https://github.com/aguncn/bookinfo-ratings.git
```

（2）进入 bookinfo-ratings 目录，后面的操作在此目录内进行：

```
cd bookinfo-ratings
```

（3）bookinfo-ratings 目录里的文档结构如下：

```
-rw-r--r--   1  root  root  1149  Mar 27 10:57  Dockerfile
-rw-r--r--   1  root  root  539   Mar 27 21:08  .gitlab-ci.yml
drwxr-xr-x  24  root  root  4.0K  Mar 27 21:49  node_modules
-rw-r--r--   1  root  root  160   Mar 27 21:48  package.json
-rw-r--r--   1  root  root  6.8K  Mar 27 21:49  package-lock.json
-rw-r--r--   1  root  root  8.5K  Mar 27 21:48  Ratings.js
-rw-r--r--   1  root  root  61    Mar 27 21:48  README.md
drwxr-xr-x   2  root  root  55    Mar 27 21:48  script
```

2. 安装 node.js 及 npm 依赖包

Node.js 由 Ryan Dahl 开发并发布于 2009 年 5 月，是一个基于 Chrome V8 引擎的 JavaScript 运行环境，使用了一个事件驱动、非阻塞式 I/O 模型，让 JavaScript 运行在服务端的开发平台，它让 JavaScript 成为与 PHP、Python、Perl、Ruby 等服务端语言平起平坐的脚本语言。

由于 CentOS 中并未预装此语言环境，所以需要自行安装。

（1）node.js 安装：

```
yum install nodejs
```

（2）设置 npm 国内下源（略）。

（3）npm 依赖包安装：

npm install

3. 运行程序

node Ratings.js 8004

4. 测试效果

在 Win10 上访问：http://192.168.1.212:8001/productpage?u=normal，正常输出如图 2-5 所示（此时，需要保证 productpages、details、reviews、ratings 服务同时启动）。

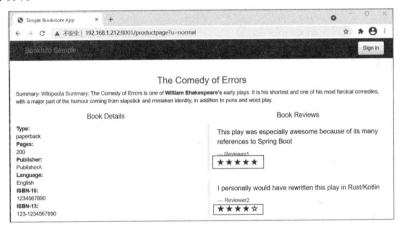

图 2-5　ratings 浏览器测试效果

至此，bookinfo 的所有组件启动完成，且浏览器显示正常。

2.5　Go 语言的 demo 组件应用

Bookinfo 项目已包含了目前 IT 界开发的一些主流语言，为了演示毕方（Bifang）自动化部署系统的更多可能性，我又做了一个基于 Go 语言开发、至简的 Web 应用，算是将主流编程语法一网打尽。此 demo 应用的 GitHub 网址：https://github.com/aguncn/bifang-go-demo。

为了精简演示，此 Go 语言代码没有依赖任何第三方库。

1. 克隆代码

（1）克隆代码后在 212 主机安装：

```
git clone https://github.com/aguncn/bifang-go-demo.git
```

（2）进入 bifang-go-demo 目录，后面的操作在此目录内进行：

```
cd bifang-go-demo
```

（3）核心代码为 main 目录下的 main.go：

网址：https://github.com/aguncn/bifang-go-demo/blob/main/main/main.go。

```go
package main

import (
    "fmt"
    "log"
    "net/http"
)

func doRequest(w http.ResponseWriter, r *http.Request) {
    fmt.Fprintf(w, "Hello Golang!")          // 这个写入 w 的是输出到客户端的
}
func main() {
    http.HandleFunc("/", doRequest)          // 设置访问的路由
    err := http.ListenAndServe(":9090", nil) // 设置监听的端口
    if err != nil {
        log.Fatal("ListenAndServe: ", err)
    }
}
```

可以看到，此 Go 代码只引用了 fmt、log、net/http 这样的内置功能库。只在网页上输出最简单的 Hello Golang! 字符，监听服务的端口为 9090。

（4）bifang-go-demo 目录里的文档结构如下：

```
drwxr-xr-x 2 root root  21 Mar 27 22:27 bin
drwxr-xr-x 2 root root  44 Mar 27 22:26 env
-rw-r--r-- 1 root root 539 Mar 27 21:08 .gitlab-ci.yml
drwxr-xr-x 2 root root  51 Mar 27 22:26 main
drwxr-xr-x 2 root root  55 Mar 27 22:26 script
-rw-r--r-- 1 root root 606 Mar 27 22:26 sonar-project.properties
```

2. Golang 环境安装及编译

Golang（又称 Go）是 Google 的 Robert Griesemer、Rob Pike 及 Ken Thompson 开发的一种静态强类型、编译型语言。Go 语言语法与 C 语言相近，但功能上有内存安全、GC（垃圾回收）、结构形态及 CSP-style 并发计算。

（1）Golang 环境安装。

由于 CentOS 中并未预装 Golang 语言环境，需要自行安装。最简单的 yum 安装命令如下：

```
yum install golang
```

（2）编译程序：

```
go build -o bin/go-demo main/main.go
```

如果正常编译完成，会在 bin 目录生成一个 go-demo 的可执行文件。

3. 运行程序

```
bin/go-demo
```

4. 测试效果

在 Win10 上访问：http://192.168.1.212:9090/，正常输出如图 2-6 所示。

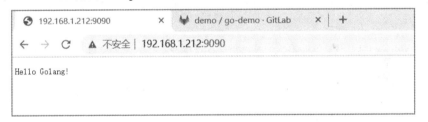

图 2-6　go-demo 浏览器测试效果

2.6　使用 GitLab 管理所有源代码

前面的章节示范了如何从 github.com 拉取 demo 项目的源代码，然后在服务器上编译、安装、启动，让读者了解了从头到尾的纯手工操作。

作为毕方（BiFang）自动化部署系统开发的前期准备工作，根据第 1 章展示的部署拓扑图，需要在本地建立一个 GitLab 服务器来管理这些 demo 项目的源代码，同时利用 GitLab 的 CI/CD 功能来实现毕方系统的 CI 持续化集成功能。

本章先讲解如何建立一个 GitLab 服务器来进行项目的源代码管理。第 3 章将讲解如何安装 gitlab-runner、编写 ".gitlab-ci.yml" 文件来实现 GitLab 的 CI/CD 功能。

限于本书篇幅和主题，本章只演示简单的安装配置过程。关于 GitLab 的更多功能，读者可以通过其官网或其他途径进行系统的学习，也可以参考《Python 3 自动化软件发布系统：Django 2 实战》和《Go 语言运维开发：Kubernetes 项目实战》，这两本书里都有 GitLab 源代码管理和 CI/CD 的更多细节。

2.6.1　安装 docker 版 GitLab 服务器

计划将 GitLab 服务器安装在 211 主机上，所以接下来的操作都是在 211 主机上进行的。由于采用 gitlab/gitlab-ce:13.3.6-ce.0 这个镜像来进行快速安装，所以需要先安装 docker。

温馨提示，此版本的 GitLab 镜像仅用于测试目的，如果想在公司企业内部使用镜像版的 gitlab-ce，最好还是采用多个镜像的方式（包括 nginx、redis、pqsql 等）以提升 GitLab 的服务性能，同时做好备份和恢复的准备工作。

（1）安装 docker：

```
yum install docker-ce
```

（2）设置 docker 镜像的国内源下载，以便加快 docker 镜像的下载速度（略），并启动 docker：

```
systemctl start docker
```

（3）下载 GitLab 镜像：

```
docker pull gitlab/gitlab-ce:13.3.6-ce.0
```

（4）建好本地项目，用于镜像的目录挂载：

```
mkdir /gitlab
```

（5）启动 GitLab 镜像：

```
docker run -itd \
    --publish 8443:443 \
```

```
--publish 8180:8180 \
--publish 8022:22 \
--name gitlab \
--restart unless-stopped \
-v /gitlab/etc:/etc/gitlab \
-v /gitlab/log:/var/log/gitlab \
-v /gitlab/data:/var/opt/gitlab \
gitlab/gitlab-ce:13.3.6-ce.0
```

可以看到，我们将 GitLab 镜像的配置、日志和数据都挂载到了本地目录，且将 GitLab 的运行端口和宿主机的端口做了相应的映射。

（6）修改 GitLab 配置并重启 GitLab 镜像。

编辑 /gitlab/etc/gitlab.rb 文件的 external_url 属性值如下：

```
external_url 'http://192.168.1.211:8180'
```

再次提醒，如果读者的测试环境与本书的环境 IP 这些不一样，请做相应更改。以后这方面就不再提醒啦！

（7）登录 GitLab。

在 Win10 上访问：http://192.168.1.211:8180/，一切正常的话，经过管理员密码的更改即可进入登录界面，这表示我们的容器版 GitLab 服务器已安装就绪，如图 2-7 所示。

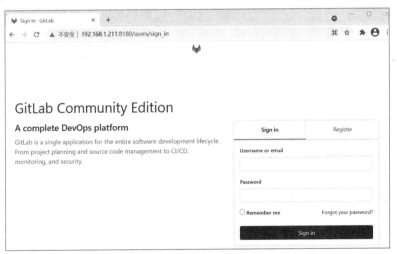

图 2-7　GitLab 服务登录界面

2.6.2 将GitHub的代码纳入GitLab管理

下面的操作可以将GitHub里的demo代码快速迁移到2.6.1小节建好的GitLab服务器中，减少了中间环节。

1. 建立示例的项目及应用

（1）在测试环境的GitLab上建好两个group，bookinfo用于存放之前的微服务组件，demo用于存放go-demo代码，如图2-8所示。

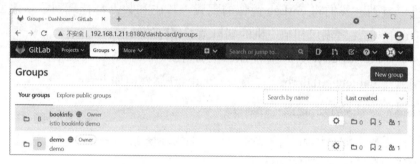

图2-8　GitLab里的Groups列表

（2）在每个Groups下面分别建好对应的项目应用。为了能在同一个目录操作，所以GitLab上的名称与GitHub的名称稍有区别，但都有语义化的含义，如图2-9所示。

图2-9　GitLab里的项目应用列表

2. 克隆两个仓库的代码到本地

本章前面已介绍过克隆GitHub里的代码，这里仅列出克隆GitLab代码的命令。这里的操作使用的是 Win10 系统。但无论是 Linux 还是 Windows 系统，其 git clone 的命令都是相同的：

```
git clone http://192.168.1.211:8180/bookinfo/productpage.git
git clone http://192.168.1.211:8180/bookinfo/details.git
git clone http://192.168.1.211:8180/bookinfo/reviews.git
git clone http://192.168.1.211:8180/bookinfo/ratings.git
git clone http://192.168.1.211:8180/demo/go-demo.git
```

3. 将 GitHub 的代码目录复制到 GitLab 的代码目录

这里不再列出命令或操作，按项目一一对应好即可。

4. 上传 GitLab 的本地代码目录到 GitLab 的代码仓库

分别进入 GitLab 复制本地的目录，按先后顺序运行如下三个命令，即可将代码上传到 GitLab 仓库（如果需要用户密码，也需要正常输入）：

```
git add .
git commit -m "first commit"
git push
```

5. 登录到 GitLab

在任何一个项目应用中均可看到，GitHub 上的代码已传输到 GitLab 中，如图 2-10 所示。

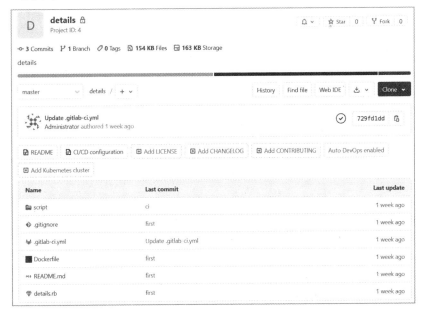

图 2-10　GitLab 中的项目应用代码

读者需要留心 GitLab 中的 ProjectID 值（图 2-10 中 details 项目应用的 ProjectID 为 4），这个 ID 值在 BiFang 系统中是用来定位这个项目应用在 GitLab 的唯一对应关系。

2.7 小　　结

本章带领读者首先熟悉了 bookinfo 这个 demo 项目的手工部署。在这一部署过程中，涉及比较多的运维知识点。希望读者能在学习的过程中逐个细化并消化，相信一定大有收获。

接下来又加入了一个基于 Go 语言的 demo 项目，希望能涵盖现在国内 IT 界更多的主流编程语言。在本章的最后，在测试环境下搭建了一个基于 docker 的 GitLab 服务器，并将 GitHub 上的 demo 代码迁移到了这个 GitLab 服务器中。

有了本章的知识铺垫，第 3 章就可以进入 GitLab 的 CI/CD 功能学习了，并且学习如何使用 Python 的 GitLab 库、触发自动化的软件构建过程，然后将生成的软件包上传到我们自制的文件服务器当中。

未来和未知，都是人生旅途中的星辰大海。保持激情，积累元气，再出发！

第 3 章 实现 GitLab 的 CI/CD 功能

学到很多东西的诀窍,就是不要一下子学很多的东西。

——约翰·洛克

第 2 章主要讲了一个示例项目——Bookinfo 微服务组件的手工部署,且把这些代码纳入了 GitLab 源代码管理里。本章将基于 GitLab 的 CI/CD 功能,实现一个半自动脚本化的持续化集成功能。在持续化集成和交付部署领域,最出名的莫过于 Jenkins。

Jenkins 是一款著名的、可扩展的、用于自动化部署的开源 CI/CD 工具。Jenkins 完全用 Java 编写,是在 MIT 许可下发布的。它有一组强大的功能,可以将软件的构建、测试、部署、集成和发布等相关任务自动化。这款用于测试的自动化 CI/CD 工具可以在 MacOS、Windows 和各种 Unix 版本(如 OpenSUSE、Ubuntu、Red Hat 等)系统上使用。除了通过本地安装包安装,它还可以在任何安装过 Java 运行时环境(Java Runtime Environment,JRE)的机器上单独安装或者作为一个 Docker 安装。

GitLab CI/CD 先前是作为一个独立项目发布的,从 2015 年 9 月发布的 GitLab 8.0 正式版开始集成到 GitLab 主软件。GitLab CI/CD 和 GitLab 是用 Ruby 和 Go 编写的,并在 MIT 许可下发布。除了其他 CI/CD 工具关注的 CI/CD 功能外,GitLab CI/CD 还提供了计划、打包、源码管理、发布、配置和审查等功能。

在使用 GitLab CI/CD 时,phase 命令包含一系列阶段,这些阶段将按照精确的顺序实现或执行。在实现后,每个作业都被描述和配置了各种选项。每个作业都是一个阶段的一部分,会在相似的阶段与其他作业一起自动并行运行。一旦你这样做,作业就被配置好了,你就可以运行 GitLab CI/CD 管道了,而且可以检查某个阶段你指定的每一个作业的状态。这也是 GitLab CI/CD 与其他用于 DevOps 测试的 CI/CD 工具的不同之处。

在作者之前出版的书中,既使用 Jenkins,也使用 GitLab 的 CI/CD。为了让本书保持一个紧凑和耦合的结构,在此选用了 GitLab 的 CI/CD 功能,且本书更重视 GitLab 的 CI 功能，从而弱化其 CD 功能。因为 CD 的大部分功能是在毕方（BiFang）自动化部署系统上实现的，而这也是本书核心的知识内容。

在当今云原生时代，Tekton、JenkinX、Argo 和 Spinnaker 等新一代的云原生 CI/CD 工具在持续地涌现，这些内容不在本书讲述的范围，但也值得有兴趣的读者不断地关注。一个典型的 CI/CD 流程图如图 3-1 所示。

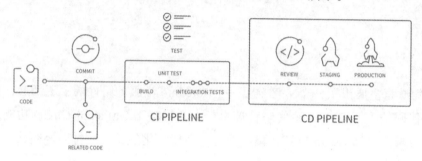

图 3-1　一个典型的 CI/CD 流程图

而毕方（BiFang）自动化部署系统测试环境的 CI 部署的流程如图 3-2 所示。

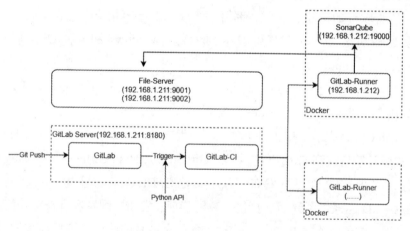

图 3-2　毕方（BiFang）的 CI 处理流程

毕方（BiFang）的 CI 处理流程、代码的编译打包及测试动作，不是在代码合并后自动发生的，而是通过 Pipeline triggers 的方式手动触发。这一触发动作是通过毕方（BiFang）调用 Python API 的方式实现的。

为了节约测试服务器，我们将 CI 的处理过程均分在两个服务器上。211 主机运行 GitLab Server 和 File-Server（用作软件仓库），212 主机运行 gitlab-Runner 和 SonarQube（用作代码质量检查），这两个服务都是运行在 Docker 环境中的。

本章最终要实现的效果是通过 Python API 触发 GitLab 的 CI/CD 过程。GitLab Server 的基本安装配置在第 2 章已讲过。所以本章的主要内容包括以下几部分：

- file-server 文件服务器的实现。
- gitlab-runner 服务的实现。
- 启用 GitLab 的 Pipeline triggers 功能。
- GitLab 的 Python API 示例。
- SonarQube 安装配置及将其集成进 GitLab CI/CD 流水线。

3.1 file-server 文件服务器的实现

file-server 文件服务器的 GitHub 网址：https://github.com/aguncn/sgfs，它在毕方（BiFang）系列里相当于一个存放构建后软件的仓库，即提供远程文件的上传，也提供软件包的下载。这个项目的代码不是本书原创，而是 https://github.com/LinkinStars/sgfs 的项目，其基本用法在 GitHub 的 README.md 文件里已有说明。

file-server 核心的用法就是 POST 请求 9001 用于上传（需要附带 token）、GET 请求 9002 用于查看和下载。本章稍后会展示。在此再次感谢开源的 IT 同行们，有了这群人，世界更有激情。

接下来将主要讲解如何编译生成 file-server 可执行文件，及相对于原版代码本书的主要更改。因为 file-server 是部署在 211 主机上的，所以本节的操作主要是在 211 主机上进行。

3.1.1 编译及运行软件

（1）如果 git 和 Golang 未安装，可使用如下命令先行安装：

```
dnf install git
dnf install golang
```

这里为了避免重复命令（有骗稿费之嫌疑），不再使用 yum，而是使用 CentOS8 默认的 dnf 命令。dnf 代表 Dandified yun，是基于 RPM 的 Linux 发行版的软件包管理器。dnf 是 yun 的下一代版本，并打算在基于 rpm 的系统中替代 yun。

dnf 功能强大且具有健壮的特征。dnf 使维护软件包组变得容易，并且能够自动解决依赖性问题。dnf 还能兼容使用 yum 的配置文件和命令。

（2）克隆代码后在 212 主机安装：

```
git clone https://github.com/aguncn/sgfs.git
```

（3）进入 sgfs 目录进行编译：

```
cd sgfs
export GOPROXY=https://goproxy.cn        --- 设置 goproxy 国内代理
./build.sh                               --- 编译，生成的文件位于 ./bin/linux/sgfs
```

编译生成的 bin 文件夹里的文件结构如下所示：

```
drwxr-xr-x  2 root root      71   Mar 28 20:24  darwin
drwxr-xr-x  4 root root      97   Mar 28 20:25  linux
-rw-r--r--  1 root root 7307594   Mar 28 20:24  sgfs-v1.0.3-2-g4bccb07-darwin.tar.gz
-rw-r--r--  1 root root 4035194   Mar 28 20:24  sgfs-v1.0.3-2-g4bccb07-linux.tar.gz
-rw-r--r--  1 root root 6976316   Mar 28 20:24  sgfs-v1.0.3-2-g4bccb07-windows.tar
drwxr-xr-x  2 root root      38   Mar 28 20:24  windows
```

Golang 支持跨平台的交叉编译，所以这次编译同时生成了支持 Mac、Win 和 Linux 平台的可执行文件，并且压缩了这 3 个目录。我们只关心 Linux 平台的文件，所以 Linux 目录和 sgfs-v1.0.3-2-g4bccb07-linux.tar.gz 压缩文件才是我们要用的。

sgfs-v1.0.3-2-g4bccb07-linux.tar.gz 即为 Linux 目录的压缩文件。接下来进入 Linux 目录，其文档结构如下所示：

```
-rw-r--r--  1 root root      423   Mar 28 20:24  conf.yml
drwxr-xr-x  2 root root       38   Mar 28 20:25  logs
-rwxr-xr-x  1 root root 10231808   Mar 28 20:24  sgfs
-rwxr-xr-x  1 root root      210   Mar 28 20:24  shutdown.sh
-rwxr-xr-x  1 root root       76   Mar 28 20:24  startup.sh
drwxr-xr-x  2 root root        6   Mar 28 20:25  upload
```

- sgfs 是编译生成的可执行文件。
- conf.yml 是 sgfs 程序的配置文件。
- upload 和 logs 分别是存放软件的目录和记录 log 的目录。
- startup.sh 和 shutdown.sh 分别为启、停脚本，是对 sgfs 可执行文件的包装。

（4）运行 file-server。

```
./startup.sh        --- 启动命令
starting sgfs...    --- 控制台输出
```

当控制台输出正常后，看看 sgfs 这个 file-server 软件仓库服务器的配置，因为在接下来的测试或是之后的 GitLab CI/CD 中，会用到其中的设置，网址：https://github.com/aguncn/sgfs/blob/master/conf.yml。

```yaml
# Maximum size of single upload file (in MB)
max_upload_size: 20

# Maximum overall request size (in MB)
max_request_body_size: 30

# root path for upload file
upload_path: upload

# File Operating Port
operation_port: 9001

# File Visit port
visit_port: 9002

# Operation certificate
operation_token: 654321

# Whether generate a file directory index (index page url is "http://127.0.0.1:9002/")
generate_index_pages: true
```

这个配置文件的注释还是很清晰的。单个文件最大支持 20MB，上传端口为 9001，下载访问端口为 9002，操作 token 为 654321，（这个重要！）读者可根据实际情况修改这些属性，本书未修改这些属性，后面的一些设置以此为准。

3.1.2 测试上传下载功能

（1）上传。

登录 212 主机，使用 curl 命令进行文件上传：

```
curl -F "file=@istio-bookinfo.zip" \
     -F "token=654321" \
```

```
    -F "uploadSubPath=/test/istio" \
    http://192.168.1.211:9001/upload-file          ---curl 的 -F 参数指定配置
```

如果一切正常则输出如下：

```
{"code":1,"message":"Save file success.","data":"//test/istio/istio-bookinfo.zip"}
```

其表示上传成功。注意要正确定位文件路径并上传 token，以及上传地址有 upload-file。

（2）浏览和下载。

在 Win10 访问：http://192.168.1.211:9002/，即可看到刚才上传的文件，如图 3-3 所示。

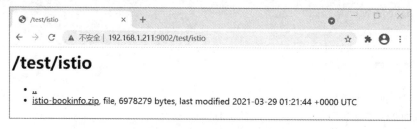

图 3-3　浏览器查看上传文件

通过 wget 命令，可测试下载：

```
wget http://192.168.1.211:9002/test/istio/istio-bookinfo.zip
```

3.1.3　代码修改说明

此处只修改了 sgfs 项目的一个文件，让其更简化。网址为 https://github.com/aguncn/sgfs/blob/master/service/upload.go，代码如下：

```
package service

import (
    //"path"
    "strings"

    "github.com/LinkinStars/golang-util/gu"
    "github.com/valyala/fasthttp"
    "go.uber.org/zap"

    "github.com/LinkinStars/sgfs/config"
```

```go
    //"github.com/LinkinStars/sgfs/util/date_util"
)

func UploadFileHandler(ctx *fasthttp.RequestCtx) {
    // Get the file from the form
    header, err := ctx.FormFile("file")
    if err != nil {
        SendResponse(ctx, -1, "No file was found.", nil)
        return
    }

    // Check File Size
    if header.Size > int64(config.GlobalConfig.MaxUploadSize) {
        SendResponse(ctx, -1, "File size exceeds limit.", nil)
        return
    }


    // authentication
    token := string(ctx.FormValue("token"))
    if strings.Compare(token, config.GlobalConfig.OperationToken) != 0 {
        SendResponse(ctx, -1, "Token error.", nil)
        return
    }


    // Check upload File Path
    uploadSubPath := string(ctx.FormValue("uploadSubPath"))
    // 注释掉,之前加了日期目录
    // visitPath := "/" + uploadSubPath + "/" + date_util.GetCurTimeFormat(date_util.YYYYMMDD)
    visitPath := "/" + uploadSubPath
    dirPath := config.GlobalConfig.UploadPath + visitPath
    if err := gu.CreateDirIfNotExist(dirPath); err != nil {
        zap.S().Error(err)
        SendResponse(ctx, -1, "Failed to create folder.", nil)
        return
    }
    // 注释掉,不要取后缀,直接取文件名
    // suffix := path.Ext(header.Filename)
    // filename := createFileName(suffix)
    filename := header.Filename

    fileAllPath := dirPath + "/" + filename

    /*
       注释掉,有同名,就报错
```

```go
    // Guarantee that the filename does not duplicate
    for {
      if !gu.CheckPathIfNotExist(fileAllPath) {
        break
      }
      filename = createFileName(suffix)
      fileAllPath = dirPath + "/" + filename
    }
*/

    // Save file
    if err := fasthttp.SaveMultipartFile(header, fileAllPath); err != nil {
      zap.S().Error(err)
      SendResponse(ctx, -1, "Save file fail.", err.Error())
    }

    SendResponse(ctx, 1, "Save file success.", visitPath+"/"+filename)
    return
}

/*
注释掉,不需要重命名文件
func createFileName(suffix string) string {
  ......
}
*/
```

在修改的地方都加了注释说明。

由于毕方(BiFang)自动化部署系统在上传时保留了发布单上的时间戳命名、上传的文件和定义好的目录层次,所以 sgfs 里的一些通用逻辑,这里不再需要。

经过如此这般一番打磨,我们为毕方准备的 file-server 就完成了。当然,这只是测试环境的建设过程。如果是应用于线上环境,那就将 file-server 作为服务自启动,上传文件的容量规则和备份策略等都需要配套跟上。

接下来,就可以进入 gitlab-runner 的配置了。

3.2 Docker 版 gitlab-runner 的安装配置

在第 2 章已安装好 GitLab Server。本节就来集成 GitLab 的 CI/CD 功能,安装 gitlab-runner，让 GitLab 帮我们持续构建软件包。那 gitlab-runner 又是什么呢？与 GitLab CI 有什么关系呢？

gitlab-runner 是配合 GitLab CI 使用的。一般来说，GitLab 里面的每一个工程都会定义一个属于这个工程的软件集成脚本（项目根目录下的 .gitlab-ci.yml 文件），用来自动化地完成一些软件集成工作。当这个工程的仓库代码发生变动时，比如有人 push 了代码，或是通过 API 触发 Pipeline triggers 后，GitLab 就会将这个变动通知 GitLab CI。这时 GitLab CI 会找出与这个工程相关联的 Runner，并通知这些 Runner 把代码更新到本地并执行预定义好的执行脚本。

所以，gitlab-runner 就是一个用来执行 ".gitlab-ci.yml" 文件脚本的东西。可以想象一下：Runner 就像一个个工人，而 GitLab CI 就是这些工人的一个管理中心，所有工人都要在 GitLab CI 里登记注册，并且表明自己是为哪个工程服务的。当相应的工程发生变化时，GitLab CI 就会通知相应的工人执行软件集成脚本。

gitlab-runner 可以分为两种类型：Shared Runner（共享型）和 Specific Runner（指定型）。

- Shared Runner：这种 Runner（工人）所有工程都能够用。只有系统管理员能够创建 Shared Runner。
- Specific Runner：这种 Runner 只能为指定的工程服务。拥有该工程访问权限的人都能够为该工程创建 Shared Runner。

本书采用的是共享型 Runner，一切以方便快速为准。按本书的规划，gitlab-runner 安装在 212 主机，所以本节执行的相关命令是在此主机上进行的。

1. 准备镜像

在 212 主机上先 pull 如下 docker 镜像：

```
docker pull gitlab/gitlab-runner
docker pull docker
```

2. 运行 docker 版 gitlab-runner

在 docker run 命令中将 gitlab-runner 的配置挂载到宿主机上，便于查看修改。同时，为了在 gitlab-runner 中执行 docker 这一命令，也将宿主机的 docker.sock 挂载进了 docker 容器中：

```
docker run -d \
    --name gitlab-runner \
```

```
--restart always \
-v /srv/gitlab-runner/config:/etc/gitlab-runner \
-v /var/run/docker.sock:/var/run/docker.sock \
gitlab/gitlab-runner:latest
```

3. 获取 GitLab Server 上的 gitlab-runner 连接信息

打开 GitLab Server 上的任意项目，导航到 Settings → Runners，可以获取到 gitlab-runner 的连接信息，如图 3-4 所示。

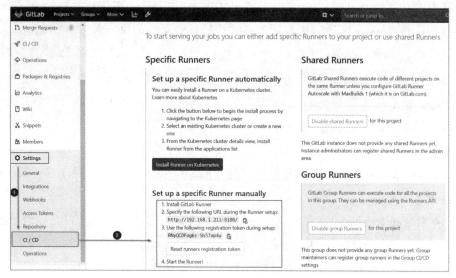

图 3-4　获取 GitLab 上的 Runner 连接信息

在我的计算机上，gitlab-runner 注册时需要连接网址：http://192.168.1.211:8180/，连接 token 为 RNyQCDFogbs-Sh57ap4u。

4. gitlab-runner 注册

当获取到注册信息，gitlab-runner 又运行起来后就可以使用如下命令进行注册：

```
docker exec gitlab-runner gitlab-runner register -n \
    --url http://192.168.1.211:8180/ \
    --registration-token RNyQCDFogbs-Sh57ap4u \
    --tag-list runInDocker \
    --executor docker \
    --docker-image docker \
    --docker-volumes /root/.m2:/root/.m2 \
    --docker-volumes /root/.npm:/root/.npm \
    --docker-volumes /var/run/docker.sock:/var/run/docker.sock \
```

第 3 章 实现 GitLab 的 CI/CD 功能

--description "runInDocker"

- executor 使用 docker。
- 通过 docker-image 指定一个 docker 镜像，这里使用的是 docker:latest。
- 通过 docker-volumns 挂载本地目录：挂载 docker.sock 是为了 docker:latest 镜像操控 Runner 服务器的 docker 服务；

  挂载 ".m2" 文件夹是为了避免 maven 每次编译项目时都重新下载 jar 包；

  挂载 ".npm" 文件夹是为了避免 Node.js 每次构建项目时都重新下载 module。

当正确注册完成后再看 GitLab Server 的界面，会在 Runners 看到刚刚注册的 gtitlab-runner，其信息显示与注册时的信息一致，如图 3-5 所示。

图 3-5 gitlab-runner 注册成功

进行到这里，我们可以做一个简单的 echo 级测试，但为了连续性，等我们配置好下一节的 Pipeline triggers 后再测试吧。

3.3 启用 GitLab 的 Pipeline triggers 功能

gitlab-runner 在上一节已配置完成，接下来看看如何触发 GitLab 的 pipeline 流水线构建交付功能。GitLab 代码提交或推送触发 CI pipeline 需要满足两个条件：① 仓库根目录下存在 ".gitlab-ci.yml" 文件；② 该项目有可用的 gitlab-runner。这样仓库每收到一次 push，runner 都会自动启动 pipeline，结果显示在项目的 pipeline 页面。

GitLab CI 使用 Yaml 文件（.gitlab-ci.yml）来管理项目配置，该文件存放于项目仓库的根目录，它定义该项目如何构建。

".gitlab-ci.yml" 文件告诉 gitlab-runner 要做什么事情，默认三个 stage：build、test、deploy。不必 3 个都用，没有 job 的节点会被忽略。

关于 .gitlab-ci.yml 的语法细节和 stage、job 等阶段的配置，读者可以参考

GitLab 的官网或其他资源。本节仅以可用的".gitlab-ci.yml"配置开始讲解,只使用了 only-triggers 这个有点不常用的语法。使用这一 pipeline 语法的目的是不希望每次运行代码时,GitLab 都触发自动构建过程,而是要手工携带 token 触发了 trigger,GitLab 才构建。对于毕方系统来说,这样的构建流水线更有可控性。

1. 获取 Pipeline triggers token

打开 GitLab Server 上的任何一个项目,导航到 Settings → CI/CD → Pipeline triggers,可以获取 Pipeline triggers 的 token。如果还没有 token,可以新建一个,如图 3-6 所示。

图 3-6 获取 Pipeline triggers 的 token

我测试环境的 go-demo 应用,其触发 token 为 559fbd3381bc39100811bd00e499a7。接下来的内容就以实现的 go-demo 项目应用的配置来讲解。

2. go-demo 的 CI/CD 文件——.gitlab-ci.yml

网址:https://github.com/aguncn/bifang-go-demo/blob/main/.gitlab-ci.yml。

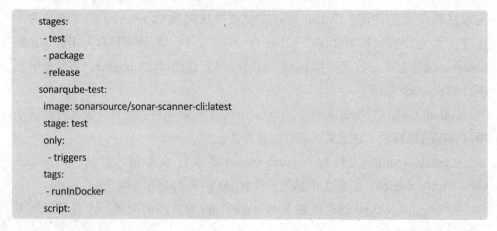

```
      - sonar-scanner
  go_package:
    image: golang
    stage: package
    only:
     - triggers
    tags:
     - runInDocker
    script:
     - echo ${APP_NAME}
     - echo ${RELEASE}
     - echo ${BUILD_SCRIPT}
     - echo ${DEPLOY_SCRIPT}
     - echo ${FILE_UP_SERVER}
     - go version
     - sh ${BUILD_SCRIPT}
     - curl -F "file=@${DEPLOY_SCRIPT}" -F "token=654321" -F "uploadSubPath=/${APP_NAME}/${RELEASE}" ${FILE_UP_SERVER}
     - curl -F "file=@${ZIP_PACKAGE_NAME}" -F "token=654321" -F "uploadSubPath=/${APP_NAME}/${RELEASE}" ${FILE_UP_SERVER}
  empty-release:
    stage: release
    only:
     - triggers
    tags:
     - runInDocker
    script:
     - echo "empty release."
```

文件说明：

- 由于两个 curl 命令行比较长，为方便编排显示，故缩小字号。
- 此 GitLab pipeline 分三个阶段：测试、构建打包和发布。
- 测试阶段的代码质量检查系统 SonarQube 的安装配置将在 3.4 节讲解，此处不提。
- 发布阶段只是一个 echo 输出，并无实际动作。
- 在打包阶段，sh ${BUILD_SCRIPT} 命令用于编译 Go 程序，然后是两个 curl 命令，一个用于上传构建压缩好的软件包，另一个用于上传部署脚本文件，而上传文件的地址就是本章开头建好的文件仓库服务器。**软件包在毕方部署时，远程 salt minion 通过 wget 命令获取；毕方在部署时通过 salt 调用部署脚本文件，传入参数去远程 minion 执行一系列命令（这是核心处理流程）。
- only 条件表示此 pipeline 只能由 triggers token 来触发构建。
- tags 条件表示此 pipeline 只能在标有 runInDocker 的 gitLab-runner 上运行，而在 3.2 节的第 4 小节中注册 gitlab-runner 时已加入此 tag（--tag-list runInDocker）。
- 运行各个阶段的 script 脚本时，从 API 一共需要传递 6 个变量值，即应用组件名称：${APP_NAME}、发布单号：${RELEASE}、软件压缩包名：{ZIP_PACKAGE_NAME}、编译脚本：${BUILD_SCRIPT}、部署脚本：${DEPLOY_SCRIPT}、软件仓库服务器：${FILE_UP_SERVER}。
- 把编译脚本和部署脚本都置于 GitLab 源代码版本管理之下，也是目前 GitOps 的一种实践。

3. go-demo 的 CI/CD 文件——sonar-project.properties

网址：https://github.com/aguncn/bifang-go-demo/blob/main/sonar-project.properties。

```
sonar.host.url=http://192.168.1.212:19000
sonar.sourceEncoding=UTF-8
sonar.login=admin
sonar.password=password
sonar.projectKey=go-demo
sonar.projectName=go-demo
sonar.projectVersion=1.0
sonar.golint.reportPath=report.xml
sonar.coverage.reportPath=coverage.xml
sonar.coverage.dtdVerification=false
sonar.test.reportPath=test.xml
sonar.sources=./main/
sonar.sources.inclusions=**/**.go
sonar.sources.exclusions=**/**_test.go,**/vendor/*.com/**,**/vendor/*.org/**, **/vendor/**
sonar.tests=./main/
sonar.test.inclusions=**/**_test.go
sonar.test.exclusions=**/vendor/*.com/**,**/vendor/*.org/**,**/vendor/**
```

文件说明：
- 项目根目录下的 sonar-project.properties 文件会被 sonar-scanner 默认调用。
- 此 sonar-project.properties 定义了 SonarQube 服务器地址、用户名、密码及扫描哪些文件。SonarQube 服务器地址的搭建将在 3.4 节讲解。

4. go-demo 的 CI/CD 文件——build.sh

网址：https://github.com/aguncn/bifang-go-demo/blob/main/script/build.sh。

```
#!/bin/sh

echo 'beging build demo-go app...'

go build -o bin/go-demo main/main.go

tar -zcvf go-demo.tar.gz env/ bin/

echo 'finish build demo-go app...'
```

文件说明：

- 在 ".gitlab-ci.yml" 的 package 阶段会调用此脚本，因为我们定义的命令是 sh ${BUILD_SCRIPT}，而之后传入 ${BUILD_SCRIPT} 的值是 script/build.sh，所以此文件会被执行。
- 由于 go-demo 比较简单，所以直接使用 go build 命令编译完成。
- 如果一些应用还有不同的环境配置的差异，可以使用配置中心来完成，也可以在这里将不同的环境打包在一起，而在部署解包时再针对环境提取文件。

5. go-demo 的 CI/CD 文件——deploy.sh

这里为了内容的完整性列出了这个部署脚本，它同样存在于每一个毕方项目的 script 目录下。

在编译阶段，deploy.sh 脚本和压缩包一起被上传到了软件仓库服务器，在 ".gitlab-ci.yml" 的 package 阶段执行脚本命令为：curl -F "file=@${DEPLOY_SCRIPT}" -F "token=654321" -F "uploadSubPath=/${APP_NAME}/${RELEASE}" ${FILE_UP_SERVER}，而之后我们传入 ${DEPLOY_SCRIPT} 的值是 script/deploy.sh，所以此文件会被上传。

在部署阶段，毕方系统通过 saltstack 调用远程执行这个部署脚本，这些都是后面的内容，所以此处暂时不再提。

3.4 GitLab Pipeline triggers 的 Python API 示例

在 GitLab Pipeline triggers 的设置页面，有说明用 curl 命令如何触发构建的过程。但这样的操作并不好用，只能做临时的用途。如果想嵌入应用程序，这条路是行不通的，最好还是使用成熟的第三方库来操作。python-gitlab 库就是在 Python 语言中专门用来操作 GitLab 的第三方库，其官方文档网址为 https://

python-gitlab.readthedocs.io/en/stable/。

在本节乃至本书都使用这个库来触发 GitLab 的构建过程，本节因为要编辑 Python 文件，建议读者在 Win10 上实践本节内容。

1. 准备好 GitLab CI/CD 的文件与配置

Go-demo 项目的相关文件已准备好，网址为 https://github.com/aguncn/bifang-go-demo。

（1）文件。
- 项目根目录下的".gitlab-ci.yml"。
- 项目根目录下的 sonar-project.properties。
- script 目录下的 build.sh。
- script 目录下的 deploy.sh。

（2）Pipeline triggers token。

token 为 3.3 节显示的 559fbd3381bc39100811bd00e499a7。

（3）GitLab Server Access token。

这个 token 用于通过 GitLab API 直接访问 GitLab Server Access。管理员通过网页右上角的下拉菜单选择 settings → Access Token 来到操作界面，通过新建一个 API Access Token 而获得。但要注意保存好这个 token，因为它只有在新建完成之后显示一次，如图 3-7 所示。

图 3-7　获取 GitLab Server Access Token

本次测试环境获取到的 Access token 为 yixczsJ6xupwpKZNvRgj。

2. 安装第三方库 python-gitlab

```
pip install python-gitlab      --- 自动安装最新版本
pip install python-gitlab==2.5.0    --- 作者写书时的版本
```

3. 编写测试代码

网址：https://github.com/aguncn/bifang/blob/main/bifangback/utils/gitlab.py（改编）。

```python
import gitlab

git_url = 'http://192.168.1.211:8180'
git_access_token = 'yixczsJ6xupwpKZNvRgj'
project_id = 1
app_name = 'go-demo'
release_name = '2021-04-01'
git_branch = 'master'
git_trigger_token = '559fbd3381bc39100811bd00e499a7'
build_script = 'script/build.sh'
deploy_script = 'script/deploy.sh'
zip_package_name = 'go-demo.tar.gz'
file_up_server = 'http://192.168.1.211:9001/upload-file'


def main():
    try:
        gl = gitlab.Gitlab(git_url, private_token=git_access_token)
        project = gl.projects.get(project_id)
        pipeline = project.trigger_pipeline(git_branch,
                    git_trigger_token,
                    variables={"RELEASE": release_name,
                               'APP_NAME': app_name,
                               'BUILD_SCRIPT': build_script,
                               'DEPLOY_SCRIPT': deploy_script,
                               'ZIP_PACKAGE_NAME': zip_package_name,
                               'FILE_UP_SERVER': file_up_server})
        print(pipeline.id)
        print(pipeline.web_url)
    except Exception as e:
        print(e)


if __name__ == '__main__':
    main()
```

代码说明：

- 通过 git_url 和 git_access_token 实例化一个 GitLab 连接实例。
- 通过 project_id 定位一个唯一的项目，id 为 1 的项目，在作者的测试环境里即为 go-demo（查看 project_id 的方法，详见 2.6.2 小节）。
- 调用项目实例的 trigger_pipeline() 方法，传入合适的参数变量即触发了一次构建。
- 如果构建开始，则输出本次构建的 pipeline_id 和 web_url。

4. 运行代码

运行上一节代码，如果一切正常则控制台输出如下内容（id 和 url 因具体而异）：

```
90
http://192.168.1.211:8180/demo/go-demo/-/pipelines/90
```

如果我查看 go-demo 的 pipelines，就可以看到此次构建的结果和具体输出，如图 3-8 所示。

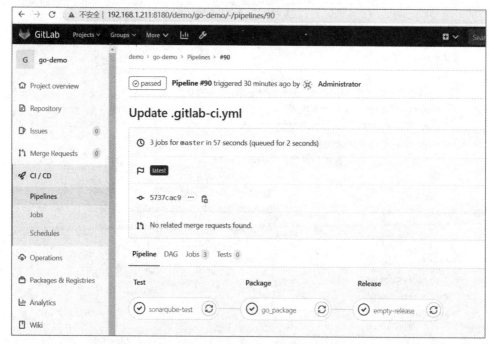

图 3-8　查看 go-demo 的 pipelines 构建结果

如果访问：http://192.168.1.211:9002/，也可以看到刚刚构建完成并上传到了软件仓库服务器上的软件，目录都是按配置组装好的，如图 3-9 所示。

第 3 章 实现 GitLab 的 CI/CD 功能

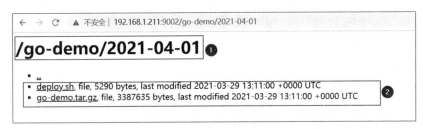

图 3-9 查看软件仓库服务器上的文件

至此，通过 Pthon API 触发 GitLab Pipeline trigger 的流程演示完成。读者如果只是看这一章，可能会感觉眼花缭乱、云里雾里。毕竟这章的知识点还是很多的。希望读者能在自己的计算机上实践起来，可能开始比较困难，也会有很多细节要理解。一旦成功一次、两次、三次之后，这一过程也就真的理解了，并且将其变成了自己的知识，获得满满的收获感。

3.5 将 SonarQube 集成进 GitLab 的 CI/CD 流程

尽管本章的知识点比较多，但还是要再增加一个知识点。它从内容上就属于这一章，那读者就再忍忍吧。

这就是在 3.3 节第 2 小节中定义的 test 阶段，使用 SonarQube 进行代码质量扫描。本章就来建一个 SonarQube 服务器，并接受 3.3 节第 3 小节中定义的 sonar-project.properties 文件的扫描配置。

SonarQube 是一个开源的代码分析平台，用来持续分析和评测项目源代码的质量。通过 SonarQube 可以检测出项目中重复的代码、潜在 bug、代码规范和安全性漏洞等问题，并通过 SonarQube web UI 展示出来。支持的语言包括：Java、PHP、C#、C、Cobol、PL/SQL 和 Flex 等。

SonarQube 作为一个工具平台，建起来是比较容易的。但它作为质量管理的工具，需要的是很多质量管理的专业技能和思考框架，这些技能就超出作者本人的能力了。所以本节仅讲解 SonarQube 的安装与 GitLab 的集成，更专业的质量分析就留给更专业的人去做吧。

按本章开始的规划，SonarQube 安装在 212 主机，所以本节的操作是在 212 主机上进行的。

3.5.1 获取并运行SonarQube镜像

（1）获取镜像：

```
docker pull sonarqube:8.6-community
docker pull sonarsource/sonar-scanner-cli:latest       --- 这个镜像将在下面一节用到
```

（2）运行镜像：

```
docker run --name sonarqube  --restart always \
  -p 19000:9000 \
  -v /sonarqube/data:/opt/sonarqube/data \
  -v /sonarqube/extensions:/opt/sonarqube/extensions \
  -v /sonarqube/logs:/opt/sonarqube/logs \
  -d sonarqube:8.6-community
```

（3）访问测试。

网址为 http://192.168.1.212:19000/，出现登录界面即表示安装成功，如图 3-10 所示。

图 3-10　SonarQube 登录界面

3.5.2 将SonarQube集成进GitLab的CI/CD流程

将 SonarQube 集成进 GitLab 的 CI/CD 流程，需要以下操作：

（1）".gitlab-ci.yml"里对代码质量扫描阶段的定义。

Go-demo 项目中 ".gitlab-ci.yml" 中，此阶段的定义如下：

```
sonarqube-test:
  image: sonarsource/sonar-scanner-cli:latest
```

在测试阶段会调用 sonarsource/sonar-scanner-cli:latest 镜像来运行脚本，而运行的命令是 sonar-scanner，sonar-scanner 默认读取项目根目录下的 sonar-project.properties 文件进行代码扫描。待扫描完成后会将扫描结果发送到 sonar-project.properties 指定的 SonarQube 服务器上。

（2）项目根目录下存在 sonar-project.properties 文件。

sonar-project.properties 的内容在前面已有列明，这里不再重复。sonar-project.properties 配置文件的编写是一个很细致的工作。有兴趣的读者可以通过其官网或其他途径了解更多。

3.5.3 查看SonarQube的扫描结果

在 3.4 节的第 3 小节中运行完那个测试脚本后，相应的 SonarQube 扫描结果也已上传。登录：http://192.168.1.212:19000，即可看到更多代码质量报告，如图 3-11 和图 3-12 所示。

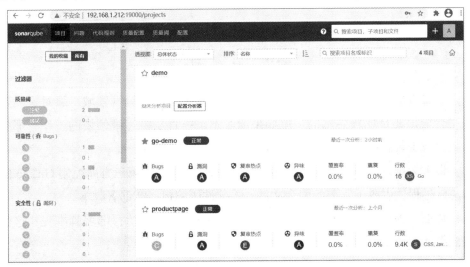

图 3-11 SonarQube 代码质量报告 1

图 3-12 SonarQube 代码质量报告 2

如果要实现代码质量与部署流程联动，可以从 SonarQube 平台中提取出每次的构建质量入库，然后设定一个质量水准，如果质量不过关，则不允许部署。之前为了控制部署流程，我曾将发布单和 jira 需求号及发布流程结合起来，也就是需求的流程没有走到指定位置，则不允许部署。

在此，读者也可以结合自己公司的实际场景，设计出更多有效的控制手段。

3.6 小　结

本章真的是知识点较多的一章。如果读者一直从事运维领域，那对这些开源平台应该还是比较熟悉的。但如果读者不是一直从事运维领域的工作，那有些概念或工具就需要不断地熟悉和了解。

世上无难事，只怕有心人。本章实现了通过 Python 第三方库控制 GitLab 的 CI/CD 过程，生成了需要的两个文件：①软件压缩包；②部署脚本。

那么，如何将软件压缩包推送到指定服务器，然后执行部署脚本呢？这就轮到第 4 章的主角——SaltStack 出场了。

第 4 章　使用 SaltStack 实现远程部署功能

故立志者，为学之心也；为学者，立志之事也。

——王阳明

本书要开发的毕方（BiFang）自动化部署系统除了是一个前后端分离的系统（前端以 Vue 框架实现，后端以 Django 框架实现）外，还是一个运维领域的自动化系统，所以需要读者具备一些运维知识才能更好地学习此系统。

毕方（BiFang）自动化部署系统的运维知识有两大块：一是第 3 章介绍的如何将源代码通过 CI/CD 流水线生成可交付软件；二是本章要介绍的如何将交付的软件快速标准地部署到服务器上。在这之后才开始学习毕方（BiFang）自动化部署系统编码开发的知识。

在运维领域，远程管理服务器和执行命令也是一个专门的自动化运维门类，能实现这些功能的常见工具平台有 SaltStack、Ansible、Chef、Fabric 和 Puppet 等。出于个人的知识积累，我在这里选用了 SaltStack 来完成此项工作。

本章最终要实现的效果是通过 Python API 调用 SaltStack 来执行部署脚本，完成 go-demo 项目的远程部署工作。本章主要内容如下：

- SaltStack 介绍。
- SaltStack 安装配置。
- Salt-API 安装配置。
- Saltypie 第三方库的安装使用。
- 通过 SaltStack 调用 deploy.sh 脚本远程部署。

4.1　SaltStack 简介

SaltStack 是一种基于 C/S 架构的服务器基础架构集中化管理平台，管理端称为 Master（主控端），客户端称为 Minion（被控端）。SaltStack 具备配置管理、远程执行、监控等功能，一般可以理解为是简化版的 Puppet 和加强版的 Func。SaltStack 本身是基于 Python 语言开发实现的，其结合了轻量级的消息队列软件 ZeroMQ 与 Python 第三方模块（Pyzmq、PyCrypto、Pyjinjia2、python-msgpack 和 PyYAML 等）构建。

通过部署 SaltStack 环境，运维人员可以在成千上万台服务器上做到批量执行命令，根据不同的业务特性进行配置集中化管理、分发文件、采集系统数据及软件包的安装与管理等。

1. SaltStack 特性

- 部署简单、方便。
- 支持大部分 UNIX/Linux 及 Windows 环境。
- 主从集中化管理。
- 配置简单、功能强大、扩展性强。
- 主控端和被控端基于证书认证，安全可靠。
- 支持 API 及自定义模块，可通过 Python 轻松扩展。

2. SaltStack 三大组件

- Grains：静态组件，Minion 启动时收集信息。
- Pillar：动态组件，定义变量、信息和密码。
- State：核心功能，通过制订好的 sls 文件对被控主机进行管理。

3. SaltStack 常用模块

- pkg 模块：包管理，包括增删更新。
- file 模块：管理文件操作，包括同步文件、设置权限和所属用户组、删除文件等操作。
- cmd 模块：在 minion 上执行命令和脚本。
- user 模块：管理系统账户操作。
- service 模块：管理系统服务操作。
- cron 模块：管理 cron 服务操作。

4. SaltStack 通信模式

SaltStack 通信模式如图 4-1 所示。

- Minion 是 SaltStack 需要管理的客户端安装组件，其主动连接 Master 端，并从 Master 端得到资源状态信息，同步资源管理信息。
- Master 作为控制中心运行在主机服务器上，负责 Salt 命令运行和资源状态的管理。
- ZeroMQ 是一款开源的消息队列软件，用于在 Minion 端与 Master 端建立系统通信桥梁。
- Daemon 是运行于每一个成员内的守护进程，承担着发布消息及通信端口监听的功能。

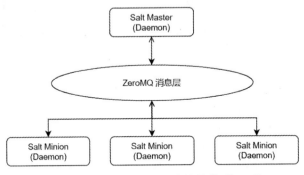

图 4-1　SaltStack 通信模式 ".png"

5. SaltStack 执行原理

SaltStack 执行原理如图 4-2 所示。

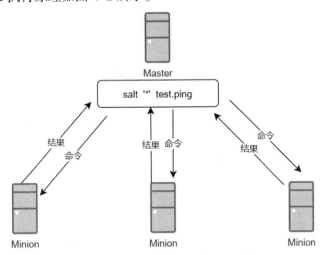

图 4-2　SaltStack 远程命令执行原理

- Minion 是 SaltStack 需要管理的客户端安装组件，其主动连接 Master 端，并从 Master 端得到资源状态信息，同步资源管理信息。
- Master 作为控制中心运行在主机服务器上，负责 Salt 命令运行和资源状态的管理。

- Master 上执行某条指令通过队列下发到各个 Minions 去执行，并返回结果。

4.2 SaltStack 的安装配置

按本书第 1 章的测试环境介绍，本章的 SaltStack 部署在两个机器上，包括一个 SaltMaster、两个 SaltMinion 和一个 Salt-API，如图 4-3 所示。

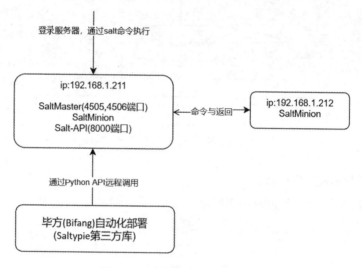

图 4-3 测试环境 SaltStack 部署图

1. 安装 SaltStack

（1）在 211 和 212 主机上，如有必要先升级系统软件包和 Python 3。是否有必要以后面的安装是否顺利为准：

```
dnf update
dnf install python3
```

（2）在 211 和 212 主机上安装 saltstack 官方的 yum 仓库 (不通过 epel)。

```
wget https://repo.saltstack.com/py3/redhat/salt-py3-repo-latest.el8.noarch.rpm
dnf -y install salt-py3-repo-latest.el8.noarch.rpm
```

一个小贴士：要是这个 rpm 在 Linux 里，不好使用 wget 命令下载的话，可以使用迅雷下载传到机器上再安装。

（3）在 211 主机上安装 SaltStack 全家桶：

```
dnf install -y salt-master salt-minion salt-ssh salt-syndic salt-cloud salt-api
```

（4）在 212 主机上安装 SaltMinon 相关应用：

```
dnf install -y salt-minion salt-ssh
```

2. 修改 SaltMinion 配置，启动服务、接受 Key

Salt 软件安装好后要经过这步操作才能正常使用 Salt 功能。由于这是测试环境，一切都以能运行最少配置为原则。

（1）修改 211 主机和 212 主机配置。

在 /etc/salt/minion 文件里找到 Master 和 id，定义好 Master 的连接地址和自身的 id 命名。为了语义统一，将每个服务的 minon id 均命名为自身的 IP 地址。当然，也可以命名为其他分类的名称，只要 id 不冲突均可：

```
master: 192.168.1.211
id: 192.168.1.211           --- 在 212 主机上，则更改为 212。
```

（2）启动服务。

执行如下命令，启动服务：

```
systemctl start salt-master     --- 只在 211 主机上执行
systemctl start salt-minion     --- 在 211 和 212 主机上均执行
```

Salt–API 在后面再配置，此时先略过。

（3）在 Master 上接受 Minon 的连接 Key。

默认设置下，Master 是手工同意 Minion 的连接请求，使用如下命令达成：

```
salt-key -A                --- 接受所有 Minion 的连接请求
salt-key -L                --- 显示当前 Master 上管理的 Key 的列表
```

如果前面的设置都正确，salt-key -L 命令的控制台结果如下：

```
Accepted Keys:
192.168.1.211
192.168.1.212
Denied Keys:
Unaccepted Keys:
Rejected Keys:
```

可以看到，211 主机和 212 主机都已纳入 Master 的管理范围内。

3. 测试 Salt 命令

Salt 命令格式为 salt [options] '<target>' <function> [arguments]，其中的 target

如果为 '*'，则表示对所有 minion 进行操作，function 一般以 "<模块.命令>" 的形式构成。这里请读者一起练练手，熟悉一下 Salt 常规的几个命令的作用。

（1）test.ping 用于测试 Minion 节点的存活性：

```
salt '*' test.ping   --- 本行为命令，后面为控制台输出，本节格式类此
192.168.1.211:
    True
192.168.1.212:
    True
```

（2）cmd.run 用于在 Minion 端执行命令：

```
salt '*' cmd.run 'uptime'
192.168.1.211:
    23:15:31 up 47 min,  1 user,  load average: 0.66, 0.56, 0.49
192.168.1.212:
    23:15:31 up 46 min,  1 user,  load average: 0.01, 0.07, 0.08
```

（3）cmd.script 用于在 Minion 执行脚本(此脚本是放于 Master 指定位置或 http 路径下)。在 /srv/salt/ 目录下新建一个 test.sh 脚本，内容如下：

```
echo 'begin'
cd /tmp
touch salt_test
echo 'finish'
```

然后运行 cmd.script 命令，就可以将存放在 Master 上的脚本传送到 Minion 上去执行。此处使用了 SaltStack 自身的 "salt://" 协议，这个协议默认读取的根目录位置就是 "/srv/salt/"，将要执行的脚本存放于此，就可以通过 "salt://" 来定位脚本位置了：

```
salt '*' cmd.script salt://test.sh
192.168.1.211:
    ----------
    pid:
        11787
    retcode:
        0
    stderr:
    stdout:
        begin
        finish
192.168.1.212:
    ----------
    pid:
```

```
    2171
retcode:
    0
stderr:
stdout:
    begin
    finish
```

如果查看 211 和 212 主机，在"/tmp"目录下会生成一个名为 salt_test 的空文件。cmd.script 模块命令不但支持自创的"salt://"协议，而且支持更大众化的"http://"协议。如果让部署脚本存放于支持"http://"协议的文件仓库服务器上，再安装一个 salt-API 来解耦与 master 的关系，那毕方（BiFang）自动化部署系统就可以远程通过 SaltMaster 指定 SaltMinion 来部署服务了。而这正是毕方（BiFang）自动化部署系统的系统构架的核心思路，希望读者能好好揣摩理解。

接下来就演示一下这种执行方式。

（4）cmd.script 使用"http://"协议远程执行脚本。

现在假定 test.sh 脚本存放于 Master 可访问的一个 http 文件仓库服务器上，其访问地址为 http://192.168.1.211:9002/test.sh。现执行如下命令：

```
salt '*' cmd.script http://192.168.1.211:9002/test.sh
```

如图 4-4 所示的输出和第（3）步上的输出完全相同，但已将执行脚本和 SaltStack 服务解耦，更值得推荐。

图 4-4　cmd.script 命令调用 http 脚本

SaltStack 的更多功能和命令，读者可以通过其官网或其他途径继续学习。其官网的 URL 为 https://docs.saltproject.io/en/latest/。

4.3 启用 Salt-API 功能

4.2 节已配置好 SaltStack Master 和 Mionion，并且测试了几个远程执行命令。本节来启用 Salt-API 功能。

SaltStack 官方提供的 REST API 格式的 Salt-API 项目，将使 SaltStack 与第三方系统集成变得尤为简单。而在毕方（BiFang）自动化部署系统中，使用 Salt-API 可以让毕方的部署与 SaltMaster 服务器解耦。读者可以想想，要是没有 Salt-API，那只能在 SaltMaster 控制 SaltMinion，相当于任何操作都会与 SaltMaster 绑定；而 Salt-API 将 SaltMaster 的功能通过 REST API 的方式暴露出来，那第三方系统（包括毕方）只要 http 可达，就能很方便地控制 SaltMinion 了。

在 4.2 节已经在 211 主机上安装好 Salt-API 了，所以只要完成配置，启动服务就好了。

1. 配置 Salt-API

（1）创建认证用户，并设置密码：

```
useradd -M -s /sbin/nologin saltapi          --- 新建用户 Salt-API
echo 'saltapipwd' | passwd --stdin saltapi   --- 设置用户 Salt-API 的密码为 saltapipwd
```

（2）生成自签名证书用于 Salt-API 的远程连接。

```
salt-call --local tls.create_self_signed_cert    --- 下面为控制台输出
local:
Created Private Key: "/etc/pki/tls/certs/localhost.key."
Created Certificate: "/etc/pki/tls/certs/localhost.crt."
```

（3）打开 SaltMaster 配置文件的包含功能，编辑"/etc/salt/master"，将 default_include 行的"#"注释取消：

```
default_include: master.d/*.conf
```

（4）创建 API 配置文件，将上面生成的证书写到配置文件。新增 /etc/salt/master.d/api.conf 文件，内容如下：

```
rest_cherrypy:
    host: 192.168.1.211
    port: 8000
    ssl_crt: /etc/pki/tls/certs/localhost.crt
    ssl_key: /etc/pki/tls/certs/localhost.key
```

（5）创建 API 认证配置文件。新增 /etc/salt/master.d/auth.conf 文件，内容如下：

```
external_auth:
    pam:
        saltapi:
            - .*
            - '@wheel'
            - '@runner'
            - '@jobs'
```

（6）重启 SaltMaster 和启动 Salt-API。重启服务让上面的配置更改生效。

```
systemctl restart salt-master
systemctl start salt-api
```

2. 测试 Salt-API 功能

简单的测试可以使用 curl 命令。

（1）用 curl 测试 token 生成：

```
curl -sSk https://192.168.1.211:8000/login \
-H 'Accept: application/x-yaml' \
-d username=saltapi \
-d password=saltapipwd \
-d eauth=pam           --- 下面为控制台输出
return:
- eauth: pam
    expire: 1608602843.72114
    perms:
    - .*
    - '@wheel'
    - '@runner'
    - '@jobs'
    start: 1608559643.721139
    token: c1df246e1cda26c337872ea336faa9bdfc203b61
    user: saltapi
```

（2）用 curl 测试 test.ping。使用上一步获取的 token 来执行远程命令：

```
curl -sSk https://192.168.1.211:8000 \
-H 'Accept: application/x-yaml' \
-H 'X-Auth-Token: c1df246e1cda26c337872ea336faa9bdfc203b61' \
-d client=local \
-d tgt='*' \
```

```
-d fun=test.ping    --- 下面为控制台输出
return:
- 192.168.1.211: true
  192.168.1.212: true
```

测试完成，可以确认 Salt-API 功能已正常。

3. 使用 saltypie 第三方库操作 Salt-API

已展示如何使用 curl 命令来操作远程 Salt-API。在之前开发类似应用的经历中，会使用 Python 的 requests 库来手工封装一个操作 Salt-API 的方法库，网上也有一些文档在做此类事。我决定换一种方式，使用 saltypie 这样的第三方库来操作 Salt-API，一来试一下这个库的稳定性，二来免得自己维护一些版本升级的事宜。

saltypie 的下载地址：https://pypi.org/project/saltypie/。

saltypie 的文件地址：https://cathaldallan.gitlab.io/saltypie/。

使用 saltypie 的步骤：

（1）安装 saltypie 库：

```
pip install saltypie
```

（2）测试 saltypie 库：

```
from saltypie import Salt

salt = Salt(
    url='https://192.168.1.211:8000',
    username='saltapi',
    passwd='saltapipwd',
    trust_host=True,
    eauth='pam'
)

exe_return = salt.execute(
    client=Salt.CLIENT_LOCAL,
    target="*",
    fun='test.ping'
)
print('exe_return:', exe_return)
```

代码说明：

● saltypie 的使用相当简单，就是用 Salt-API 的配置初始化一个 salt 操作实例。
- 调用 salt 实例的 execute() 方法，传入要执行的命令即可获取返回值。

执行上面的代码，正常输出如下，和直接在 SaltMaster 执行 test.ping 返回的值一样，只需要稍候解析一下数据结构而已：

```
exe_return: {'return': [{'192.168.1.211': True, '192.168.1.212': True}]}
```

4.4 结合部署脚本、软件包和 saltypie 来实现远程部署

经过本章前面知识点的讲解，实现远程部署已没有什么障碍。本节结合所有的工作，以 go-demo 项目为例将远程部署的步骤串联起来。

4.4.1 测试环境

这里先列一下当前测试环境的服务器及运行服务，如表 4-1 所列。

表 4-1 测试环境的服务器及运行服务

服 务	IP 及端口	位 置
文件仓库服务器	192.168.1.211:9002	—
Go-demo 软件包	192.168.1.211:9002	/go-demo/20210320104629503523FC/go-demo.tar.gz
Go-demo 部署脚本	192.168.1.211:9002	/go-demo/20210320104629503523FC/depoy.sh
SaltMaster	192.168.1.211	—
SaltMinion	192.168.1.211 192.168.1.212	
Salt-API	192.168.1.211:8000	—
测试机器 (saltypie)	192.168.1.10	

再加一个测试环境的部署服务运行流程图，辅助理解，如图 4-5 所示。

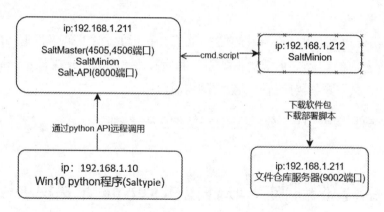

图 4-5 测试环境的部署服务运行流程

网址为 http://192.168.1.211:9002/go-demo/20210320104629503523FC，在服务目录下已存在对应的压缩包和部署脚本：

```
..
deploy.sh,      file, 5237 bytes, last modified 2021-03-20 05:06:29 +0000 UTC
go-demo.tar.gz, file, 3387634 bytes, last modified 2021-03-20 02:47:49 +0000 UTC
```

4.4.2 deploy.sh 部署脚本

本节将讲解 deploy.sh 脚本的内容。这个脚本是为了 demo 而写的，很多脚本健壮性的判断细节并没有体现。如果读者要采用这种模式，部署脚本需要不断加强，加入更多的判断以保证每一步都有正常的执行和回馈。

deploy.sh 网址：https://github.com/aguncn/bifang-go-demo/blob/main/script/deploy.sh。此服务编写的核心思路就是为脚本传入不同的流程参数，就可以实现相应服务的软件拉取、服务停止、软件部署或回滚、服务启动、状态检测等过程。

（1）部署脚本接受的参数。

此部署脚本目前接收如下 7 个参数：

```
# APP 名称
APP_NAME=$1
# RELEASE 发布单参数
RELEASE=$2
# ENV 环境参数
ENV=$3
# APP 压缩包名参数
ZIP_PACKAGE_NAME=$4
```

```
# APP 压缩包的网络下载地址
ZIP_PACKAGE_URL=$5
# PORT 服务端口参数
PORT=$6
# ACTION 服务启停及部署参数
ACTION=$7
```

脚本本身有注释，含义不用过多解释。这里主要关注的是最后一个参数 ACTION($7)。在毕方（Bifang）自动化部署系统的每一次具体的部署过程中，前面6个参数都是固定的；最后一个参数每次调用的都是不同的参数，比如一次正常的部署要先后传递如下参数：

'fetch', 'stop', 'stop_status', 'deploy', 'start', 'start_status', 'health_check'。

现在不理解这些参数不要紧，后面会说明，现在只是留下一个初步印象。

（2）部署脚本的变量及常量：

```
# APP 可执行文件名或包名
PACKAGE_NAME="go-demo"
# APP 部署根目录
APP_ROOT_HOME="/app"
# 可执行文件所有目录
BIN_PATH="./bin"
# APP 软件包保存根目录
LOCAL_ROOT_STORE="/var/ops"

# 由前面的变量组合起来的目录
# APP 安装路径
APP_HOME="$APP_ROOT_HOME/$APP_NAME"
# APP 当前运行版本的压缩包保存路径
LOCAL_STORE="$LOCAL_ROOT_STORE/$APP_NAME/current"
# APP 上一个版本的压缩包保存路径，用于一次性回滚
LOCAL_BACK="$LOCAL_ROOT_STORE/$APP_NAME/backup"
```

这些变量也都有注释。读者要注意的是，对于每一个具体的服务，这些变量有可能需要一些细微调整。比如 PACKAGE_NAME 和 BIN_PATH 这两个变量，如果在后面的函数里用得少，完全可以不用。又或是部署复杂，也可以增加新的变量。其他几个变量如 LOCAL_ROOT_STORE、APP_HOME、LOCAL_STORE 和 LOCAL_BACK 等，最好统一标准化。

将这部分定义在此，并集成进开发同事的代码中，也是 DevOps 的一种体现。让每个开发维护自己服务的启停，以后迁移到云集群上时，开发同事就可

以更顺利地控制自己的服务。

（3）公共函数：

```
#解决对命令的执行结果重复判断问题
shopt -s expand_aliases

alias CHECKRETURN='{
    ret=${?}
    if [ ${ret} -ne 0 ]; then
        read errmsg
        echo ${errmsg}
        return ${ret}
    fi
}<<<'

#获取应用进程 PID，供后面的函数判断之用
PID=$(ps aux |grep "${PACKAGE_NAME}"|grep -v "salt"|grep -v "grep"|awk '{print $2}')
CHECKRETURN "ERROR: 获取应用进程 ID 失败 "
echo $PID
```

shell 编程相对于其他高级语言，对执行错误或意外的处理比较烦琐。这里做一个统一的返回值意外处理。在毕方系统调用 Salt-API 后执行是否成功的重要判断，是返回结果里一定要有"success"关键字，所以在执行函数时，成功了一定要 echo 说明。在 shell 执行完后，要经常判断服务进程是否存在，这里也提取出来，统一处理。

（4）ACTION 参数与对应服务的处理：

```
case "$ACTION" in
    fetch)
        fetch
        ;;
    deploy)
        deploy
        ;;
    rollback)
        rollback
        ;;
    start)
        start
        ;;
    stop)
        stop
```

```
            ;;
        stop_status)
            stop_status
            ;;
        start_status)
            start_status
            ;;
        health_check)
            health_check
            ;;
        restart)
            stop
            start
            ;;
        *)
            echo $"Usage: $0 {7 args}"
esac
```

在部署脚本里,我们将根据 ACTION 不同的参数,调用不同的函数。比如,如果传入的 ACTION 参数为 fetch,则调用 fetch 这个函数,实现软件包的获取;如果传入的 ACTION 参数为 stop,则调用 stop 这个函数,停止服务。其他类似,下面会具体说明。

(5)获取及备份软件包:

```
#先建立相关目录,备份上次部署的软件包,再从构件服务器上获取软件包,保存到指定目录
#只支持一次回滚,若想回滚多次,最好再重新部署之前的发布单
fetch() {
    #判断之前是否存在相关目录,如果没有,再新建
    if [ ! -d $APP_HOME ];then
        mkdir -p $APP_HOME
        CHECKRETURN "ERROR: 建立 $APP_HOME 目录失败 "
    fi
    if [ ! -d $LOCAL_STORE ];then
        mkdir -p $LOCAL_STORE
        CHECKRETURN "ERROR: 建立 $LOCAL_STORE 目录失败 "
    fi
    if [ ! -d $LOCAL_BACK ];then
        mkdir -p $LOCAL_BACK
        CHECKRETURN "ERROR: 建立 $LOCAL_BACK 目录失败 "
    fi
    #删除上上次备份的软件包 ( 无多次回滚 )
    if [ -f "$LOCAL_BACK/$ZIP_PACKAGE_NAME" ];then
```

```bash
            mv $LOCAL_BACK/$ZIP_PACKAGE_NAME /tmp/
            CHECKRETURN "ERROR: 移动上次备份的软件包失败 "
    fi
    # 备份上次的软件包
    if [ -f "$LOCAL_STORE/$ZIP_PACKAGE_NAME" ];then
            mv $LOCAL_STORE/$ZIP_PACKAGE_NAME
            $LOCAL_BACK/$ZIP_PACKAGE_NAME
            CHECKRETURN "ERROR: 备份上次的软件包失败 "
    fi
    # 获取本次的部署包
    wget -q -P $LOCAL_STORE $ZIP_PACKAGE_URL
    CHECKRETURN "ERROR: 获取本次的部署包失败 "

    echo "APP_NAME: $APP_NAME prepare success."
}
```

在每次拉取软件包时先建好目录，再备份好上一次的软件包到 $LOCAL_BACK，再通过 wget 命令下载本次要部署的软件包，保存在本地的 LOCAL_STORE 目录下。因为通过 Salt-API 传过来的参数里已定位好 $ZIP_PACKAGE_URL 软件压缩包的路径，所以这里是可以自动化部署的。

（6）回滚：

```bash
# 回滚，从 BACKUP 目录解压恢复
rollback() {
    rm -rf $APP_HOME/*
    CHECKRETURN "ERROR: 删除当前 APP 应用的目录文件失败 "
    tar -xzvf $LOCAL_BACK/$ZIP_PACKAGE_NAME -C $APP_HOME
    CHECKRETURN "ERROR: 从 $LOCAL_BACK 回滚目录中解决部署包失败 "
    echo "APP_NAME: $APP_NAME rollback success."
}
```

回滚比较简单，因为仅支持一次回滚，所以把 $LOCAL_BACK 目录里的原软件包解压到运行目录即可。

（7）部署：

```bash
# 清除目录已有文件，将 CURRENT 解压到运行目录
deploy() {
    rm -rf $APP_HOME/*
    CHECKRETURN "ERROR: 删除当前 APP 应用的目录文件失败 "
    tar -xzf $LOCAL_STORE/$ZIP_PACKAGE_NAME -C $APP_HOME
    CHECKRETURN "ERROR: 从 $LOCAL_STORE 下载目录中解决部署包失败 "
    # 此处还可以根据传过来的 $ENV,$PORT,$RELEASE 等参数，做一个性化的部署及配置处理
```

```
#甚至可以传更多的参数，进行增量全量，软件包或配置的个别部署处理
echo "APP_NAME: $APP_NAME deploy success."
}
```

部署过程就是将 wget 到 $LOCAL_STORE 目录的软件包解压到服务运行目录 $APP_HOME 中即可。

如果我们在压缩包中包含了关于环境的配置或是其他个性化的部署，都可以在这个函数里实现，传入脚本的 7 个参数都可以使用，如果不够还可以再在脚本里增加更多变量。

（8）启动：

```
#启动应用，传递了 port 和 env 参数
start() {
    #判断是否该 APP 还在运行，如果运行则不启动，报错并返回
    if [ -n "$PID" ]; then
        echo "Project: $APP_NAME is running, kill first or restart, failure start."
        return 1
    fi

    #此处为真正启动命令 (不同的应用，必须重写此处，不能统一处理)
    cd "$APP_HOME"
    nohup $BIN_PATH/$PACKAGE_NAME >/dev/null 2>&1 &
    CHECKRETURN "ERROR: 启动 APP 应用的命令失败 "
    sleep 2
    echo "APP_NAME: $APP_NAME is start success . "
}
```

为了保险起见，在启动之前先查看是否还有以前的进程存在。如果之前进程都已停止，再调用 nohup $BIN_PATH/$PACKAGE_NAME >/dev/null 2>&1 & 命令启动新的服务进程。

此处，不同的应用启动服务也是不同的写法 (如 Java-jar、Python 和 Ruby 等)，所以这个脚本需要开发和运维共同维护好，但以开发为主。在脚本调优或标准化脚本时，才需要运维介入。

（9）停止：

```
stop() {
    if [ -n "$PID" ]; then
        #这里可以加入程序的自然停止命令或是拉出服务集群，实现优雅地停止命令
        #但如果一个程序能直接被 kill -9，但依然能保证数据一致性，服务不受影响，也是鲁
         棒性（Robust）表现，自己权衡
```

```
        kill -9 $PID
        CHECKRETURN "ERROR: 杀死 APP 应用的命令失败 "
        sleep 2
    fi
    echo "APP_NAME: $APP_NAME is success stop."
}
```

此处停止脚本是直接 kill-9。中心思路：部署脚本的维护是个不断完善和提高的过程。

（10）启动状态判断：

```
start_status() {
    if [ -n "$PID" ]; then
        echo "APP_NAME: $APP_NAME is success on running."
    else
        echo "APP_NAME: $APP_NAME is failure on running."
    fi
}
```

服务的启停是以服务进程是否存在为判断依据。

（11）停止状态判断：

```
stop_status() {
    if [ -n "$PID" ]; then
        echo "APP_NAME: $APP_NAME is failure on stop."
    else
        echo "APP_NAME: $APP_NAME is success on stop."
    fi
}
```

（12）服务健康检查：

```
health_check() {
    #此处可以加一些进程是否存在、端口是否开启或是访问指定的 URL 是否存在的功能
    echo "APP_NAME: $APP_NAME is success health."
}
```

作为一个测试环境的 demo 项目，此处直接输出成功。

4.4.3　测试deploy.sh功能

（1）测试 fetch action 功能。

```
from saltypie import Salt
```

```python
salt_url = 'https://192.168.1.211:8000'
salt_user = 'saltapi'
salt_pwd = 'saltapipwd'
eauth = 'pam'
target_list = ['192.168.1.212']
script_url = 'http://192.168.1.211:9002/go-demo/20210320104629503523FC/deploy.sh'
app = 'go-demo'
release = '20210320104629503523FC'
env = 'dev'
zip_package_name = 'go-demo.tar.gz'
zip_package_url = 'http://192.168.1.211:9002/go-demo/20210320104629503523FC/go-demo.tar.gz'
service_port = '9090'
action = 'fetch'


def salt_cmd():

    salt = Salt(
        url=salt_url,
        username=salt_user,
        passwd=salt_pwd,
        trust_host=True,
        eauth=eauth
    )

    arg_list = [script_url, '{} {} {} {} {} {}'.format(app, release, env,
                                                        zip_package_name, zip_package_url,
                                                        service_port, action)]
    exe_return = salt.execute(
        client=Salt.CLIENT_LOCAL,
        target=target_list,
        tgt_type='list',
        fun='cmd.script',
        args=arg_list,
    )
    return exe_return['return']


# 使用 saltypie 来获取返回值, 不自己写 http 请求, 更容易解析结果
ret = salt_cmd()
# salt 找不到服务器, 则返回列表里字典中有一个必为空, 这一步要提前判断
if any(not item for item in ret):
```

```
            print(' 找不到 salt minion 客户端：' + str(ret))
    for server in ret:
        for ip, detail in server.items():
            print('ip: ', ip)
            print('retcode: ', detail['retcode'])
            print('stdout: ', detail['stdout'])
            print('stderr: ', detail['stderr'])
            print('pid: ', detail['pid'])
```

代码解释：

- 先人工定义好连接 Salt-API 连接信息及需要传递给 deploy.sh 脚本的 7 个参数。
- 当执行完 salt.execute() 方法后再解析出返回值输出。

控制台输出结果如下：

```
ip: 192.168.1.212
retcode: 0
stdout:
APP_NAME: go-demo prepare success.
stderr:
pid: 3386
```

同时查看 212 主机上的 /var/ops/go-demo 目录，会发现 deploy.sh 脚本已新建好存储和备份目录，且已将软件包下载到指定的目录，如图 4-6 所示。

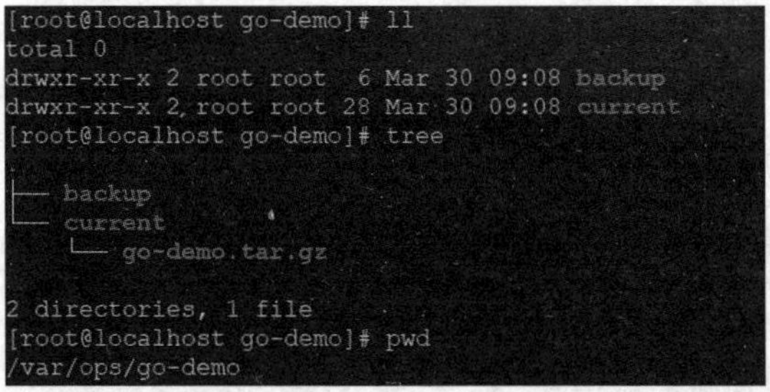

图 4-6　deploy.sh 的 fetch action 参数获取软件包

（2）测试 deploy action 功能。

代码和第（1）步测试一样，只是将 "action = 'fetch'" 更改为 "action = 'deploy'"。执行完本次脚本之后，控制台输出结果如下：

```
ip: 192.168.1.212
retcode: 0
stdout:
```

APP_NAME: go-demo deploy success.
stderr:
pid: 3437

同时,查看 212 主机上的 /app/go-demo 目录,会发现 deploy.sh 脚本已将压缩软件包解压到指定目录内,如图 4-7 所示。

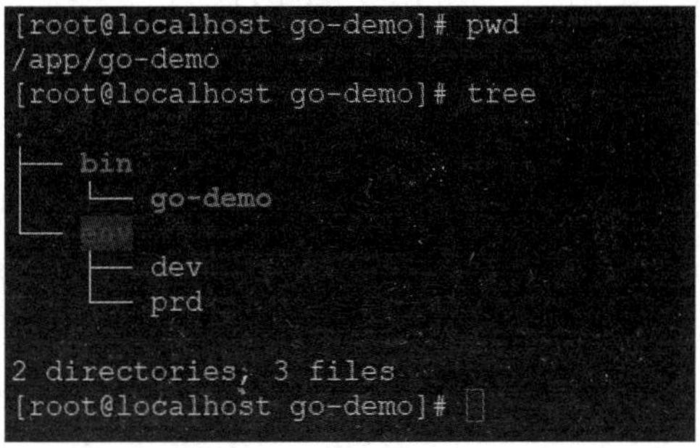

图 4-7 deploy.sh 的 deploy action 参数部署软件

(3)测试 start action 功能。

代码和第(1)步测试一样,只是将"action = 'fetch'"更改为"action = 'start'"。执行完本次脚本之后,控制台输出结果如下:

ip: 192.168.1.212
retcode: 0
stdout:
APP_NAME: go-demo is start success .
stderr:
pid: 3472

同时访问:http://192.168.1.212:9090/,发现 go-demo 服务已启动,说明脚本功能基本正常,如图 4-8 所示。

图 4-8 deploy.sh 的 start action 参数启动服务

4.5 小　结

本章主要讲解了 SaltStack 的知识，以及如何在测试环境部署 SaltStack 的 Master、Minion 及 API 服务；还讲解了如何通过 saltypie 这个第三方库操作 Salt-API 提供的服务。使用 saltypie，结合 cmd.script 使用 "http://" 协议，我们完全地将 Salt 提供的服务和毕方系统调用它的服务实现了解耦。最后通过分析毕方系统提供的较通用的 deploy.sh 脚本的代码讲解和测试实操，进一步强化本章的学习内容，也为后面的毕方系统的开发打下良好的底层知识基础。

到本章为止，从通过 GitLab 进行 CI/CD 到通过 SaltStack 进行自动化远程脚本部署，开发毕方自动化部署系统的前置知识和环境准备已全部完成。从第 5 章开始将进入开发毕方后端服务的内容。

更精彩，不容错过！

第 5 章　Python、Django 与 DRF 的开发环境

哲学家只是用不同的方式解释世界，而问题在于改变世界。

——马克思

前 4 章已将毕方（BiFang）自动化部署系统的功能需求、Demo 项目和脚本化 API 调用达到 CI/CD 的过程讲解清楚。本章将介绍如何搭建起毕方（BiFang）自动化部署系统的后端开发环境，以便进入下一章的代码编写。至于前端开发环境的搭建和代码编写，将在学习完后端后进行。

搭建毕方系统的后端开发环境，主要分为四大块内容：Python、Django、DRF（Django REST framework，本书简称为 DRF）和 IDE（PyCharm）。本章就围绕这四块内容展开，希望读者能跟上学习进度。

在此提醒一点，如果读者前面 4 章的内容还没有完全消化吸收，最好暂时不要进行此章节的学习，回头再实操一下前面 4 章的内容，毕竟"磨刀不误砍柴工"嘛！

5.1　Python 环境安装

这里主要介绍 Windows 环境下的 Python 环境安装。在 Linux 环境下，因为本书测试环境使用的是 CentOS 8，本身自带了 Python 3.6.8 的语言环境，所以省略此步骤。

话说回来，CentOS 被红帽收购后，自 2020 年起改变 CentOS 开源发行版的目的，将其逐渐纳入了 Fedora 到 Redhat 发行的中间环节。这一定位的改变，社区预测之后很多公司将会重新选择 CentOS 8 替代版。关于这一事件，作为 IT 人士保持关注吧。要相信，道路曲折，但明天会更好。

尽管 CentOS 3 已自带 Python 3 语法环境，但在做 Python 的虚拟化环境和 pip 第三方库安装时，Windows 和 Linux 还是有一些差别的，到时会一一说明。

1. Windows 环境下的 Python 安装

（1）下载。

Python 3.8.6 的下载网址：https://www.python.org/downloads/release/python-386/。

Windows 可执行安装包下载网址：https://www.python.org/ftp/python/3.8.6/python-3.8.6.exe。

（2）安装。

下载 python-3.8.6.exe，双击此文件，然后一直单击下一步就可以了。相信会 Python 语言编写的人安装应该不成问题，作者不在此赘述。只建议在选择安装目录时不要用默认的目录，而是选择 C 盘或 D 盘专门的 Python38 目录存放，图 5-1 是我自己的安装目录，供参考。

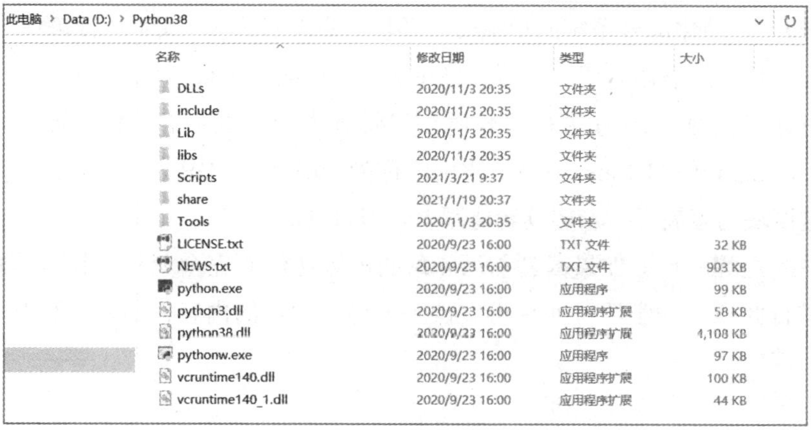

图 5-1 Windows 选择独立的 Python 安装目录

（3）测试。

在 Windows 下进入 cmd 命令模式，输入 python 命令，测试安装是否成功。

```
C:\Users\ccc>python   --- 下面为控制台输出及简单测试
Python 3.8.6 (tags/v3.8.6:db45529, Sep 23 2020, 15:52:53) [MSC v.1927 64 bit (AMD64)] on win32
Type "help", "copyright", "credits" or "license" for more information.
>>> print("hello, python")
hello, python
>>> 1 + 2
3
>>>exit()
```

如果是在 Linux 下，不用安装 Python，可直接输入 python3 命令，测试是否已安装。注意，我安装的 CentOS 8 里没有 python 命令，这里可能需要明确一下 Python 的版本，少犯经验错误吧。如果读者习惯了直接输入 python，也可以通过建软链的方式来解决这个小问题：

```
[root@localhost ~]# python3   --- 下面为控制台输出及简单测试
Python 3.6.8 (default, Aug 24 2020, 17:57:11)
[GCC 8.3.1 20191121 (Red Hat 8.3.1-5)] on linux
Type "help", "copyright", "credits" or "license" for more information.
>>> print('hello, linux python')
hello, linux python
>>> 1 + 5
6
>>>exit()
```

2. Python 虚拟环境的安装

Python 虚拟环境的主要目的是给不同的工程创建互相独立的运行环境。在 Python 虚拟环境下，每个工程都有自己的依赖包，而与其他的工程无关。不同的虚拟环境中同一个包可以有不同的版本，且虚拟环境的数量没有限制。我们之前可以用 virtualenv 或者 pyenv 等工具来创建多个虚拟环境，现在可以轻松地使用 Python 内置的 venv 模块来创建一个 Python 虚拟环境。

网上关于自哪一个 Python 版本开始内置 venv 模块的说法不一，从 3.3 到 3.6 都有。我是实用主义者，我检验的标准：如果使用 venv 内置模块命令报错，那版本就不支持。另外，在安装 Python 虚拟环境时，我的建议是将一个主机上所有的虚拟环境放在同一个目录里进行管理，而不要安装在各自的项目代码下面，因为这会增加源代码管理的难度。

（1）建立 Python 虚拟环境。

Windows 和 Linux 系统中安装 Python 虚拟环境的命令一致，不需要分别说明。输入如下命令，即可建立一个虚拟环境：

```
C:\Users\ccc>python -m venv bifangback_py_venv
```

上面的命令会在当前目录下新建一个名为 bifangback_py_venv 的虚拟环境文件夹，这个文件夹里的文件都是专门为 bifangback_py_venv 这个虚拟环境服务的。

在下面激活这个虚拟环境的步骤之前，先输入 cd 进入这个 bifangback_py_venv 目录。

（2）激活 python 虚拟环境。

Windows（通过运行 Scripts 目录下的脚本进行）：

```
C:\Users\ccc\bifangback_py_venv>Scripts\activate              --- 激活虚拟环境
(bifangback_py_venv) C:\Users\ccc\bifangback_py_venv>         --- 开头的括号表明环境
```

Linux（通过导入 bin/ 下脚本进行）：

```
[root@localhost bifangback_py_venv]# source bin/activate            --- 激活虚拟环境
(bifangback_py_venv) [root@localhost bifangback_py_venv]#           --- 开头的括号表明环境
```

（3）退出 Python 虚拟环境。

Windows（通过运行 Scripts 目录下的脚本进行）：

```
(bifangback_py_venv) C:\Users\ccc\bifangback_py_venv>Scripts\deactivate.bat    --- 退出
C:\Users\ccc\bifangback_py_venv>                                               --- 正常环境
```

Linux（独立命令）：

```
(bifangback_py_venv) [root@localhost bifangback_py_venv]# deactivate           --- 退出
[root@localhost bifangback_py_venv]#                                           --- 正常环境
```

本书后面部分如果没有特殊说明，针对 Python 环境的第三方库安装都在此 Python 虚拟环境下进行，文字或截图都不再另行说明，望读者谨记。

5.2　Django 及 DRF 库安装

通过 5.1 节的学习，假定读者已在自己的开发环境上安装好了 Python，接下来进行 Django 及 DRF 的第三方库的安装，要记得在虚拟环境下进行哟。

Python 的第三方库的安装，一般都是通过 pip 命令进行的（之前还有一种 easy_install.py 安装）。pip 是 Python 包管理工具，该工具提供了对 Python 包的查找、下载、安装和卸载的功能。目前版本的 Python 安装包已自带该工具。

pip 的安装是相当灵活的，它既可以从 pypi（国外）直接下载安装，也可以设置镜像源从国内加速下载安装；它既可以从目录下有 setup.py 的安装，也可以从一个 tar.gz 或 zip 压缩包安装；它既可以直接从一个编译好的 whl 文件安装，也可以在安装过程中进行二进制编译再安装；它既可以从一个 requirements.txt 文件列表中安装，也可以在这个文件中区分不同的操作系统进行安装；它既可以指定版本进行安装，也可以在安装的过程中将依赖的其他第三方库一并安装。总之，非常好！

5.2.1 设置pip国内镜像源

如果不设置 pip 的安装源，其默认是从 https://pypi.org/ 网站上拉取下载。但这个网站在国外，于是国内很多人就自动同步了这个仓库内容。我们将安装源设置为国内的话，下载速度会提高很多。

现在各行各业都在使用这个方法，在国内建同步镜像站来加速国外的资源。君不见，maven、npm、goproxy、docker 等都在使用相应的对策。条件好的公司甚至可以在公司内部也各搞一套镜像源，私家花园加游泳池，才是人生赢家。

（1）临时使用，在 pip 命令里跟上镜像源地址：

```
pip install -i https://pypi.tuna.tsinghua.edu.cn/simple numpy
```

（2）Windows 中，%APPDATA%\pip\pip.ini 文件内容：

```
[global]
index-url = https://pypi.tuna.tsinghua.edu.cn/simple
[install]
trusted-host = https://pypi.tuna.tsinghua.edu.cn
```

（3）Linux 中，~/.pip/pip.conf 文件内容：

```
[global]
index-url = https://pypi.tuna.tsinghua.edu.cn/simple
[install]
trusted-host = https://pypi.tuna.tsinghua.edu.cn
```

（4）当前国内可能 pip 镜像源：

阿里云：http://mirrors.aliyun.com/pypi/simple/。

豆瓣：http://pypi.douban.com/simple/。

清华大学：https://pypi.tuna.tsinghua.edu.cn/simple/。

中国科学技术大学：http://pypi.mirrors.ustc.edu.cn/simple/。

华中科技大学：http://pypi.hustunique.com/。

5.2.2 Django安装

Django 是 Python 编程语言驱动的一个开源模型—视图—控制器 (model

view controller，MVC）风格的 Web 应用程序框架。使用 Django 可以在几分钟内就创建一个高品质、易维护、数据库驱动的应用程序。

目前，国内使用 Python 开发 Web 的框架，最主要使用的就是 Django 和 Flask，其各有特征。Django 体系完善，但学习曲线陡峭一些。Flask 小巧灵活，但在大一点的项目中也要不断整合各种第三方库。

毕方（BiFang）自动化部署系统选择的是 Django 框架，本书选用的版本是写代码时的最新版本 3.1.4，其官网：https://www.djangoproject.com/。Django 的安装很简单，使用如下 pip 命令，即可完成安装：

```
pip3 install django==3.1.4   --- 安装 Django 指定版本，后面为控制台输出
Collecting django==3.1.4
Downloading https://pypi.tuna.tsinghua.edu.cn/packages/08/c7/7ce40e5a5cb47ede081b9fa8a3dd93d101c884882ae34927967b0792f5fb/Django-3.1.4-py3-none-any.whl (7.8MB)
100% |████████████████████████████████| 7.8MB 245kB/s
Requirement already satisfied: pytz in /usr/lib/python3.6/site-packages (from django==3.1.4)
Collecting sqlparse>=0.2.2 (from django==3.1.4)
Downloading https://pypi.tuna.tsinghua.edu.cn/packages/14/05/6e8eb62ca685b10e34051a80d7ea94b7137369d8c0be5c3b9d9b6e3f5dae/sqlparse-0.4.1-py3-none-any.whl (42kB)
100% |████████████████████████████████| 51kB 7.1MB/s
Collecting asgiref<4,>=3.2.10 (from django==3.1.4)
Downloading https://pypi.tuna.tsinghua.edu.cn/packages/89/49/5531992efc62f9c6d08a7199dc31176c8c60f7b2548c6ef245f96f29d0d9/asgiref-3.3.1-py3-none-any.whl
Installing collected packages: sqlparse, asgiref, django
Successfully installed asgiref-3.3.1 django-3.1.4 sqlparse-0.4.1
```

从控制台输出的信息可以看到，pip 使用了国内清华大学的镜像源下载。并且在安装 Django 的过程把它所依赖的其他第三方库也一并下载安装完成了。

5.2.3　Django REST framework（DRF）的安装

在 5.2.2 小节已经安装完成了 Django 框架，按照通常的想法，那就应该可以写毕方的后台应用了，那为什么还要安装 DRF 呢？

绝对观点来说，我们可以不安装 DRF，只使用 Django 来进行后端的开发。但再想想另一层，在做企业开发时，我们希望使用的工具不能只是完成，还要快速规范地完成！也就是除了要质量，还要效率！如果每一个后端 API 都需要自己来实现序列化、反序列化、请求用户的获取、请求 token 的解析，还有 API 的限流、Models 表字段的封装等工作，会使我们的开发进度很慢且自己写这些容易出错和重复，不一定是最佳实践。

在这样的场景下，DRF 就很有必要了。DRF 是在 Django 框架的基础上进行了二次开发，植入了更多开箱即用的功能，是一个用于构建 Web API 的强大而又灵活的工具。其主要特点如下：

- 提供了定义序列化器 Serializer 的方法，可以快速根据 Django ORM 或者其他库自动序列化 / 反序列化。
- 提供了丰富的类视图、Mixin 扩展类，简化视图的编写。
- 丰富的定制层级：函数视图、类视图、视图集合到自动生成 API，满足各种需要。
- 多种身份认证和权限认证方式的支持。
- 内置了限流系统。
- 直观的 API web 界面。
- 可扩展性，插件丰富。

开发毕方（BiFang）自动化部署系统，主要用的就是 DRF 的功能。当然，在一个具体场景下也有使用 Django 本身提供的功能写的 API。本书选用的版本是写代码时的最新版本 3.12.2，其官网：https://www.django-rest-framework.org/。

DRF 的安装很简单，使用如下 pip 命令即可完成安装：

```
pip3 install djangorestframework==3.12.2            --- 安装 DRF 指定版本，后面为控制台输出
Collecting djangorestframework==3.12.2
Downloading https://pypi.tuna.tsinghua.edu.cn/packages/8e/42/4cd19938181a912150e55835109b19
33be26b776f3d4fb186491968dc41d/djangorestframework-3.12.2-py3-none-any.whl (957kB)
100% |████████████████████████████████| 962kB 1.8MB/s
Requirement already satisfied: django>=2.2 in /usr/local/lib64/python3.6/site-packages (from
djangorestframework==3.12.2)
Requirement already satisfied: sqlparse>=0.2.2 in /usr/local/lib/python3.6/site-packages (from
django>=2.2->djangorestframework==3.12.2)
Requirement already satisfied: asgiref<4,>=3.2.10 in /usr/local/lib/python3.6/site-packages (from
django>=2.2->djangorestframework==3.12.2)
Requirement already satisfied: pytz in /usr/lib/python3.6/site-packages (from django>=2.2-
>djangorestframework==3.12.2)
Installing collected packages: djangorestframework
Successfully installed djangorestframework-3.12.2
```

从控制台看到 Successfully 字样，便知 DRF 已成功安装。

5.3　安装开发毕方（BiFang）所有的第三方库

在习得 5.2 节的技巧后，相信读者可以把一般的 Python 第三方库通过 pip 命令安装到虚拟环境中了。"百尺竿头须进步，十方世界是全身。"现在把难度再升一级，把 pip 的安装向最佳实践推进一步，看看如何以 requirements 文件的形式进行安装。

以 requirements 文件的形式进行安装第三方库，这个需求的场景是什么呢？

想象一下这个场景：我在自己的开发环境开发完了毕方 (BiFang) 系统，其中使用了很多第三方库，那么，如何将我使用的第三方库和每个库特定的版本很好地传递给读者或是我的其他开发伙伴呢？如果我换电脑进行开发了呢？这不能每次都口传心授、大脑记忆、运维程序、差什么再安装什么或是发邮件说明吧？

这时，有一个 requirements.txt 文件无疑是很好的沟通手段，在这个文件中会列明毕方 (BiFang) 系统开发过程中使用的所有第三库和相应的版本。有了这个文件，它可以让读者或开发伙伴在自己的电脑上构建出和我完全相同的开发环境，最大限度地保持开发的连贯性和运行的稳定性。

当然，如果读者有其他语言的开发经验，自然知道 Java 开发当中的 pom.xml 文件、Node.js 开发过程中的 package.json 文件以及 Golang 开发当中的 go.mod 文件。它们这些文件的作用和 Python 开发当中的 requirements.txt 文件相同。这说明，这种依赖的固定，已是软件开发行业中的一种最佳实践。

不过，我得在这里吐槽一下 Python 的依赖管理，自动化程度没有其他开发语言效率高。安装依赖的时候不会自动维护 requirements.txt 文件 (Node.js)。并且依赖管理文件的机制没有强制性，我个人认为最好的管理模式还是先加入依赖列表文件，才可以自动下载再在代码中导入使用 (maven 可以)。

1. 生成开发环境的第三方库依赖列表文件

（1）导出依赖库列表。

freeze 命令可以冻结当前环境的依赖包。下面的命令可以将当前 Python 环境的第三方库依赖列表导入 requirements.txt 文件中：

```
pip3 freeze > requirements.txt
```

（2）查看当前 requirements.txt 内容。

如果只安装 5.2 节的 Django 及 DRF，当前 requirements.txt 文件的内容如下：

```
asgiref==3.3.1
Django==3.1.4
djangorestframework==3.12.2
pytz==2021.1
sqlparse==0.4.1
```

读者从这个文件里可以了解 requirements 的格式。可以看到文件的每一行都标明了依赖库的名称和版本，中间使用"=="表明固定具体版本。

2. 按 requirements.txt 文件导入第三方库

有了上一节生成的 requirements.txt 文件，在其他的计算机上可以使用如下命令，轻松地安装这个文件中的所有依赖库列表：

```
pip3 install -r requirements.txt
```

这时如果再使用 pip3 freeze 命令查看(不用在末尾导入文件)，就会发现 Django、DRF 以及它们的依赖库已经安装在计算机上了，如果我们将 requirements.txt 纳入 Git 源代码管理系统，开发环境的重建也是比较方便的。

3. 按操作系统类型分别安装 requirements.txt

按前两节介绍的 requirements.txt 文件的操作，这个知识点就讲完了吗？并没有！还有一种情况是我们在安装 Python 依赖库时会遇到的，就是同一个依赖库在不同的操作系统里，安装方式是不一样的。比如 mysqlclient 这个依赖库是 python3 后连接 mysql 或 mariadb 数据库的推荐使用库，这个库在 Windows 环境推荐使用 whl 单文件安装，而在 Linux 环境中需要正常的编译安装。那这种文件能在一个 requirements.txt 文件内解决吗？还是需要维护两个不同的文件？

针对这种情况，requirements.txt 文件也是可以解决的，但需要一些特殊的写法，按如下写法编写的 requirements.txt 文件就可以解决 mysqlclient 库的安装问题：

```
mysqlclient @ file:///D:/mariadb/mysqlclient-1.4.6-cp38-cp38-win_amd64.whl; platform_system == "Windows"
mysqlclient==1.4.6; platform_system == "Linux"
```

当然，这样的写法需要在 Windows 里先将 mysqlclient-1.4.6-cp38-cp38-win_amd64.whl 文件放在指定的位置。在 Linux 里使用 dnf 命令将 python3-devel、 mysql-devel 和 gcc 对应系统库依赖先安装上。

当使用这种写法之后，pip3 install -r requirements.txt 命令就会在 Windows 和 Linux 控制上产生如下输出：

> Ignoring mysqlclient: markers 'platform_system == "Linux"' don't match your environment
> ===================== 上面为 windows 输出，下面为 linux 输出 =================
> Requirement 'mysqlclient @ file:///D:/mariadb/mysqlclient-1.4.6-cp38-cp38-win_amd64.whl' looks like a filename, but the file does not exist
> Ignoring mysqlclient: markers 'platform_system == "Windows"' don't match your environment

4. 毕方（BiFang）的 requirements 文件

在开发毕方（BiFang）自动化部署系统的过程中，作者使用了很多第三方依赖库，每个库都有自己特定的作用。这里先给出一个总的列表，让读者有个总体印象，在本书后面使用到时再介绍这个库的作用。

网址：https://github.com/aguncn/bifang/blob/main/bifangback/requirements.txt。

```
asgiref==3.3.1
certifi==2020.12.5
cffi==1.14.5
chardet==4.0.0
click==7.1.2
colorama==0.4.4
colorclass==2.2.0
coreapi==2.3.3
coreschema==0.0.4
Django==3.1.4
django-cors-headers==3.6.0
django-filter==2.4.0
django-simple-history==2.12.0
djangorestframework==3.12.2
djangorestframework-jwt==1.11.0
drf-yasg==1.20.0
gevent==21.1.2
greenlet==1.0.0
gunicorn==20.1.0
h11==0.12.0
httpcore==0.12.2
httpx==0.16.1
idna==2.10
inflection==0.5.1
itypes==1.2.0
Jinja2==2.11.2
```

```
MarkupSafe==1.1.1
mysqlclient @ file:///D:/mariadb/mysqlclient-1.4.6-cp38-cp38-win_amd64.whl; platform_system == "Windows"
mysqlclient==1.4.6; platform_system == "Linux"
packaging==20.8
pycparser==2.20
PyJWT==1.7.1
pyparsing==2.4.7
python-gitlab==2.5.0
pytz==2020.5
requests==2.25.1
rfc3986==1.4.0
ruamel.yaml==0.16.12
ruamel.yaml.clib==0.2.2
saltypie==0.14.0
six==1.15.0
sniffio==1.2.0
sqlparse==0.4.1
termcolor==1.1.0
terminaltables==3.1.0
uritemplate==3.0.1
urllib3==1.26.2
uvicorn==0.13.3
zope.event==4.5.0
zope.interface==5.3.0
```

5.4 PyCharm 安装配置

当前面安装步骤完成后，开发环境的准备只差一个 IDE 就全部到位了。本节将讲解 PyCharm（IDE）社区版的安装。至于另一个流利的 Python 编辑器 vscode（Visual Studio Code），读者可以自行学习了解。

1. PyCharm 安装

下载地址：https://www.jetbrains.com/pycharm/download/other.html。

本书开发环境使用的版本为 Community Edition（社区版）2020.2。读者选择新的版本也不会有问题，一般版本越高功能越强。

安装流程在此不表，安装完成后查看菜单 Help → About，可看到 PyCharm 版本信息：

```
PyCharm 2020.2.3 (Community Edition)
Build #PC-202.7660.27, built on October 6, 2020
Runtime version: 11.0.8+10-b944.34 amd64
VM: OpenJDK 64-Bit Server VM by JetBrains s.r.o.
Windows 10 10.0
GC: ParNew, ConcurrentMarkSweep
Memory: 1902M
Cores: 12
```

2. 建立毕方（BiFang）项目

在继续操作 PyCharm 前，先使用 Django 建立好毕方（BiFang）项目的目录，然后再继续操作使之更有针对性。

（1）新建项目。

进入 Python 虚拟环境，到 D:\Code\bifang 目录（此处以我的计算机的实际目录演示）下运行如下命令，生成 bifangback 项目目录：

```
django-admin startproject bifangback
```

这行代码将会在当前 D:\Code\bifang 目录下创建一个 bifangback 目录。来看看 startproject 创建了什么：

```
bifangback/
    manage.py
    bifangback/
        __init__.py
        asgi.py
        settings.py
        urls.py
        wsgi.py
```

这些目录和文件的用处：

- 最外层的 bifangback/ 根目录：只是项目的容器，Django 不关心它的名字，可以将它重命名为任何你喜欢的名字。
- manage.py：一个可以用各种方式管理 Django 项目的命令行工具。
- 里面一层的 bifangback/ 目录：包含你的项目，它是一个纯 Python 包。它的名字就是当你引用它内部任何东西时需要用到的 Python 包名（比如 bifangback.urls）。
- bifangback/__init__.py：一个空文件，告诉 Python 这个目录应该被认为是一个 Python 包。如果你是 Python 初学者，可以阅读官方文档中的更多关于包的知识。

- bifangback /settings.py：Django 项目的配置文件。
- bifangback /urls.py：Django 项目的 URL 声明就像你的网站"目录"。
- bifangback/wsgi.py：作为项目运行在 WSGI 兼容的 Web 服务器上的入口。在本书附录里使用 nginx 和 gunicorn 部署毕方（BiFang）项目，将需要指定向文件的配置。
- bifangback/asgi.py：Django3 版本新增的文件作为项目运行在 ASGI 兼容的 Web 服务器上的入口。它用于服务 Django 项目中的异步处理代码，在毕方（BiFang）项目中没有用到这类功能，暂不关注。

（2）启动项目。

当建好 bifangback 项目后，使用下面的命令启动服务：

```
python manage.py runserver                    --- 下面为控制台输出
Watching for file changes with StatReloader
Performing system checks...

System check identified no issues (0 silenced).

You have 18 unapplied migration(s). Your project may not work properly until you apply the migrations
           for app(s): admin, auth, contenttypes, sessions.
Run 'python manage.py migrate' to apply them.
April 01, 2021 - 16:50:40
Django version 3.1.4, using settings 'bifangback.settings'
Starting development server at http://127.0.0.1:8000/
Quit the server with CTRL-BREAK.
```

在浏览器中输入：http://127.0.0.1:8000，即可访问此服务。如果在网页里看到一个待飞的小火箭，说明服务启动成功，如图 5-2 所示。

3. PyCharm 配置 Python 虚拟环境

我们已安装好 PyCharm，且已建好 bifangback 项目，现在就可以启动 PyCharm 并打开 bifangback 所在的目录，在这个目录下编辑已有的代码文件，或是新建自己的代码文件进行编辑工作了。Hold on! 还有点细活儿没做完。我们还需要为 PyCharm 里 bifangback 项目配置 Python 虚拟环境。配置好了这个虚拟环境才可以让这个 IDE 不仅有常见的自动缩进、空行提醒和错误检查等功能，还可以进入依赖库检查，提示依赖库里的方法参数和调用规则等更多高级功能，从而加快开发速度，减少错误。

图 5-2　启动 Django 服务

（1）查看或修改环境配置。

单击 PyCharm 的菜单 File → settings，定位到 bifangback 项目的 Python 解释器，如果其解释器已是虚拟环境的 Python，说明此配置正确，如图 5-3 所示。

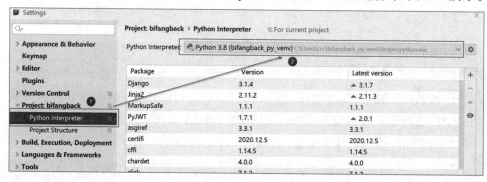

图 5-3　查看 bifangback 的 Python 虚拟环境设置

如果 Python 解释器配置得不对，则需要在这个界面里编辑配置，将之前存在的 Python 虚拟环境加进来，如图 5-4 所示。

第 5 章　Python、Django 与 DRF 的开发环境

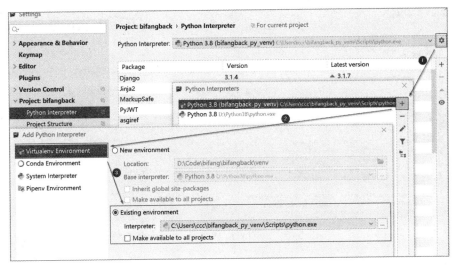

图 5-4　新增 bifangback 的 Python 虚拟环境

（2）让 PyCharm 管理 bifangback 的启停服务。

当 PyCharm 里有 Python 虚拟环境时，IDE 就帮我们在以后的编码中提高了很多工作效率。现在做个小小的进步，让 PyCharm 也管理起 bifangback 项目的服务启停命令吧，这样我们就不用在 Windows 的 command 终端控制台和 PyCharm 之间切来切去了。

单击菜单 Run → Run…，选择 Edit Configurations…，在跳出来的窗口中定义好 manage.py 的路径和运行参数，再给它命名（runserver）就完成了，如图 5-5 所示。

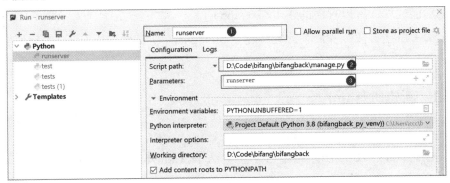

图 5-5　PyCharm 管理 bifangback 服务启停

经此设置，以后就可以通过 PyCharm 里的按钮进行 bifangback 项目的服务启停了，如图 5-6 所示。

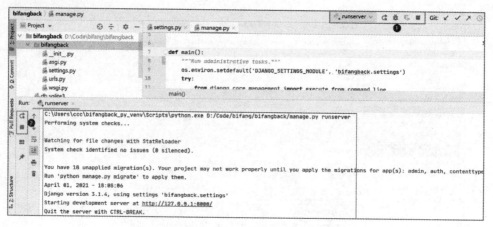

图 5-6　PyCharm 操作 bifangback 启停

5.5　新建 bifangback 项目的 app

在 Django 开发中有两个主要的结构概念：一是 project，二是 app。我们在 5.4 节中通过 django-admin startproject 命令新建了一个 project。而本节就来新建 bifangback 项目所需要的 app。在层级上，project 是 Django 的顶层，在一个 project 里面，可以有多个 app。

从区别上来看，project 包含一些全局配置，这些配置构成一个全局的运行平台，各个 app 都运行在这个全局的运行平台上，而 app 代表的是一个相对独立的功能模块，所以程序的逻辑都在 app 中。

为了讲述方便，我们在此统一建立 bifangback 所需要的所有 app，后面的章节再来填充每个 app 的具体内容。

5.5.1　新建所有app

（1）命令行新建所有 app。

在 windows command 命令行中，进入 D:\Code\bifang\bifangback 目录运行如下命令：

```
django-admin startapp account        ---bifangback 的账号相关功能
django-admin startapp app            ---bifangback 的应用组件增删查改
django-admin startapp project        ---bifangback 的项目增删查改
```

django-admin startapp cmdb	---bifangback 的数据库 models 管理
django-admin startapp deploy	---bifangback 的部署
django-admin startapp env	---bifangback 的环境流转
django-admin startapp gitapp	---bifangback 的 git 代码管理器列表显示
django-admin startapp history	---bifangback 的服务器历史和发布单历史功能
django-admin startapp release	---bifangback 的发布单相关功能
django-admin startapp server	---bifangback 的服务器增删查改
django-admin startapp permission	---bifangback 的权限管理
django-admin startapp stats	---bifangback 的数据统计功能

运行完这些命令后，文件目录结构如图 5-7 所示，其他 11 个 app 目录的文件结构和图 5-7 中展开的 account 应用的结构相同。

图 5-7　bifangback 新建所有 app 后的文件目录结构

每一个 app（恰巧，我们也新建了一个名称为 bifangback app 的 Django app，两个 app 的使用场景不要搞混）目录下的文件作用如下：

- -init-.py：初始化文件，同样也标志 app 可以被引用。
- admin.py：它是我们的后台管理工具，后期可以通过它管理我们的 model 和数据库。由于在 bifangback 项目里统一了 models 数据库表的生成，均放在了 cmdb 这个 app 里，所以其他 app 里的这个 admin.py 文件均为空。
- apps.py：这个是 Django 生成的 app 名称的文件，这个名称会在下面

settings.py 文件里的 INSTALLED_APPS 段使用。
- models.py：模型文件里面放的都是数据库表的映射。由于在 bifangback 项目里统一了 models 数据库表的生成，并都放在了 cmdb 这个 app 里，所以其他 app 里的 models.py 文件均为空。
- tests.py：放置测试文件。
- views.py：放置视图函数文件。
- migrations：数据迁移目录，它负责迁移文件、生成数据库表数据，后期要使用它去结合 models 生成数据库表。在 bifangback 项目里统一了 models 数据库表的生成，并都放在了 cmdb 这个 app 里，其他 app 目录里的 migrations 目录内容均为空，方便集中管理。

关于 Django 的 app，我也有话要说。这些生成的 app 目录才是一种脚手架快速生成功能。如果理解了 Django 的文件结构，手工一个个建好文件和目录是可以的。

另外，app 目录里的 views.py 文件是 Django 预命名的。在很多复杂应用中，我们需要新建一些类似 views 命名的文件，以区分同一个 app 里很多不同的子功能模块。只要在 urls 路由文件里能定位到这些 views 文件即可。

现在讲这些可能有些虚，当是立此存照，等读者学完本书，再来重温这里的内容吧。

（2）将新建的应该注册到 project 的配置文件中。

打开 D:\Code\bifang\bifangback\bifangback 目录下的 settings.py 文件，这个文件是设置 bifangback 全局性变量的。将 INSTALLED_APPS=[…] 这个列表内容更新为如下内容：

```
INSTALLED_APPS = [
    'account.apps.AccountConfig',
    'app.apps.AppConfig',
    'project.apps.ProjectConfig',
    'cmdb.apps.CmdbConfig',
    'deploy.apps.DeployConfig',
    'env.apps.EnvConfig',
    'gitapp.apps.GitappConfig',
    'history.apps.HistoryConfig',
    'release.apps.ReleaseConfig',
    'server.apps.ServerConfig',
    'permission.apps.PermissionConfig',
```

```
    'stats.apps.StatsConfig',
    'django.contrib.admin',
    'django.contrib.auth',
    'django.contrib.contenttypes',
    'django.contrib.sessions',
    'django.contrib.messages',
    'django.contrib.staticfiles',
    'rest_framework',
    'django_filters',
    'corsheaders',
    'drf_yasg',
    'simple_history',
]
```

其中，新建的 12 个 app 在列表的开头，列表是以 Django 开头的 app，是 Django 框架内置的 app，很多功能可以开箱即用。而列表中最后 5 个 app 是我们安装了几个第三方依赖库后，需要注册的（在 requirementes.txt 文件列表中存在）。

rest_framework：DRF 安装后需要在这里注册。

django_filters：用于实现更灵活的基于 URL 参数的过滤功能。

corsheaders：处理跨域请求的依赖库。

drf_yasg：为 DRF 提供一个类似 swagger 的 API 自动文档功能的依赖库。

simple_history：能自动记录所有 models 操作的历史。

（3）更新 project 的配置文件中的中间件配置。

Django 中的中间件（middleware）是一个镶嵌到 Django 的 request/response 处理机制中的一个 hooks 框架，是一个修改 Django 全局输入/输出的底层插件系统。由于新增了 corsheaders 和 simple_history 两个依赖库，它们也要求更新 Django 配置文件中的中间件。

将 settings.py 文件中 MIDDLEWARE =[…] 这个列表内容更新为如下内容：

```
MIDDLEWARE = [
    'django.middleware.security.SecurityMiddleware',
    'django.contrib.sessions.middleware.SessionMiddleware',
    'corsheaders.middleware.CorsMiddleware',
    'django.middleware.common.CommonMiddleware',
    'django.middleware.csrf.CsrfViewMiddleware',
    'django.contrib.auth.middleware.AuthenticationMiddleware',
    'django.contrib.messages.middleware.MessageMiddleware',
```

```
'django.middleware.clickjacking.XFrameOptionsMiddleware',
'simple_history.middleware.HistoryRequestMiddleware',
]
```

相对于默认的中间件处理流程，多了 CorsMiddleware 和 HistoryRequestMiddleware。

（4）修改时区和语言。

这一步是为了对中文及东八区更好地支持，避免出现中文乱码。修改 settings 文件中的 LANGUAGE_CODE 和 TIME_ZONE 即可：

```
LANGUAGE_CODE = 'zh-hans'
TIME_ZONE = 'Asia/Shanghai'
```

5.5.2 整合bifangback项目的urls路由

通过前面的操作，新建的 app 及一些依赖库的设置已注册到 bifangback 的配置中去了。现在可以在这里了解一下 Django 框架的组件和处理流程，把前面的一些知识串起来，再看看还有哪些知识需要去补全。

Django 框架——整体使用一种称为 MTV 模式、保持各组件之间松耦合关系。

- Model（模型）：负责业务对象与数据库的对象（ORM）。
- Template（模版）：负责如何把页面展示给用户，由于 bifangback 是前后端分离的开发模式，Django 后端只提供数据流，不提供网页渲染，所以本书不涉及这块内容。
- View（视图）：负责业务逻辑，并在适当的时候调用 Model 和 Template。同样，基于上面的理由，View 在本书中只会调用 Model，而不会调用 Template。

此外，Django 还有一个 URL 分发器，它的作用是将一个个 URL 页面请求分发给不同的 view 请求，view 再调用相应的 Model。Django 框架的处理流程如图 5-8 所示。

用户访问首先通过 urls 进行访问路径的匹配，然后转到匹配到的方法或函数，再转到 views 进行逻辑处理。若需要访问数据库，则通过 models 访问，获取需要的数据后返回给 views。views 处理完后，若是 API 调用方式直接返回数据给用户（这是 bifangback 项目的处理方式），否则通过 template 对指定的模板进行渲染，然后将对应的 html 返回给页面。

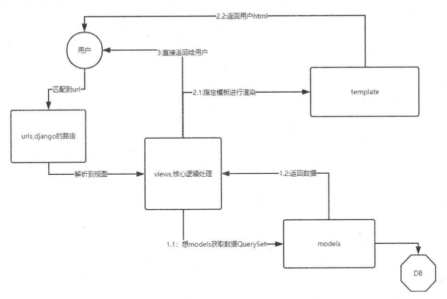

图 5-8　Django 框架处理流程

那么，Django 框架是如何定义 urls 路由的呢？在 settings.py 配置文件中，有如下一项配置：

ROOT_URLCONF = 'bifangback.urls'

这个配置表示，在默认情况下处理路由的就是 D:\Code\bifang\bifangback\bifangback 目录里的 urls.py 文件。当项目扩大时，使用一个 urls.py 文件来管理所有的路由，会显得臃肿和混乱。此时一般采用的策略是分而治之，也就是让每个 app 目录里面都有一个自己的 urls.py 文件，它负责这个 app 里的路由；然后，再将所有 app 目录里的 urls.py 文件汇总给 D:\Code\bifang\bifangback\bifangback 目录里的 urls.py 文件，从而达到了分而治之的目的。这个总的 urls.py 文件里只用维护一些第三方依赖库的路由，及包括各 app 的子路由文件即可。

下面就来实现这样一个路由文件的整合吧。

（1）建立各个 app 自身的 urls.py 文件。

以 cmdb 这个 app 为例，在 cmdb 目录下新建一个 urls.py 文件，其文件内容暂时如下：

```
from django.urls import path
app_name = "cmdb"
```

```
urlpatterns = [

]
```

其他 app 目录下依次建立。可以看到这个 urls.py 相当于一个占位的文件，里面的路由暂时为空，待以后再来填充这个文件。

（2）总的 urls.py 包含各个 app 的子路由文件。D:\Code\bifang\bifangback\bifangback\urls.py，如下所示：

```
from django.contrib import admin
from django.urls import path
from django.urls import re_path
from django.urls import include
from rest_framework import permissions
from drf_yasg.views import get_schema_view
from drf_yasg import openapi

schema_view = get_schema_view(
    openapi.Info(
        title=" 毕方 (Bifang) 自动化部署平台 API",
        default_version='v1',
        description=" 毕方 (Bifang) 自动化部署平台，引导你进入 devops 开发的领域。",
        terms_of_service="https://github.com/aguncn/bifang",
        contact=openapi.Contact(email="aguncn@163.com"),
        license=openapi.License(name="BSD License"),
    ),
    public=True,
    permission_classes=(permissions.AllowAny,),
)

urlpatterns = [
    path('admin/', admin.site.urls),
    path('api-auth/', include('rest_framework.urls')),
]

urlpatterns += [
    re_path(r'^swagger(?P<format>\.json|\.yaml)$', schema_view.without_ui(cache_timeout=0), name='schema-json'),
    path('swagger/', schema_view.with_ui('swagger', cache_timeout=0), name='schema-swagger-ui'),
    path('redoc/', schema_view.with_ui('redoc', cache_timeout=0), name='schema-redoc'),
]

urlpatterns += [
```

```
    path('account/', include('account.urls')),
    path('app/', include('app.urls')),
    path('project/', include('project.urls')),
    path('cmdb/', include('cmdb.urls')),
    path('deploy/', include('deploy.urls')),
    path('env/', include('env.urls')),
    path('gitapp/', include('gitapp.urls')),
    path('history/', include('history.urls')),
    path('release/', include('release.urls')),
    path('server/', include('server.urls')),
    path('permission/', include('permission.urls')),
    path('stats/', include('stats.urls')),
]
```

可以看到在总的 urls.py 文件中，除了依赖库的 DRF、swagger 的路由设置，其他都只是包含了各个子 app 里的 urls.py 文件。后期开发时增加各自 app 下的路由即可。

（3）测试。

由于下一章要讲解毕方（BiFang）自动化部署系统的数据库设计和对 Django Model 模型的 ORM 操作，这里先将 Django 系统自身的数据库合并吧。运行下面两条命令，将完成此任务：

```
python manage.py makemigrations     --- 将 model 变化记录成文件
python manage.py migrate            --- 将记录的文件合进数据库，下面为控制台输出
Operations to perform:
  Apply all migrations: admin, auth, contenttypes, sessions
Running migrations:
  Applying contenttypes.0001_initial... OK
  Applying auth.0001_initial... OK
  Applying admin.0001_initial... OK
  Applying admin.0002_logentry_remove_auto_add... OK
  Applying admin.0003_logentry_add_action_flag_choices... OK
  Applying contenttypes.0002_remove_content_type_name... OK
  Applying auth.0002_alter_permission_name_max_length... OK
  Applying auth.0003_alter_user_email_max_length... OK
  Applying auth.0004_alter_user_username_opts... OK
  Applying auth.0005_alter_user_last_login_null... OK
  Applying auth.0006_require_contenttypes_0002... OK
  Applying auth.0007_alter_validators_add_error_messages... OK
  Applying auth.0008_alter_user_username_max_length... OK
  Applying auth.0009_alter_user_last_name_max_length... OK
```

Applying auth.0010_alter_group_name_max_length... OK
Applying auth.0011_update_proxy_permissions... OK
Applying auth.0012_alter_user_first_name_max_length... OK
Applying sessions.0001_initial... OK

此时，再启动 bifangback 服务（通过 IDE 或控制台）。如果一切配置均完成，就可以看到 Django admin 管理后台和 swagger 的文档界面，如图 5-9 和图 5-10 所示。而"http://127.0.0.1:8000/ 首页"还没有实现，所以是一个 debug 的 404 页面，如图 5-11 所示。尽管现在还不能见到更多效果，但我们一直在进步，是不是？

网址：http://127.0.0.1:8000/admin。

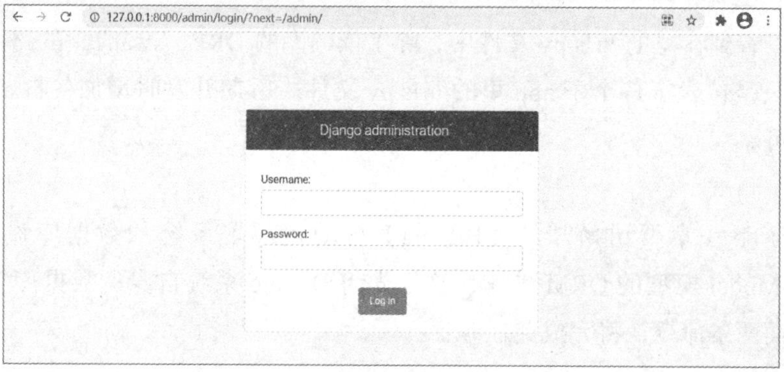

图 5-9 Django admin 后台管理界面

网址：http://127.0.0.1:8000/swagger/。

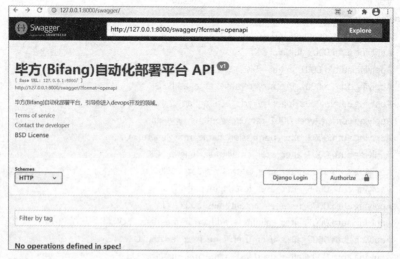

图 5-10 Django swagger 文档界面

第 5 章　Python、Django 与 DRF 的开发环境

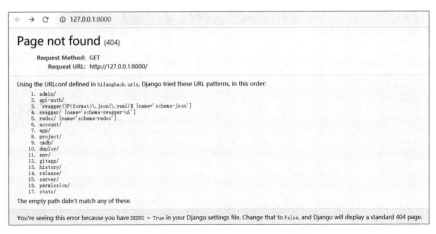

图 5-11　bifangback 404 首页

5.6　本章 GitHub 代码拉取运行

为了方便读者实践，已将本章代码放于 GitHub 上单独的仓库中，如果我在本章的讲解中有任何遗漏，代码也可以说明问题，读者可以自行下载运行。这里讲一下如何操作。以后每章节都有类似环境，会在每章前面列明，不再一一讲解。

本章代码 GitHub 地址：

https://github.com/aguncn/bifang-book/tree/main/bifang-ch5。

（1）在 Windows 上安装好 GitHub 工具。

我选择的工具：https://www.git-scm.com/download/win。

（2）在本地磁盘中，选择一个目录下载保存。

此处选择的目录是 D:\Code 目录。

（3）运行 git 命令，下载仓库代码：

```
git clone https://github.com/aguncn/bifang-ch5.git
```

（4）运行 PyCharm，打开 D:\Code\bifang-ch5\bifangback 作为项目目录，如图 5-12 所示。

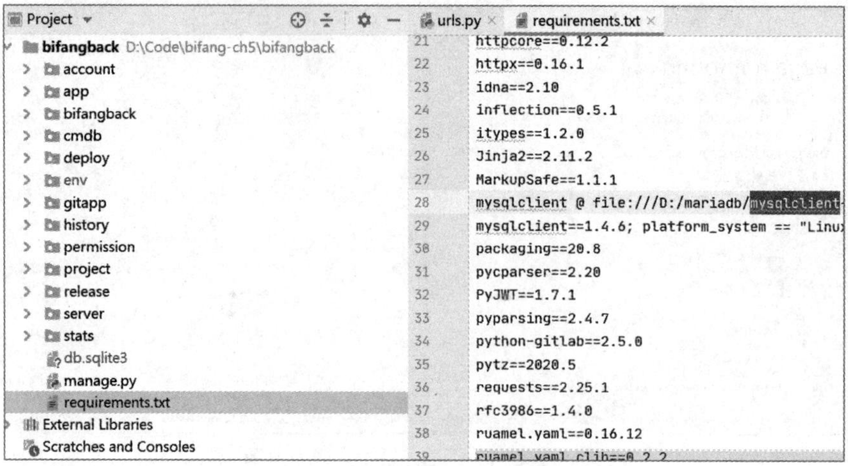

图 5-12　bifangback ch5 项目打开

（5）安装配置在 Windows 命令行控制台，安装好 Python 虚拟环境，在虚拟环境里安装好 requirements.txt 里的第三方依赖库（注意系统软件要先满足，mysqlclient 的 whl 文件要先下载到指定目录）。本章前面已说明，不再讲解。

（6）最后，在 IDE 里指定 Python 虚拟环境，启动 bifangback 服务即可打开浏览器测试。本章前面已说明，不再讲解。

5.7　小　结

本章主要讲解了建立 bifangback 项目的开发环境。涉及的知识包括安装 Python、Django 和 DRF，Python 虚拟环境的安装配置、requirements.txt 文件的生成和使用以及 PyCharm 这个 IDE 的安装配置。接着，使用 django-admin 命令建立 bifangback 项目的骨架目录，预建了所有的 app，设计开发了总分结构的 urls 路由系统。

本章以实践的内容为主，理论的东西比较少。我希望读者在反复练习的过程中，深入学会如何自己开发一个系统的技能。且 Python 语法、Django 框架以及牵扯到的细节知识大多都需要专著才能讲解到方方面面的内容。书写那样一本本专著，在本书的计划之外，也在本人的能力之外，还请读者谅解。所以也请有兴趣的读者继续找各种好的文档资料深入学习。当然，也可以随时和我沟通交流。闻道有先后，术业有专攻，三人行，则必有我师也。

我此次就反其道而行之，先像学徒闷头苦干，把一个个能眼见为实的代码"码"出来，知道了在具体的一个项目中能用到哪些知识后，再潜心研究去把握一些全局系统性的知识。相信这样，学习才更有针对性。

　　这一章我们搭建好了戏台，莫忙唱戏。下一章我会带领读者再挖深坑打好地基，把 bifangback 的数据库设计好。下章基础牢固之后，就可以放手去写一些自动化部署的功能逻辑了。

第 6 章　毕方（BiFang）数据库设计

每天反复做的事情造就了我们，然后你会发现，优秀不是一种行为，而是一种习惯。

——亚里士多德

本章 GitHub 代码地址：https://github.com/aguncn/bifang-book/tree/main/bifang-ch6。

本书到此已建立好了毕方（BiFang）自动化部署系统的后端——backfangback 项目的开发环境，Django 框架里的 Project、app、urls 等都已配置到位。本章就系统地讲解毕方系统所用到的数据库表，这在 Django 框架里是以 Models 模型的形式来呈现的。

到本章结尾，我们会建立好毕方系统需要使用的所有数据库，并为这些数据库增加一些模拟的数据，以免后面的开发数据显得太空旷。

6.1　Django model 与 ORM

本章还是以实战操作为主，理论讲解为辅，让读者的知识点以刚刚够用为限。更全面的知识还需要读者平时多学习积累。

1. model 与 ORM 简介

在 Django 中，与数据库相关的模块是 model 模块，它提供了一种简单易操作的 API 方式与数据库交互，它通过 ORM（Object Relational Mapping）映射的方式来操作数据库。其实现了数据模型与数据库的解耦，即数据模型的设计不需要依赖于特定的数据库，通过简单的配置就可以轻松更换数据库。

一个类对应数据库一张表，一个类属性对应该表的一个字段，一个实例化的类对象就是一个表中的一行数据信息。在开发的阶段，工程师只需要就 Python 语言本身进行代码设计，而不用太过于分散注意力去操作 SQL 原生操作语句，提供了一个自动化生成访问数据库的 API。

ORM 的作用如图 6-1 所示。

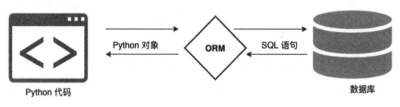

图 6-1　ORM 的作用

ORM 的优点：
- 实现了代码与数据库的解耦合。
- 开发者不需要操作太多的原生 SQL，可以提高开发效率。
- 防止 SQL 注入，通过对象操作的方式，默认就是防止 SQL 注入。

ORM 的缺点：
- 牺牲性能——对象转换到 SQL 会存在一定的消耗。
- 当需要操作较复杂的语句时，用 ORM 对象操作的方式很难实现。

ORM 的常用字段如下：
- AutoField：int 自增列，必须填入参数 primary_key=True。当 model 中如果没有自增列，则自动创建一个列名为 id 的列。
- IntegerField：一个整数类型，范围为 –2147483648 ～ 2147483647。
- CharField：字符类型，必须提供 max_length 参数，max_length 表示字符长度。
- DateField：日期字段，日期格式为 YYYY-MM-DD，相当于 Python 中的 datetime.date() 实例。
- DateTimeField：日期时间字段，格式为 YYYY-MM-DD HH:MM[:ss[.uuuuuu]] [TZ]，相当于 Python 中的 datetime.datetime() 实例。

如果读者之前没有接触过 java，那么现在讲这些 Golang 语言的 ORM，可能会使读者觉得比较虚无缥缈。稍有耐心一些，本章后面就会用到这些字段来生成数据库。

2. bifangback 项目的数据库连接设置

Django 的数据库连接配置也在 settings.py 文件里进行。settings.py 文件里 DATABASES 列表的默认设置如下：

```
DATABASES = {
```

```
    'default': {
        'ENGINE': 'django.db.backends.sqlite3',
        'NAME': BASE_DIR / 'db.sqlite3',
    }
}
```

这表示 Django 默认使用的是 SQLite 数据库。SQLite 是一个相当小巧的数据库，它实现了自给自足的、无服务器的、零配置的、事务性的 SQL 数据库引擎。据称，SQLite 是在世界上最广泛部署的 SQL 数据库引擎。

前面说过，当我们使用了 Django 的 ORM 后，数据库的迁移是很方便的。如果后期要使用功能一些更强的数据，比如 Mysql 或 Mariadb，那应该如何操作呢？

当需要这类数据库迁移时，我们不需要更改任何 ORM 语句，只需要先建好数据库，然后将 settings.py 里 DATABASES 的列表变量更新为数据库实际连接信息即可：

```
# 安装好 mysqlclient 之后 ( 使用 whl 文件安装 )，可以切换为 mysql 数据库
DATABASES = {
    'default': {
        'ENGINE': 'django.db.backends.mysql',
        'NAME': 'bifang',
        'USER': 'root',
        'PASSWORD': 'root_password',
        'HOST': '127.0.0.1',
        'PORT': '3306',
        'OPTIONS': {
            'charset': 'utf8mb4',
            # 'init_command': "SET sql_mode='STRICT_TRANS_TABLES'",
        }
    }
}
```

现在，我们暂时不用如此操作，所以代码里加上了注释。在本书中，直到所有开发代码完成，SQLite 都可以满足开发所需，且使用 SQLite 重建和删除数据库等也十分方便。在附录里，当需要将毕方系统部署在线上环境时，再来使用这样的数据库连接吧。

3. 使用 DB Browser for SQLite 查看 SQLite 数据库

由于 Django 默认使用的是 SQLite 数据库，在第 5 章运行 python manage.py migrate 时，在 SQLite 数据库中生成了 Django 默认内置的一些数据库。为了方便调试和测试，我们怎么查看 SQLite 中的数据库内容呢？

第 6 章　毕方（BiFang）数据库设计

查看 SQLite 数据内容的软件有不少，我在这里介绍一下自己使用过的 DB Browser for SQLite（https://sqlitebrowser.org/）。当下载完成并安装好 DB Browser for SQLite 后，运行此软件。打开 bifangback 项目生成的 db.sqlite3 文件（此文件默认生成在 Django 项目根目录下），就可以查看此数据库的结构或是各个数据表的具体内容了，如图 6-2 所示。

图 6-2　使用 DB Browser for SQLite 查看 SQLite 数据库

在第 5 章讲过，我们将 bifangback 项目的所有 app 应用到的 model，都集中存放在 cmdb 这个 app 里。所以，所有 model 文件都在此目录下创建。

6.2　git model

本章从本节开始逐一讲解 bifangback 项目里的 model 设计。讲解的顺序先从简单的、无依赖的数据表讲起，再过滤到有外键的、复杂的数据表。这样更连贯自然，前面的数据表可以为后面的数据表作铺垫。

git 数据表存放的是 bifangback 连接 GitLab 服务器的信息，如果企业只有一个 GitLab 服务器，可以将这些信息配置在 settings.py 文件里，可以达到同样的效果。为了灵活，bifangback 将这些配置信息放在了数据库中。

1. bifangback models 抽象基类

因为在所有的 bifangback 数据表中，有一些共同的字段是每个数据表都需要存在的。我们将这部分字段提取出来形成一个 models 抽象基类，后面所有的 model 都继承自这个基类，这种写法使得代码质量更好，且更容易维护。

在 cmdb 目录下新建一个 base_models.py 文件，内容如下：

```python
from django.db import models
from simple_history.models import HistoricalRecords
from django.contrib.auth.models import User


# 基础虚类，所有 model 的共同字段，其他 model 由此继承，包括记录 ORM 操作历史的 history 字段。
class BaseModel(models.Model):
    # unique=True 用于保证名称不重复
    name = models.CharField(max_length=100,
                            unique=True,
                            verbose_name=" 名称 ")
    description = models.CharField(max_length=100,
                                   null=True,
                                   blank=True,
                                   verbose_name=" 描述 ")
    # 为了避免删除关联记录，bigang 里所有外键都是 on_delete=models.SET_NULL
    create_user = models.ForeignKey(User,
                                    blank=True,
                                    null=True,
                                    on_delete=models.SET_NULL,
                                    verbose_name=" 用户 ")
    # auto_now 用于 ORM 更新记录时，每次自动更新此字段时间
    update_date = models.DateTimeField(auto_now=True)
    # auto_now_add 只用于第一次新增时，自动更新此字段时间
    create_date = models.DateTimeField(auto_now_add=True)
    # 用于扩展
    base_status = models.BooleanField(default=True)
    # Django 的 simple_history 库加此字段，用于自动保存每个表的操作历史
    history = HistoricalRecords(inherit=True)

    # property 用于为 ORM 查询增加一个计算型字段
    @property
    def username(self):
        return self.create_user.username
```

```
# 记录的默认显示值
def __str__(self):
    return self.name

class Meta:
    # abstract 关键字，表示此 class 不会生成数据表，只能被继承使用
    abstract = True
    # 默认排序规则，按更新时间降序
    ordering = ('-update_date',)
```

- models 抽象基类不会创建实际的数据表，在模型的 Meta 类里有 abstract=True 元数据项，每个数据表的名称、描述、创建者和时间等都集中在此类里了。History 字段是为了让系统自动记录每次数据库的变更历史。
- models.CharField() 中的 CharField 字符型字段、models.DateTimeField() 时间型字段、models.BooleanField() 布尔型字段，就是本节第 1 小节里列过的字段类型，把这些类型和下面在数据库里实际生成的字段类型做比较，就应该可以把握这些字段的使用方法了。
- 这个抽象基类只是个一次性操作，后面的 model 都继承自它。

2. git model 类设计

为了避免已有的第三方依赖库也有叫 git 从而发生名称冲突，所以 git model 类的名称在此处更改为 GitTb（取 Git Table 的意思）。

在 cmdb 目录下新建一个 git_models.py 文件，内容如下：

```
from django.db import models
from .base_models import BaseModel


# GitLab 代码仓库的地址，可能有的公司有多个 git 仓库，所以独立出一个数据表
# 如果只有一个代码仓库，当然也可以直接在 Django 的 settings 文件里定义
class GitTb(BaseModel):
    # gitlab 的 URL
    git_url = models.URLField(verbose_name="Git API 地址 ")
    # 用于 Python 的 GitLab 库去进行 API 认证时需要
    git_token = models.CharField(max_length=64,
                                 default='no_token',
                                 verbose_name="Git API 认证 token")
    # GitLab 的版本仅用于展示记录，无实际用途
    git_ver = models.CharField(max_length=16,
```

```
                        default='12.10',
                        verbose_name="Git 版本 ")

    class Meta:
        # 用于定义数据表名称，而不使用系统自动生成的
        db_table = 'GitTb'
        # 用一起定义 admin 后台显示名称 ( 规则为数据表名及简短中文 )
        verbose_name = 'GitTb 代码仓库 '
        verbose_name_plural = 'GitTb 代码仓库 '
```

- from .base_models import BaseModel：表示导入刚刚创建的 BaseModel 基类。
- class GitTb（BaseModel）：表示 GitTb 类继承自 BaseModel。
- 这个数据表主要作为 app 应用组件的外键。
- 其他字段都有注释，不再解释。

3. git model 模型注册并迁移数据库

Django 的数据库迁移主要扫描的是每个 app 下面的 models.py 文件，所以需要将 GitTb 类注册到这个模型文件中。在 cmdb 目录下的 models 文件里加入如下内容：

```
from django.db import models

from .git_models import GitTb
```

在 Windows 的 command 命令控制台，使用如下命令进入 Python 虚拟环境：

```
call C:\Users\ccc\bifangback_py_venv\Scripts\activate
```

然后，运行如下命令，生成数据库迁移文件：

```
python manage.py makemigrations        --- 生成数据库迁移文件，下面为控制台输出
Migrations for 'cmdb':
  cmdb\migrations\0001_initial.py
    - Create model HistoricalGitTb
    - Create model GitTb
```

数据库迁移文件位于 cmdb 目录的 migrations 目录下，此时如果读者去查看此目录的话会发现有一个 0001_initial.py 的文件。以后每一次 model 模型文件的变更都会新生成一个序号增量的文件，用于反映其变化。

最后，运行如下命令，将数据库迁移文件的变化，更新到实际的数据库中：

```
python manage.py migrate
```

```
Operations to perform:
    Apply all migrations: admin, auth, cmdb, contenttypes, sessions
Running migrations:
    Applying cmdb.0001_initial... OK
```

Django 框架会持续跟踪数据库的变化，并将变化的过程和文件名记录在其自身维护的数据表中。所以这两个命令可以反复执行，且不会产生破坏性后果。当 model 的变化已反映到数据库中后，这两个命令的输出会显示"No changes detected"和" No migrations to apply."。

4. 生成 git model 模拟数据

模拟数据的生成对于我们来说，还是有不好少处的。这一过程可以让我们更加熟悉数据表中的字段，学习 ORM 的一些技能，快速重建数据库环境，减少重复工作，还可以学习如何在 Django 框架中自定义操作命令。

（1）在 bifangback 项目目录下的 account 目录（也可以是其他 app，此处选择了此目录）下新建一个 management 目录，在 management 下再建一个 commands 目录。

（2）在 commands 目录下新建一个 mockdb.py 文件（这样的流程相当于为 Django 定义了一个新的管理命令：python manage.py mockdb 命令，而此命令将执行 mockdb.py 文件里的 python 代码），内容如下：

```python
from django.core.management.base import BaseCommand, CommandError
from django.contrib.auth.models import Group
from cmdb.models import *
import string
import random
import time
import datetime
from django.contrib.auth import get_user_model

User = get_user_model()

username = 'admin'
group_name = 'admin'


# 自定义命令，用于建立测试数据，很多 ORM 语句会使用
class Command(BaseCommand):
    help = 'create test data for BiFang back server.'
```

```python
    def add_arguments(self, parser):
        self.stdout.write(self.style.SUCCESS(' 没有额外参数，重建全部模拟测试数据 '))

    def handle(self, *args, **options):
        self.add_user()
        self.add_git()
        self.stdout.write(self.style.SUCCESS(' 数据重建完成，一把梭哈 ~~~'))
        # raise CommandError('Ok！')

    # 新建一个用户
    def add_user(self):
        User.objects.all().delete()
        Group.objects.all().delete()
        print('delete all user and group data')
        User.objects.create_user(username='tom', email='user@demo.com', password="password")
        User.objects.create_user(username='mary', email='user@demo.com', password="password")
        admin = User.objects.create_superuser(username, 'admin@demon.com', 'password')
        root = User.objects.create_superuser('root', 'root@demon.com', 'password')
        admin_group = Group.objects.create(name=group_name)
        Group.objects.create(name='test')
        Group.objects.create(name='dev')
        Group.objects.create(name='operate')
        admin_users = [admin, root]
        admin_group.user_set.set(admin_users)
        self.stdout.write(' 用户和用户组重建完成。')

    # 新建一个 git 仓库
    def add_git(self):
        GitTb.objects.all().delete()
        print('delete all GitTb data')
        create_user = User.objects.get(username=username)
        GitTb.objects.create(name='MainGit',
                             description=' 主要 git 库 ',
                             create_user=create_user,
                             git_url='http://192.168.1.211:8180',
                             git_token='RbCcuLssPekyVgy24Nui')
        self.stdout.write('GitTb 重建完成。')
```

- 因为 GitTb 数据表里有从 Base_models 继承的 create_user 的外键，所以在模拟数据里先将 Django 的用户和用户组建好，再建 GitTb 时，create_user 才有外键关联。新建用户时既建立了普通用户，也建立了管理员用户，并将管理员加入了 admin 用户组。

第 6 章　毕方（BiFang）数据库设计

- [Model].objects.create() 语法是 ORM 用来新建一条记录的语句。
- [Model]..objects.all().delete() 语法是 ORM 用来清空一个表中所有记录的语句。
- [Model].objects.get() 语法是 ORM 用来在数据表中查找一条特定记录的语句。
- Git_url 和 git_token 是作者自己测试环境的 GitLab 的真实值，读者在新建自己的 GitLab 时应该修改此处的值。
- GitTb 里 create_user 这种外键需要对外对应的实例赋值。
- handle() 方法里再包含真正的增加模拟数据的方法，更好管理。

（3）运行如下命令，将模拟数据迁移到数据库中：

```
python manage.py mockdb   --- 运行 mockdb 这个自定义管理命令，下面为控制台输出
没有额外参数，新建全部模拟测试数据，删除所有旧记录
delete all user and group data
用户和用户组重建完成
delete all GitTb data
GitTb 重建完成
数据重建完成，一把梭哈 ~~~
```

（4）使用 DB Browser for SQLite 查看 bifangback 的数据库结构和数据库内容，会发现 GitTb 数据表已建成，其字段和在 git_models.py 里定义一致。GitTb 数据表中的内容也和我们在 mockdb.py 文件中加入的一致，如图 6-3 和图 6-4 所示。

图 6-3　SQLite 中 GitTb 数据表结构

图 6-4　SQLite 中 GitTb 数据表内容

（5）之后新建模拟数据时只列出真正新建的函数，至于将此函数加入 handle() 的过程都和此一样，不再赘述。

5. git model 后台管理注册

Django 为人称道的还有一个设计，就是自带数据库管理后台。当建好了 model 文件后，只要在 app 目录里的 admin.py 里稍加配置，就可以通过管理后台增删查改 model 的所有数据了。

此处就来为 GitTb 增加后台管理功能吧。由于其他 model 的后台管理注册都差不多，所以只演示这一次，后面的 model 注册可以看本章的 GitHub 仓库代码。

（1）在 cmdb 的 admin.py 中加入如下内容：

```python
from django.contrib import admin
from simple_history.admin import SimpleHistoryAdmin

from .models import *

admin.site.site_header = ' 毕方 (BiFang) 自动化部署系统 '
admin.site.site_title = ' 登录毕方 (BiFang) 系统后台 '
admin.site.index_title = ' 毕方 (BiFang) 后台管理 '

class GitTbHistoryAdmin(SimpleHistoryAdmin):
    list_display = ['id', 'name', 'git_url', 'git_ver']
    history_list_display = ["status"]
    search_fields = ['name', 'git_url']

admin.site.register(GitTb, GitTbHistoryAdmin)
```

- GitTbHistoryAdmin 类用于定义 GitTb 的界面显示。继承自 SimpleHistory-Admin 类，可以在后台查看记录更改的历史。
- 使用 admin.site.register() 函数来注册 GitTb 数据表。

（2）登录 http://127.0.0.1:8000/admin/（用户名密码在建用户模拟数据时已

生成：admin/password），可以看到 GitTb 数据表已出现在 cmdb 栏目下，如图 6-5 所示。

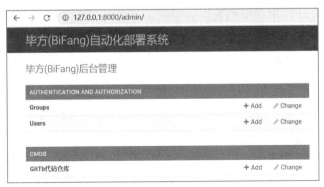

图 6-5　admin 后台管理界面

（3）进入 GitTb 可以看到这里显示的内容，和在 DB Browser for SQLite 软件中看到的一样，同时，左上角的 History 可以显示此数据表的所有操作历史，如图 6-6 所示。

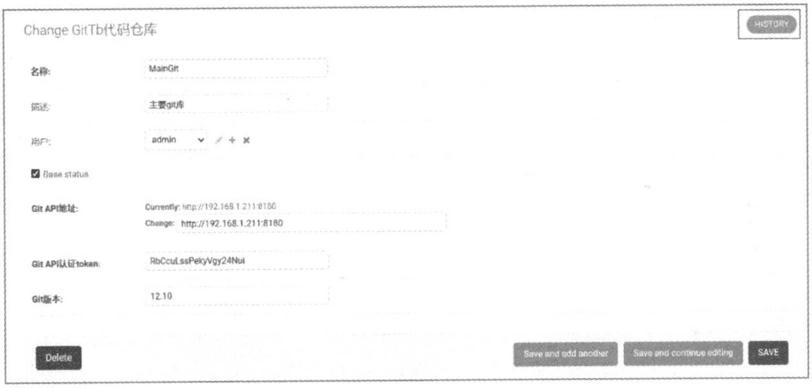

图 6-6　GitTb 数据表内容及历史查看

6.3　SaltStack model

在 6.2 节中完整地演示了如何在 Django 中设计 git model 数据表，并将其迁移到真正的数据库中。此外，还进行了模型流程，生成模拟数据和后台管理注册。

如果读者掌握了这一流程，本章余下的工作就是很多流水线一样的重复工作了。所以我在本章其他的 model 讲解时会省掉模型注册、迁移数据库和后台管理注册的流程步骤，这些流程步骤都只需换关键字就可以了。我们只专注于 model 字段的设计作用和模拟数据的生成。

省略的代码可以依据本章开头的 GitHub 仓库地址查看。

1. salt model 类设计

salt 数据表存放的是 bifangback 连接 SaltMaster 服务器的信息。如果企业只有一个 SaltMaster 服务器，可以将这些信息配置在 settings.py 文件里，达到同样的效果。但为了防止误操作，还是将测试开发环境和线上环境的 SaltMaster 服务分开部署比较安全妥当。

为了避免与有的叫 salt 的第三方依赖库名称冲突，所以 salt model 类的名称更改为 SaltTb（取 Salt Table 的意思）。在 cmdb 目录下新建一个 salt_models.py 文件，内容如下：

```python
from django.db import models
from .base_models import BaseModel


# SaltAPI 地址，不同环境可能需要不同的 salt master 隔离
# 同样，如果只有一个 salt master，只需要在 Django settings 文件里定义
class SaltTb(BaseModel):
    # BiFang 项目主要使用了 saltypie 这个第三方库操作 SaltAPI，
    # 这些字段主要满足 saltypie 的认证 SaltAPI 的参数
    salt_url = models.URLField(verbose_name="SaltAPI 地址 ")
    salt_user = models.CharField(max_length=64, verbose_name="SaltAPI 用户 ")
    salt_pwd = models.CharField(max_length=64, verbose_name="SaltAPI 密码 ")
    eauth = models.CharField(max_length=64, default='pam', verbose_name="SaltAPI 用户认证 ")
    trust_host = models.BooleanField(default=True, verbose_name="SaltAPI 安全认证 ")
    # 同样，这个版本只用于可能的展示，现在没有使用
    salt_ver = models.CharField(max_length=12, default='2019.3010', verbose_name="Salt 版本 ")

    class Meta:
        db_table = 'SaltTb'
        verbose_name = 'SaltTb 远程执行工具 '
        verbose_name_plural = 'SaltTb 远程执行工具 '
```

- 这个数据表主要作为环境的外键。因为一个 SaltMaster 可以管理多个服务，所以以环境为维度建立 SaltMaster 比较合理。

2. 生成 salt model 模拟数据

在 mockdb.py 文件中新增 salt model 的模拟数据：

```python
# 新建一个 Salt API
def add_salt(self):
    SaltTb.objects.all().delete()
    print('delete all SaltTb data')
    create_user = User.objects.get(username=username)
    SaltTb.objects.create(name='MainSalt',
                    description=' 主要 SaltApi',
                    create_user=create_user,
                    salt_url='https://192.168.1.211:8000',
                    salt_user='saltapi',
                    salt_pwd='saltapipwd',
                    eauth='pam',
                    trust_host=True)
    self.stdout.write('SaltTb 重建完成。')
```

- 每次运行 python manage.py mockdb 命令时都先清除之前数据库的所有记录。
- salt_url 和 salt_user、salt_pwd 为作者测试环境的 Salt API 的认证信息，读者如果在自己的电脑建立 Salt API，则需要自行修改此处的值。
- [Model]..objects.all().delete() 语法是 ORM 用来清空一个表中所有记录的语句。
- [Model].objects.get() 语法是 ORM 用来在数据表中查找一条特定记录的语句。
- GitTb 里的 create_user，这种外键需要对外对应的实例赋值。

3. salt model 生成的数据表结构与内容

如图 6-7 和图 6-8 所示，本章余下的数据结构和内容请读者在 DB Browser for SQLite 软件和管理后台查看，此处不再列明。

SaltTb		CREATE TABLE "SaltTb" ("id" integer NOT NULL
id	integer	"id" integer NOT NULL
name	varchar(100)	"name" varchar(100) NOT NULL UNIQUE
description	varchar(100)	"description" varchar(100)
update_date	datetime	"update_date" datetime NOT NULL
create_date	datetime	"create_date" datetime NOT NULL
base_status	bool	"base_status" bool NOT NULL
salt_url	varchar(200)	"salt_url" varchar(200) NOT NULL
salt_user	varchar(64)	"salt_user" varchar(64) NOT NULL
salt_pwd	varchar(64)	"salt_pwd" varchar(64) NOT NULL
eauth	varchar(64)	"eauth" varchar(64) NOT NULL
trust_host	bool	"trust_host" bool NOT NULL
salt_ver	varchar(12)	"salt_ver" varchar(12) NOT NULL
create_user_id	integer	"create_user_id" integer

图 6-7 SaltTb 数据结构

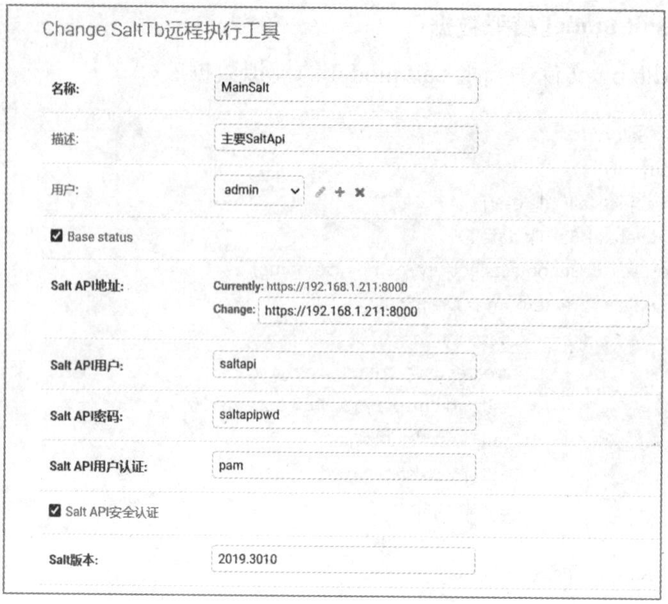

图 6-8　SaltTb 数据内容

6.4　environment model

environment model 即环境数据表，是用来区分服务器的一个维度信息（数据表名缩写为 Env）。在毕方系统中，环境信息一来为服务器分组分类，二来可以在 DevOps 的 CI/CD 流程中串联起测试部门的角色。一个发布单只有经过一个对应的环境流转操作之后，才可以在对应的环境里进行部署。

1. salt model 类设计

bifangback 管理每个环境的服务器是通过 SaltMaster 进行的，这些技术细节在第 4 章已讲解明白。Env 数据表存放的是 bifangback 管理的环境信息。一般 IT 企业为保证软件上线质量的稳定，至少会分为测试环境和线上环境。

在 cmdb 目录下新建一个 env_models.py 文件，内容如下：

```
from django.db import models
from .base_models import BaseModel
from .salt_models import SaltTb


#开发环境、线上环境等
```

第 6 章 毕方（BiFang）数据库设计

```python
class Env(BaseModel):
    # 因为继承自 BaseModel，所以 name、description、user、date 这些字段都有了，不用重复
    # 使用 id，可以在必要时减少一些对数据表自增 id 的依赖，做数据库迁移方面还是有好处的
    env_id = models.IntegerField(default=0, verbose_name=" 环境 Id")
    # 一般 SaltMaster 都是分环境建立的，这样能达到批量管理的目的
    # 多套 SaltMaster 可以隔离安全网络
    salt = models.ForeignKey(SaltTb,
                             related_name="ra_env",
                             blank=True,
                             null=True,
                             on_delete=models.SET_NULL,
                             verbose_name="Salt 地址 ")

    class Meta:
        db_table = 'Env'
        verbose_name = 'Env 环境 '
        verbose_name_plural = 'Env 环境 '
```

- 字段中的 default=0 用于设置该字段的默认生成值。
- 字段中的 verbose_name，类似于注释，又显示为后台管理界面的字段名称。
- models.ForeignKey（定义外键关联）。
- 外键定义中的 related_name 用于指定从 SaltTb 模型返回到 Env 模型的反向关系的名称。
- 外键定义中的 on_delete 是当子表中的某条数据删除后关联的外键操作。允许的选择有 SET_NULL、CASCADE、DO_NOTHING、PROTECT、SET_DEFAULT 等。此处设置为空，但数据库的外键被删除后确实会带来很多问题，最好小心规划这些操作。
- Meta 中的 db_table 不用系统默认自动生成的数据表名，而是自定义指定数据表名称。
- Meta 中的 verbose_name 和 verbose_name_plural 用于指定在后台管理界面中显示的名称。

2. 生成 environment model 模拟数据

mockdb.py 文件中新增的 Env model 的模拟数据：

```python
# 新建一个环境
def add_env(self):
    Env.objects.all().delete()
```

```python
print('delete all Env data')
create_user = User.objects.get(username=username)
salt = SaltTb.objects.order_by('?').first()
env_list = ['dev', 'prd']
for index, env in enumerate(env_list):
    Env.objects.create(name=env,
                       description=env,
                       create_user=create_user,
                       env_id=index,
                       salt=salt)
self.stdout.write('Env 重建完成。')
```

- 在模拟数据中只生成两个环境：dev 开发环境和 prd 线上环境，对于本书开发已够用。
- SaltTb.objects.order_by('?').first() 这种 ORM 语法，实现了从 SaltTb 中随机选出一条记录的目的。

6.5 project model

project model 即项目数据表。BiFang 系统中的项目是用来组织不同的 app 组件应用的（读者应将 Django 框架本身的 project 和 app 概念，与 BiFang 系统中的这两个概念区分开来。当然，在概念设计上它们有相同之处）。

一个 project 可以包含多个 app，组成一个类似微服务的系统。而一个 app 只能从属于一个 project。所以在数据库结构设计上，每一个 app 数据表里有一个外键指向 project。

1. project model 类设计

在 cmdb 目录下新建一个 project_models.py 文件，内容如下：

```python
from django.db import models
from .base_models import BaseModel




# 项目，可表示由多个相关微服务组成的项目
# 比如，istio 里的 bookin 就算是 bifang 中的一个 project
# 而其中的 productpage,reviews,details,ratings,go-demo 这些应用
# 就相当于 bifang 中的 app 应用
class Project(BaseModel):
```

```python
# 只是为了一个中文显示及统一的项目 id 加的，不解释
cn_name = models.CharField(max_length=255, verbose_name=" 中文名 ")
project_id = models.IntegerField(default=0, verbose_name=" 项目编号 ")

class Meta:
    db_table = 'Project'
    verbose_name = 'Project 项目 '
    verbose_name_plural = 'Project 项目 '
```

- 字段中的 max_length 设置该字段最多能容纳多少个字符。

2. 生成 project model 模拟数据

mockdb.py 文件中新增的 project model 的模拟数据：

```python
# 新建 demo 项目
def add_project(self):
    Project.objects.all().delete()
    print('delete all Project data')
    create_user = User.objects.get(username=username)
    project_name_list = ['User', 'Service', 'Store', 'Card', 'Support', 'BookInfo', 'Demo']
    project_cn_name_list = [' 用户管理 ', ' 服务 ', ' 库存 ', ' 购物车 ', ' 客服 ', ' 书店 ', 'Go 项目 ']
    for project_name, project_cn_name in zip(project_name_list, project_cn_name_list):
        Project.objects.create(name=project_name,
                               description=project_name,
                               cn_name=project_cn_name,
                               create_user=create_user,
                               project_id=random.randint(1000, 10000))
    self.stdout.write('Project 重建完成。')
```

- 通过 zip 循环多个列表，同时建立多个 project。
- 出于模拟，project_id 在 1000 ~ 10000 随机取值。

6.6　app model

app 是 BiFang 系统设计的一个主要数据结构，它需要和一个具体的 GitLab 上的项目代码一一对应（这里有个比较绕的概念，GitLab 上所有代码的组织，是以 groups 和 project 做两级分类的，而不是以 project 和 app 做两级分类的。所以 BiFang 系统中的 app 对应于 GitLab 中的 project。读者要注意使用上下文，辨析各自含义）。

1. app model 类设计

在 cmdb 目录下新建一个 app_models.py 文件（注意和同目录下的 apps.py 区分开来），内容如下：

```python
from django.db import models
from .base_models import BaseModel
from .git_models import GitTb
from .project_models import Project


# 应用服务，相当于一个可独立部署的微服务
# 如 istio bookinfo 中，可独立部署且使用不同语言开发的 4 个应用
# (productpage, details, reviews, ratings)
class App(BaseModel):
    # cn_name 和 app_id 意义和 project 中的一样。
    # 但要注意，这个 app_id 和一些外键表中的 app_id 有雷同，这是高级技巧
    # 这是一个坑，但本来这样命名也是最自然的
    cn_name = models.CharField(max_length=255, verbose_name="中文名")
    app_id = models.IntegerField(default=0, verbose_name="应用编号")
    # 每个 app 应用与一个代码关联
    git = models.ForeignKey(GitTb,
                            related_name="ra_app",
                            blank=True,
                            null=True,
                            on_delete=models.SET_NULL,
                            verbose_name="Git 实例")
    # 这里单独定义一个 git 中这个 app 的 id，可能有优化的空间，也可能没有
    # git 库在定义具体的代码仓库时就是要这个参数
    git_app_id = models.IntegerField(default=0, verbose_name="Git 应用 ID")
    # 为了加强控制，git 中每一个 CI/CD 功能，当代码提交后都不会自动运行
    # 而要通过一个 trigger token 去人工触发
    # 这个 trigger token 的生成在 gitlab 的 CI/CD 里很容易生成
    git_trigger_token = models.CharField(max_length=64,
                                         blank=True,
                                         null=True,
                                         verbose_name="git trigger token")
    # 将 app 与 project 进行关联，方便数据统计、关联显示
    project = models.ForeignKey(Project,
                                related_name='ra_project',
                                blank=True,
                                null=True,
                                on_delete=models.SET_NULL,
                                verbose_name='项目')
```

```
# 按 devops 及 gitops 的理念，开发维护编译脚本和部署脚本，并进行版本管理
# build_script 用于定义通过构建，生成软件包的脚本
build_script = models.CharField(max_length=255, default='build.sh',
                                verbose_name=" 编译脚本名 ")
# deploy_script 用于定义服务部署、启停、备份、回滚、健康状态检测等功能
# 如果为开发提供了模板，是很容易作为代码的一部分管理起来的，配置即代码！
# 如果 bifang 本身来存储这些脚本，反而增加管理难度，不透明度及中心化
deploy_script = models.CharField(max_length=255, default='bifang.sh',
                                 verbose_name=" 部署脚本名 ")
# 定义软件包的名称，这里也有纠结，是自定义还是强约定
# 如果规定了软件包名必须与 app 名称相同，会少很多事，但又显得过于霸道
# 这里留个小口吧，另外在部署脚本时还会有几个类似的软件包、软件压缩包名、软件部
  署目录
zip_package_name = models.CharField(max_length=255, default='demo.zio',
                                    verbose_name=" 应用压缩包 ")
# 如果 app 有脚本端口，则可能用于状态检测，增加部署的成功判断率
service_port = models.IntegerField(default=0, verbose_name=" 服务端口 ")
# 使用哪个用户名和用户组启动
service_username = models.CharField(max_length=24,
                                    blank=True,
                                    null=True,
                                    verbose_name=" 执行用户名 ")
service_group = models.CharField(max_length=24,
                                 blank=True,
                                 null=True,
                                 verbose_name=" 执行用户组 ")
# 如果 BiFang 增加独立的服务器启停功能，日志单独数据表保存
# 但这里可以保存最近启停次数，用于定义日志
op_no = models.IntegerField(default=0, verbose_name=" 启停日志次数 ")

class Meta:
    db_table = 'App'
    verbose_name = 'App 应用 '
    verbose_name_plural = 'App 应用 '
```

- 字段中的 blank=True 针对表单，你的表单可以不填写该字段，比如 admin 界面下增加 model 一条记录的时候，直观地就看到该字段不是粗体。
- 字段中的 null =True，针对数据库而言，表示数据库的该字段可以为空，那么在新建一个 model 对象的时候是不会报错的。
- 代码里已有几乎每个字段的含义和解释。

2. 生成 app model 模拟数据

mockdb.py 文件中新增的 app model 的模拟数据：

```python
# 新建 demo 应用
def add_app(self):
    App.objects.all().delete()
    print('delete all App data')
    create_user = User.objects.get(username=username)
    app_name_list = ['User-Login', 'Service-724', 'Store-Address', 'Card-Adjust', 'Support-Admin']
    app_cn_name_list = ['用户登录', '全天服务', '库存地址', '购物车调配', '客服后管']
    for app_name, app_cn_name in zip(app_name_list, app_cn_name_list):
        git = GitTb.objects.order_by('?').first()
        project = Project.objects.order_by('?').first()
        App.objects.create(name=app_name,
                           description=app_name,
                           cn_name=app_cn_name,
                           create_user=create_user,
                           app_id=random.randint(10000, 100000),
                           git=git,
                           git_app_id=1,
                           git_trigger_token='559fbd3381bc39100811bd00e499a7',
                           project=project,
                           build_script='script/build.sh',
                           deploy_script='script/deploy.sh',
                           zip_package_name='go-demo.tar.gz',
                           service_port=9090,
                           service_username='sky',
                           service_group='operate')
    app_name_list = ['ProductPage', 'Details', 'Reviews', 'Ratings', 'go-demo']
    app_cn_name_list = ['产品页', '详情页', '评论页', '评级页', 'golang 演示组件应用']
    git_app_list = [3, 4, 5, 6, 1]
    git_trigger_token_list = ['b843f743187a6632e0440a716cf038',
                              '1b57c3f005caf21f90fb3cc833eeb6',
                              '4abe06d297ded3da79ab0ff0c67f7e',
                              'c14b9eead3021965918e187eca16a0',
                              '559fbd3381bc39100811bd00e499a7']
    service_port_list = [8001, 8002, 8003, 8004, 9090]
    for app_name, app_cn_name, git_trigger_token, service_port, git_app_id in \
            zip(app_name_list, app_cn_name_list, git_trigger_token_list, service_port_list, git_app_list):
        git = GitTb.objects.order_by('?').first()
        project = Project.objects.get(name='BookInfo')
        App.objects.create(name=app_name,
                           description=app_name,
```

```
                    cn_name=app_cn_name,
                    create_user=create_user,
                    app_id=random.randint(10000, 100000),
                    git=git,
                    git_app_id=git_app_id,
                    git_trigger_token=git_trigger_token,
                    project=project,
                    build_script='script/build.sh',
                    deploy_script='script/deploy.sh',
                    zip_package_name='{}.tar.gz'.format(app_name),
                    service_port=service_port,
                    service_username='root',
                    service_group='root')
        self.stdout.write('App 重建完成。')
```

- 为了力求逼真，这里多建一些 app。
- 第一个 for 循环里的 app 纯模拟，不能操作；第二个 for 循环里的所有 app 都是我在前面讲过的，在我的测试环境里真实的 app。app_nam、trigger_token、service_port 和 gitlab project id 字段值也是真实的；如果读者在自己的电脑上与这些值不同，请自行修改。

6.7　server model

BiFang 系统的服务器数据表主要存放的是每个服务器的信息，包括 ip 端口等信息。为了支持在同一个主机部署多个 app 组件的需求，BiFang 支持同一个 ip 录入多个端口，只要端口间彼此不冲突即可。

此外，为了能追踪服务器的变化，回滚发布单，在服务器数据表里还有主备发布单字段。为了能更新发布单状态的变化，服务器里还有一个自身的状态字段，此字段是一个外键，指向一个单独的服务器状态数据表。

1. server model 类设计

在 cmdb 目录下新建一个 server_models.py 文件，内容如下：

```
from django.db import models
from .base_models import BaseModel
from .app_models import App
from .release_models import Release
from .env_models import Env
```

```python
SYSTEM_CHOICES = (
    ('WINDOWS', 'WINDOWS'),
    ('LINUX', 'LINUX'),
)


# 为了能动态管理服务器状态，我觉得加个单独的表很有必要
class ServerStatus(BaseModel):
    # ['Ready', 'Ongoing', 'Success', 'Failed']
    # [' 准备部署 ',' 部署中 ',' 部署成功 ',' 部署失败 ']
    status_value = models.CharField(max_length=1024,
                                    blank=True,
                                    verbose_name=" 状态值 ",
                                    default="Ready")

    class Meta:
        db_table = 'ServerStatus'
        verbose_name = 'ServerStatus 服务器状态 '
        verbose_name_plural = 'ServerStatus 服务器状态 '


# 为外键设置默认值
def get_server_status():
    return ServerStatus.objects.get_or_create(name='Ready')[0].id


# 部署服务器保证 ip 和 port 结合起来的唯一性，可以在一个服务器上部署多个应用
#（但不可以在同一个服务器的不同端口部署同一个应用
# 想想 salt 在执行同一批次执行到同一个服务器上的情况，会执行多次
# 也可以只一次详细考察 target_list 同为一个机器的情况吧）
# 当然，能在同一个服务器上，部署多个相同的应用，这得益于将部署脚本让开发自己维护
# 真正的 devops 团队是需要同时具有开发和运维的跨界实力的啦 ~
class Server(BaseModel):
    # GenericIPAddressField 的字段，让这里只存储 ip 地址
    ip = models.GenericIPAddressField(max_length=64, verbose_name=" 服务器 Ip")
    # 服务端口，其实这里需要优化
    # 如果有的服务启动不提供端口服务呢？乱写吗？如何保证多个无端口服务不冲突？
    port = models.IntegerField(verbose_name=" 服务器端口 ")
    system_type = models.CharField(max_length=16,
                                   choices=SYSTEM_CHOICES,
                                   default='LINUX',
                                   verbose_name=" 操作系统 ")
```

第 6 章 毕方（BiFang）数据库设计

```python
# 此服务器与哪一个 app 关联
app = models.ForeignKey(App,
                        related_name='ra_server',
                        blank=True,
                        null=True,
                        on_delete=models.SET_NULL,
                        verbose_name=' 应用服务 ')
# 环境关联，在部署时根据环境关联服务器
env = models.ForeignKey(Env,
                        related_name="ra_server",
                        blank=True,
                        null=True,
                        on_delete=models.SET_NULL,
                        verbose_name=" 环境 ")
# 保存在此服务器正在运行的 app 的最近发布单
main_release = models.ForeignKey(Release,
                        related_name='ra_server_main',
                        blank=True,
                        null=True,
                        on_delete=models.SET_NULL,
                        verbose_name=' 主发布单 ')
# 保存在此服务器正在运行的 app 的次新发布单，主要用于回滚
# BiFang 只支持最近一次回滚，不支持无限回滚
back_release = models.ForeignKey(Release,
                        related_name='ra_server_back',
                        blank=True,
                        null=True,
                        on_delete=models.SET_NULL,
                        verbose_name=' 备份发布单 ')
# 如果一个发布单部署了多次，或是分批在服务器部署，就有这个记录的必要性了
# 用于判断服务器部署是否完成
deploy_no = models.IntegerField(blank=True,
                        null=True,
                        default=0,
                        verbose_name=" 部署次数 ")
# 记录各种状态用于前端显示
# 或发布进行中或完成的判断 ( 主发布单和是否完成部署 )
server_status = models.ForeignKey(ServerStatus,
                        related_name='ra_server',
                        # default=get_server_status,
                        blank=True,
                        null=True,
                        on_delete=models.SET_NULL,
```

```
                            verbose_name="服务器状态")

class Meta:
    db_table = 'Server'
    # 一个服务器上部署多个应用，保证 ip 加 port 的唯一性
    unique_together = ('ip', 'port')
    verbose_name = 'Server 服务器 '
    verbose_name_plural = 'Server 服务器 '
```

- SYSTEM_CHOICES 字典用于让字段只包含特定的枚举值。
- 注释完整，无须多解释。

2. 生成 server model 模拟数据

mockdb.py 文件中新增的 Server model 及 ServerStatus 的模拟数据：

```
# 新建服务器状态
def add_server_status(self):
    ServerStatus.objects.all().delete()
    print('delete all ServerStatus data')
    create_user = User.objects.get(username=username)
    status_list = ['Ready', 'Ongoing', 'Success', 'Failed']
    status_value_list = [' 准备部署 ', ' 部署中 ', ' 部署成功 ', ' 部署失败 ']
    for status_name, status_value_name in zip(status_list, status_value_list):
        ServerStatus.objects.create(name=status_name,
                                    description=status_name,
                                    create_user=create_user,
                                    status_value=status_value_name)
    self.stdout.write('ServerStatus 重建完成。')

# 新建 server 服务器
def add_server(self):
    Server.objects.all().delete()
    print('delete all Server data')
    create_user = User.objects.get(username=username)
    server_status = ServerStatus.objects.get(name="Ready")
    for number in range(50):
        ip = '192.168.1.{}'.format(number)
        port = random.randint(10000, 100000)
        app = App.objects.order_by('?').first()
        env = Env.objects.order_by('?').first()
        Server.objects.create(name='{}_{}'.format(ip, port),
                              description=app.name,
                              create_user=create_user,
                              ip=ip,
```

```
                        port=port,
                        app=app,
                        env=env,
                        server_status=server_status,
                        system_type=random.choice(['WINDOWS', 'LINUX']))
app_name_list = ['ProductPage', 'Details', 'Reviews', 'Ratings']
service_port_list = [8001, 8002, 8003, 8004]
ip_list = ['192.168.1.211', '192.168.1.212', '192.168.1.213', '192.168.1.214']
env = Env.objects.get(name='dev')
for app_name, service_port in zip(app_name_list, service_port_list):
    app = App.objects.get(name=app_name)
    for ip in ip_list:
        Server.objects.create(name='{}_{}'.format(ip, service_port),
                        description=app_name,
                        create_user=create_user,
                        ip=ip,
                        port=service_port,
                        app=app,
                        env=env,
                        server_status=server_status,
                        system_type='LINUX')
app_name_list = ['go-demo']
service_port_list = [9090]
ip_list = ['192.168.1.211', '192.168.1.212']
env = Env.objects.get(name='dev')
for app_name, service_port in zip(app_name_list, service_port_list):
    app = App.objects.get(name=app_name)
    for ip in ip_list:
        Server.objects.create(name='{}_{}'.format(ip, service_port),
                        description=app_name,
                        create_user=create_user,
                        ip=ip,
                        port=service_port,
                        app=app,
                        env=env,
                        server_status=server_status,
                        system_type='LINUX')
self.stdout.write('Server 重建完成。')
```

- 'Ready'、'Ongoing'、'Success'、'Failed' 与 '准备部署'、'部署中'、'部署成功'、'部署失败' 这 4 个服务器状态值一一对应。同时，要注意区别服务器状态值和后面介绍的发布单状态值。

- 第一个 for 循环是为了制造一个纯虚构的数据，不可正常使用。
- 第二个和第三个 for 循环是我的计算机上真实的服务器数据，要是读者的测试环境与作者的不一样，请自行修改相关信息。

6.8 release model

release 是 BiFang 系统设计的一个主要数据结构。从核心作用上来说，其他的数据表设计主要是为了满足发布单的周边数据完整性。它与环境相关，也与 app 应用相关，与 GitLab 的代码分支相关。而在新建发布单、构建发布单、流转发布单和部署发布单的一系列操作流程中，都会更新发布单的状态值。所以发布单也还有一个外键关联的发布单状态数据表。

1. release model 类设计

在 cmdb 目录下新建一个 release_models.py 文件，内容如下：

```python
from django.db import models
from .base_models import BaseModel
from .app_models import App
from .env_models import Env


# 为了能动态管理发布单状态，我觉得加个单独的表很有必要
class ReleaseStatus(BaseModel):
    # ['Create', 'Building', 'BuildFailed', \
    #  'Build', 'Ready', 'Ongoing', 'Success', 'Failed']
    # [' 创建 ',' 编译中 ',' 编译失败 ', \
    #  ' 编译成功 ',' 准备部署 ',' 部署中 ',' 部署成功 ',' 部署失败 ']
    status_value = models.CharField(max_length=1024,
                                    blank=True,
                                    verbose_name=" 状态值 ")

    class Meta:
        db_table = 'ReleaseStatus'
        verbose_name = 'ReleaseStatus 发布单状态 '
        verbose_name_plural = 'ReleaseStatus 发布单状态 '


# 发布单是 BiFang 部署平台中的灵魂和纽带
# 各种配置经由它作融合，各种动态数据经由它启动发散
```

```python
class Release(BaseModel):
    # app 关联
    app = models.ForeignKey(App,
                            related_name='ra_release',
                            blank=True,
                            null=True,
                            on_delete=models.SET_NULL,
                            verbose_name=" 应用 ")
    # 环境关联，这个在新建和编译发布单过程中是没有数据的，在环境流转之后才有
    env = models.ForeignKey(Env,
                            related_name="ra_release",
                            blank=True,
                            null=True,
                            on_delete=models.SET_NULL,
                            verbose_name=" 环境 ")
    # 如果一个发布单部署了多次，或是分批在服务器部署，就有这个记录的必要性了
    deploy_no = models.IntegerField(blank=True,
                                    null=True,
                                    default=0,
                                    verbose_name=" 部署次数 ")
    # 自定义需要部署的 git 分支
    git_branch = models.CharField(max_length=255,
                                  blank=True,
                                  null=True)
    # pipeline_id 和 pipeline_url 在编译软件包的过程中生成
    # 用于获取编译状态及定位编译输出
    pipeline_id = models.IntegerField(default=0)
    pipeline_url = models.URLField(default='http://www.demo.com')
    # 这个部署脚本在 git 代码中一般会有自己独立的目录
    # 而在制品仓库时，可能就有软件包并列在同一个目录了，很清晰
    deploy_script_url = models.URLField(default=None,
                                        blank=True,
                                        null=True,
                                        verbose_name=" 部署脚本路径 ")
    # 这里生成软件包之后直接记录 url 软件包路径
    # 这样在部署脚本中就可以直接使用 wget 下载了
    # 为什么不使用 salt？其实也行，但 wget 不是更稳定和简单么？
    zip_package_url = models.URLField(default=None,
                                      blank=True,
                                      null=True,
                                      verbose_name=" 压缩制品库路径 ")
    # 记录各种状态用于前端显示
    release_status = models.ForeignKey(ReleaseStatus,
```

```
                                    related_name='ra_release',
                                    blank=True,
                                    null=True,
                                    on_delete=models.SET_NULL,
                                    verbose_name=" 发布单状态 ")

    class Meta:
        db_table = 'Release'
        verbose_name = 'Release 发布单 '
        verbose_name_plural = 'Release 发布单 '
```

- 'Create'、'Building'、'BuildFailed'、'Build'、'Ready'、'Ongoing'、'Success'、'Failed' 这几个值与 ' 创建 '、' 编译中 '、' 编译失败 '、' 编译成功 '、' 准备部署 '、' 部署中 '、' 部署成功 '、' 部署失败 ' 这 8 个发布单状态需要一一对应。同时，也要注意区分服务器状态值与发布单状态值。

2. 生成 release model 模拟数据

mockdb.py 文件中新增的 app model 的模拟数据：

```
# 新建发布单状态
def add_release_status(self):
    ReleaseStatus.objects.all().delete()
    print('delete all ReleaseStatus data')
    create_user = User.objects.get(username=username)
    status_list = ['Create', 'Building', 'BuildFailed',
                   'Build', 'Ready', 'Ongoing', 'Success', 'Failed']
    status_value_list = [' 创建 ', ' 编译中 ', ' 编译失败 ',
                         ' 编译成功 ', ' 准备部署 ', ' 部署中 ', ' 部署成功 ', ' 部署失败 ']
    for status_name, status_value_name in zip(status_list, status_value_list):
        ReleaseStatus.objects.create(name=status_name,
                                     description=status_name,
                                     create_user=create_user,
                                     status_value=status_value_name)
    self.stdout.write('ReleaseStatus 重建完成。')


# 新建 demo 发布单
def add_release(self):
    Release.objects.all().delete()
    print('delete all Release data')
    create_user = User.objects.get(username=username)
    for number in range(100):
        app = App.objects.order_by('?').first()
        env = Env.objects.order_by('?').first()
```

```
            release_status = ReleaseStatus.objects.order_by('?').first()
            random_letter = ''.join(random.sample(string.ascii_letters, 2))

            name = datetime.datetime.now().strftime("%Y%m%d%H%M%S%f") + random_letter.upper()
            Release.objects.create(name=name,
                           description=name,
                           create_user=create_user,
                           app=app,
                           env=env,
                           git_branch='master',
                           pipeline_id=0,
                           pipeline_url='http://www.demo.com',
                           deploy_script_url='http://1.2.3.4/a/b/bifang.sh',
                           zip_package_url='http://1.2.3.4/a/b/go-demo.zip',
                           release_status=release_status)
        self.stdout.write('Release 重建完成。')
```

- 这些虚拟数据只是为了展示，都不可真的使用。测试时还是要新建发布单。
- 发布单的命名为时间戳再加上两个随机字母。

6.9 permission model

BiFang 系统权限粗分为三个：新建和构建权限、流转权限以及部署权限，这个表是独立的。在权限应用列表中，将权限与 app 应用组件、操作用户关联起来，就形成了总体的权限列表。

1. permission model 类设计

在 cmdb 目录下新建一个 permission_models.py 文件，内容如下：

```
from django.db import models
from .base_models import BaseModel
from django.contrib.auth.models import User
from .app_models import App


# 创建编译发布单、环境流转、部署发布单三个大的权限
# BiFang 暂不支持基于各个具体环境的细致权限
# 大家可以自己基于书上教授的技能自行实现
class Action(BaseModel):
```

```python
    action_id = models.IntegerField(unique=True, verbose_name="权限序号")

    class Meta:
        db_table = 'Action'
        verbose_name = 'Action 权限'
        verbose_name_plural = 'Action 权限'


# 具体的权限数据表
# 如果要获取某个服务的所有权限或是某一应用的指定权限的用户列表，都是可以的
class Permission(BaseModel):
    # 权限与 app 应用级别关联
    app = models.ForeignKey(App,
                            related_name="ra_permission",
                            blank=True,
                            null=True,
                            on_delete=models.SET_NULL,
                            verbose_name="App 应用")
    # 权限与具体的权限动作（创建编译、环境流转、部署发布）关联
    action = models.ForeignKey(Action,
                               related_name="ra_permission",
                               blank=True,
                               null=True,
                               on_delete=models.SET_NULL,
                               verbose_name="操作权限")
    # 权限关联到用户
    pm_user = models.ForeignKey(User,
                                related_name="ra_permission",
                                blank=True,
                                null=True,
                                on_delete=models.SET_NULL,
                                verbose_name="权限用户")

    class Meta:
        db_table = 'Permission'
        verbose_name = 'Permission 应用权限'
        verbose_name_plural = 'Permission 应用权限'
```

- 'Create'、'Building'、'BuildFailed'、'Build'、'Ready'、'Ongoing'、'Success'、'Failed' 这几个值与 '创建'、'编译中'、'编译失败'、'编译成功'、'准备部署'、'部署中'、'部署成功'、'部署失败' 这 8 个发布单状态需要一一对应。同时，也要注意区分服务器状态值与发布单状态值。

2. 生成 release model 模拟数据

mockdb.py 文件中新增的 permission model 的模拟数据：

```
# 新建权限
def add_action(self):
    Action.objects.all().delete()
    print('delete all Action data')
    create_user = User.objects.get(username=username)
    Action.objects.create(name='Create',
                          description=' 创建编译权限 ',
                          create_user=create_user,
                          action_id=100)
    Action.objects.create(name='Env',
                          description=' 环境权限 ',
                          create_user=create_user,
                          action_id=1000)
    Action.objects.create(name='Deploy',
                          description=' 部署权限 ',
                          create_user=create_user,
                          action_id=10000)
    self.stdout.write('Action 重建完成。')

# 新建 demo 应用权限用户表
def add_permission(self):
    Permission.objects.all().delete()
    print('delete all Permission data')
    create_user = User.objects.get(username=username)
    for number in range(5):
        app = App.objects.order_by('?').first()
        action = Action.objects.order_by('?').first()
        pm_user = User.objects.order_by('?').first()
        name = '{}-{}-{}'.format(app.name, action.name, pm_user.username)
        Permission.objects.create(name=name,
                                  description=name,
                                  create_user=create_user,
                                  app=app,
                                  action=action,
                                  pm_user=pm_user)
    self.stdout.write('Permission 重建完成。')
```

- 三个权限的英文字符分别为 'Create'、'Env' 和 'Deploy'。
- 随机组合了一些操作权限。因为测试时使用的用户属于 admin 用户组，它拥有管理权限，所有 app 应用都能操作，所有权限都拥有，不受这些权限列表限制。

6.10 history model

BiFang 系统的系统记录表主要记录发布单的操作历史和服务器的操作历史，用于追踪每一个发布单和第一次对服务器的操作，可用于显示历史或是判断部署状态之用。历史数据表的模拟数据无意义，故没有增加模拟数据环节。

1. history model 类设计

在 cmdb 目录下新建一个 history_models.py 文件，内容如下：

```python
from django.db import models
from .base_models import BaseModel
from .env_models import Env
from .release_models import Release, ReleaseStatus
from .server_models import Server

# 用于在发布单的部署代码层次内的记录是部署新代码，还是回滚老代码？
# 用于发布单历史
DEPLOY_CHOICES = (
    ('deploy', '部署'),
    ('rollback', '回滚'),
)


# 用于大的方向，是在部署代码还是单纯在维护服务器启停？
# 用于服务器操作历史，区分发布单和服务器操作历史是有意义的，维护不一样，但也可以一
  追到底
OP_CHOICES = (
    ('deploy', '部署'),
    ('maintenance', '启停维护'),
)


# 对在服务器上操作的第一步做记录，后期根据需要有可能做维护改动
# 它主要能包括所有 deploy 过程中的 action_list 列表项目
ACTION_CHOICES = (
    ('fetch', '获取软件'),
    ('stop', '停止'),
    ('stop_status', '停止状态检测'),
    ('deploy', '部署'),
    ('rollback', '回滚'),
    ('start', '启动'),
    ('start_status', '启动状态检测'),
    ('health_check', '服务健康检测'),
```

```python
)

# 发布单历史记录发布单的生命周期，新建、编译、流转、部署、回滚部署
class ReleaseHistory(BaseModel):
    release = models.ForeignKey(Release,
                                related_name='ra_release_history',
                                blank=True,
                                null=True,
                                on_delete=models.SET_NULL,
                                verbose_name=' 发布单 ')
    env = models.ForeignKey(Env,
                            related_name="ra_release_history",
                            blank=True,
                            null=True,
                            on_delete=models.SET_NULL,
                            verbose_name=" 环境 ")
    release_status = models.ForeignKey(ReleaseStatus,
                                       related_name='ra_release_history',
                                       blank=True,
                                       
                                       ET_NULL,
                                       verbose_name=" 发布单状态 ")
    deploy_type = models.CharField(max_length=255,
                                   choices=DEPLOY_CHOICES,
                                   blank=True,
                                   null=True,
                                   verbose_name=" 部署类型 ")
    # 发布单的次数和这里匹配，即可找到合适的历史记录
    log_no = models.IntegerField(blank=True,
                                 null=True,
                                 default=0,
                                 verbose_name=" 部署批次 ")
    log = models.TextField(verbose_name=" 日志内容 ")

    class Meta:
        db_table = 'ReleaseHistory'
        verbose_name = 'ReleaseHistory 发布单历史 '
        verbose_name_plural = 'ReleaseHistory 发布单历史 '


# 服务器变更历史，记录服务器上的部署、停止、回滚
class ServerHistory(BaseModel):
```

```python
        server = models.ForeignKey(Server,
                                   related_name='ra_server_history',
                                   blank=True,
                                   null=True,
                                   on_delete=models.SET_NULL,
                                   verbose_name=' 服务器 ')
        release = models.ForeignKey(Release,
                                    related_name='ra_server_history',
                                    blank=True,
                                    null=True,
                                    on_delete=models.SET_NULL,
                                    verbose_name=' 发布单 ')
        env = models.ForeignKey(Env,
                                related_name="ra_server_history",
                                blank=True,
                                null=True,
                                on_delete=models.SET_NULL,
                                verbose_name=" 环境 ")
        op_type = models.CharField(max_length=255,
                                   choices=OP_CHOICES,
                                   blank=True,
                                   null=True,
                                   verbose_name=" 操作类型 ")
        action_type = models.CharField(max_length=255,
                                       choices=ACTION_CHOICES,
                                       blank=True,
                                       null=True,
                                       verbose_name=" 服务器操作类型 ")
        # 发布单或 app 应用的次数和这里匹配，即可找到合适的历史记录
        log_no = models.IntegerField(blank=True,
                                     null=True,
                                     default=0,
                                     verbose_name=" 部署启停批次 ")
        log = models.TextField(verbose_name=" 日志内容 ")

        class Meta:
            db_table = 'ServerHistory'
            verbose_name = 'ServerHistory 服务器历史 '
            verbose_name_plural = 'ServerHistory 服务器历史 '
```

尽量周全地记录所有的操作日志，log_no 用于定位每一次的变更操作，以便在前端界面上展示部署的详细记录。

第 6 章 毕方（BiFang）数据库设计

2. 数据库总汇

当所有数据表建立完，DB Browser for SQLite 和 admin 后台管理的数据表如图 6-9 和图 6-10 所示。

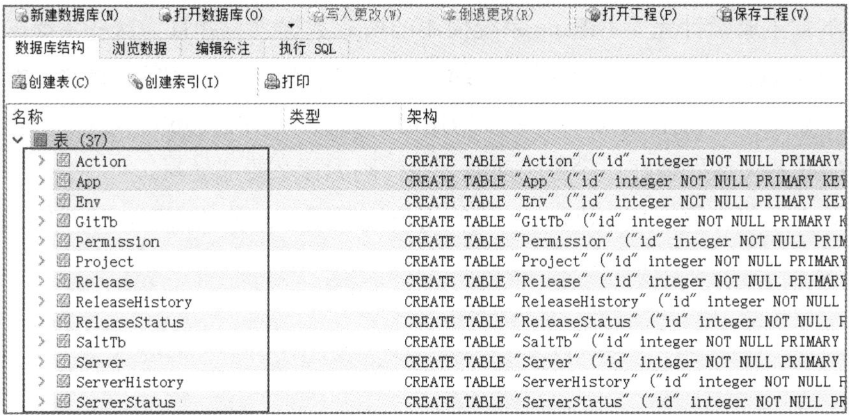

图 6-9　DB Browser for SQLite 查看所有数据表

图 6-10　admin 管理后台查看所有数据表

6.11 小　结

本章主要设计完了 BiFang 系统使用的所有数据表，并且为这些主要的表建立了模拟数据。在这个过程中也介绍了一些 ORM 的写法及字段属性。希望读者能了解每一个库表及每一个字段的用途，在接下来的代码学习过程中才能让 API 的联系关系处理好，才更能理解一些代码为什么要那样写。

第 7 章将主要介绍 BiFang 系统的用户注册登录部分的内容。让我们来写些代码吧！

第 7 章　后端用户模块

> 真正的无知不是缺少知识,而是拒绝获取知识。
>
> ——卡尔·波普尔

本章 GitHub 代码地址:https://github.com/aguncn/bifang-book/tree/main/bifang-ch7。

目前为止,已建好 GitLab 服务器,编写好 Demo 的 Bookinfo 组件代码、SaltStack 远程管理系统、File-Server 软件仓库服务器,安装好了 Python 语言虚拟环境、Django 及 DRF 框架库,建好了 bifangback 项目,并且为 bifangback 后端项目设计好了所有的数据库。

接下来,就要正式开始写 bifang 系统的后端代码了。本章主要来实现后端的用户模块,主要包括用户的注册和登录验证,附带实现一些小小的功能。

7.1　bifangback 默认首页

在第 5 章结束时,bifangback 后端项目的默认首页还只是一个 404 页面。第 6 章全都在设计数据库,对默认页面没有动过,所以还是 404。本节就来实现一个最简单的默认首页,让读者可以学到如何在 Django 里增加一个网页的常规套路。

7.1.1　新增一个默认首页的函数视图

在 bifangback 项目根目录的 bifangback 子目录,与 settings.py 同级的目录下新增一个 views.py 文件,内容如下:

```
from django.http import HttpResponse


def index(request):
    return HttpResponse('Hello, Bifang!')
```

在 Django 中，视图就是一个标准的 Python 函数或类，它的输入参数是 Web 请求对象 HttpRequest，输出返回值是 Web 应答对象 HttpResponse，大致处理流程如图 7-1 所示。

图 7-1　Django 视图

- 首先，从 django.http 模块导入了 HttpResponse 类。
- 接着，定义 index 函数，它就是视图函数。每个视图函数都使用 HttpRequest 对象作为第一个参数，并且通常称为 request。（注意，视图函数的名称并不重要，我们将其命名为 index，是因为这个名称能够比较准确地反映出它实现的功能。）
- 这个视图会返回一个 HttpResponse 对象，其中包含生成的响应。每个视图函数都负责返回一个 HttpResponse 对象。我们在这里只是生成了一个简单的 "Hello, Bifang!" 字符号，证明后端网站已成功启动。
- Django 中的 HttpRequest 对象和 HttpResponse 都包括很多属性和方法，读者可以从其官网中获取更细节的信息。

7.1.2　新增一个默认首页的路由

在 bifangback 项目根目录的 bifangback 子目录，与 settings.py 同级的目录下（也与上一小节创建的 views 同级），在项目级汇总 urls.py 文件中加入如下内容：

```
from .views import index

urlpatterns = [
    path('', index, name='index'),
    ...
]
```

路由就是用户访问网站资源的路径，在 Django 中路由要和视图函数对应起来，路由和视图的对应关系有一对一和多对一。静态路由是比较常见的路由，不使用尖括号和正则表达式定义的路由，与视图函数是一对一关系。使用尖括号定义的路由是动态的，可以定义传入变量并且指定变量的类型。
- from .views import index 表示从当前目录的 views 文件中导入 index 函数。
- path 中的第一个参数表示在访问网站时不带任何二级目录，即访问默认首页需要对应的函数。
- 省略号表示维持其他内容不变。

Django 路由方法 path 的原则可带四个参数，即 path（route、view、name、kwargs），四个参数详解如下：
- route：包含 url 模式的字符串。在处理请求时，Django 从第一个模式开始，urlpatterns 沿列表向下移动，将请求的 url 与每个模式进行比较，直到找到匹配的 url。
- view：当 Django 找到匹配的模式时，它将使用 HttpRequest 对象作为第一个参数，并将路线中所有"捕获"的值作为关键字参数来调用指定的 view 函数。
- name：命名 url 可以使你在 Django 中的其他地方（尤其是在 html 中 a 标签或者 action 响应 url 时）明确地引用它。这项强大的功能可让你修改单个文件即可对项目的 url 模式进行全局更改。
- kwargs：可以在字典中将任意关键字参数传递给目标视图，不是必需项。

7.1.3　测试默认首页

启动 bifangback 服务，在浏览器中访问：http://127.0.0.1:8000/，得到如图 7-2 所示页面，即表示默认页面没有问题。新增一个页面，在 Django 里就是这么简单；实现一个视图，再定义一个路由指向它即可。复杂点的页面还需要在视图里定义网页渲染模板，调用 ORM 从数据库中获取数据。但其操作的基本流程，并没有发生改变。

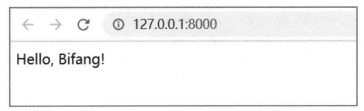

图 7-2　bifangback 默认首页

7.2 bifangback 用户与用户组的 API

本节使用 DRF 来快速实现用户与用户组的 API，而此 API 在本书后面的章节也会被前端调用。通过本节学习，希望读者可以了解 DRF 的 ModelViewSet 的序列化写法，使用它可以很快实现那些标准化的基于 Model 的 API。

关于用户账号，我们已专门新建一个 account 的 app 用来实现这些功能。所以，本节的文件操作主要是在这个目录下操作。

1. 新增 DRF 设置

因为从本节开始要开始大量使用 DRF 来实现 API 及 JWT 认证，所以在这里提前把配置设置到 settings.py 中，新增内容如下：

```
REST_FRAMEWORK = {
    'DEFAULT_PAGINATION_CLASS': 'utils.pagination.PNPagination',
    'DEFAULT_FILTER_BACKENDS': ('django_filters.rest_framework.DjangoFilterBackend',),
    'DEFAULT_PERMISSION_CLASSES': (
        'rest_framework.permissions.IsAuthenticated',
    ),
    'DEFAULT_AUTHENTICATION_CLASSES': (
        'rest_framework_jwt.authentication.JSONWebTokenAuthentication',
        'rest_framework.authentication.SessionAuthentication',
        'rest_framework.authentication.BasicAuthentication',
    ),
}

# 自定义 obtain_jwt_token 登录参数验证
AUTHENTICATION_BACKENDS = (
    'account.jwt_views.CustomJwtBackend',
)
JWT_AUTH = {
    'JWT_EXPIRATION_DELTA': datetime.timedelta(days=365),
    'JWT_AUTH_HEADER_PREFIX': 'Bearer',
    'JWT_ALLOW_REFRESH': True,
}
```

- DEFAULT_PAGINATION_CLASS 用于自定义分页。本书后面将讲解。
- DEFAULT_FILTER_BACKENDS 用于自定义过滤。本书后面将讲解。
- DEFAULT_PERMISSION_CLASSES 表示所有 API 应该需要用户验证才

可访问。
- DEFAULT_AUTHENTICATION_CLASSES 支 持 JSONWebToken（JWT）方式认证。
- JWT_AUTH 定义 JWT 的公共属性。

2. 定义用户与用户组的序列化文件

在 account 目录下新建一个 serializers.py 文件，其内容如下：

```python
from django.contrib.auth.models import Group
from django.contrib.auth import get_user_model
from rest_framework import serializers

User = get_user_model()


class GroupsReadOnly(serializers.ModelSerializer):
    class Meta:
        model = Group
        fields = ['id', 'name']


class UserSerializer(serializers.ModelSerializer):
    # many-to-many 的字段序列化
    groups_names = GroupsReadOnly(source='groups', many=True, read_only=True)

    class Meta:
        model = User                                              # 要序列的 model
        fields = ('id', 'url', 'username', 'email', 'groups', 'groups_names')   # 数据字段


class GroupSerializer(serializers.ModelSerializer):
    class Meta:
        model = Group                                             # 要序列的 model
        fields = ('url', 'name', 'id')                            # 数据字段
```

- 此文件完成了 DRF 中对用户与用户组字段的序列化和反序列化。
- 使用 get_user_model() 来获取 User 模型，在这里和 django.contrib.auth.models import User 表现相同。
- 由于我们想在用户的 API 输出中包含用户所属的用户组（一对多），所以在序列化文件中，多出一个 GroupsReadOnly 的类定义（作为 groups_names 字段，加入用户 API 中）。

- 序列化类继承自 serializers.ModelSerializer。
- fields 配置 API 中输出哪些字段。

DRF 序列化会把模型对象转换成 Python 字典，经过 response 响应以后变成 json 字符串。反序列化把客户端发送过来的数据，经过 request 以后变成字典，序列化器可以把字典转成模型。

如果想为 Django 存在的模型类快速创建序列化器，可以使用 DRF 框架封装好的 ModelSerializer 模型类序列化器，帮助我们快速创建一个 Serializer 类。它有额外的好处：基于模型类的字段条件创建约束；自动生成序列化器（serializer）；内部已经封装了 create 与 update 方法，序列化器对象在调用 save() 方法时，会自动对数据库进行操作。

3. 定义用户与用户组的视图文件

在 account 目录下新建 user_group_views.py 文件，内容如下：

```python
from django.contrib.auth.models import User, Group
from rest_framework import viewsets
from .serializers import UserSerializer, GroupSerializer


class UserViewSet(viewsets.ModelViewSet):
    """
    用户接口
    """
    queryset = User.objects.all().          # 指定 queryset
    serializer_class = UserSerializer       # 指定 queryset 对应的 serializers


class GroupViewSet(viewsets.ModelViewSet):
    """
    用户组接口
    """
    queryset = Group.objects.all()          # 指定 queryset
    serializer_class = GroupSerializer      # 指定 queryset 对应的 serializers
```

- 注释已足够，无须再解释。
- 与前一类的函数视图定义不一样，此处为类视图定义（Ddjango 中的视图定义就分为这两类）。
- order_by('-date_joined')，这种 ORM 写法表示按 date_joined 字段降序排序。

- 一般 DRF 的 API 生成时有两个数据需要满足，即序列化文件和数据库记录。另外可能还会包括过滤条件、自定义返回值和权限类等。

4. 增加路由

定义好视图之后就可以加入路由了。此处的路由定义分为两处，一处定义 DRF 的登录 API，另一处定义 account 应用的用户与用户组 API。定义 DRF 的登录 API，在汇总的 urls.py 中增加如下内容：

```
urlpatterns = [
    path('api-auth/', include('rest_framework.urls')),
    ...
]
```

定义 account 应用的用户与用户组 API，在 account 目录中的 urls.py 中增加如下内容：

```
from rest_framework import routers
from .user_group_views import UserViewSet, GroupViewSet
from .views import UserRegisterView


# 使用 router 注册 view，绑定 url 映射关系
# 关于什么时候使用 router，什么时候不能使用，看 API 的集成程度
router = routers.DefaultRouter()
router.register(r'users', UserViewSet)          # 绑定 view 到 users 路由下
router.register(r'groups', GroupViewSet)        # 绑定 view 到 groups 路由下
```

对于 DRF 中的视图集 ViewSet，我们除了可以自己手动指明请求方式与动作 action 之间的对应关系外（普通路由），还可以使用 routers 来帮助我们快速实现路由信息。

5. 测试用户与用户组 API

在本节前面的步骤操作完成后，启动 bifangback 应用。浏览器访问 http://127.0.0.1:8000/account/，默认出现如图 7-3 所示页面，单击右上角登录后，就会出现如图 7-4 所示页面，表示所有 API 都需要登录用户才能访问。

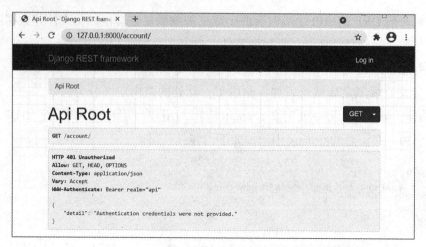

图 7-3　bifangback 用户与用户组 API 默认页面

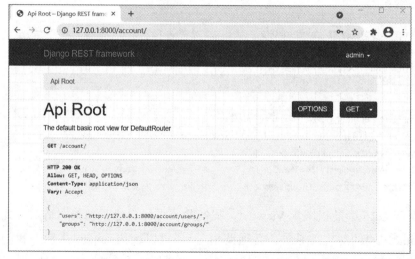

图 7-4　bifangback 用户与用户组 API 登录后页面

7.3　bifangback 用户注册

实现 bifangback 项目的用户注册，本节的 API 实现需要自定义返回，且和 swagger 文档做了一点集成，这些知识点将在本节一并学习。

1. 新增用户注册视图

在 account 目录下的 views.py 文件中（这个文件在使用 django-admin startapp account 这个脚手架命令时已自动生成），新增如下内容：

```python
from rest_framework.views import APIView
from django.contrib.auth import get_user_model
from utils.ret_code import *
from drf_yasg.utils import swagger_auto_schema

User = get_user_model()


# 用户注册
class UserRegisterView(APIView):
    """
    用户注册

    参数：
    {
        "username": "first",
        "password": "passwd",
        "passwordConfirm": "passwd",
        "email": "dem@sa.com"
    }
    """
    authentication_classes = []
    permission_classes = []

    @swagger_auto_schema(
        tags=['Users']
    )
    def post(self, request, *args, **kwargs):
        req_data = request.data
        username = req_data['username']
        password = req_data['password']
        password_confirm = req_data['passwordConfirm']
        email = req_data['email']
        if password != password_confirm:
            return_dict = build_ret_data(THROW_EXP, '两次输入密码不一致！')
            return render_json(return_dict)
        if User.objects.filter(username=username):
            return_dict = build_ret_data(THROW_EXP, '用户已经存在，请重新选择用户名！')
            return render_json(return_dict)
        try:
            User.objects.create_user(username=username,
                                    password=password,
                                    email=email,
```

```
                          is_active=True)
        return_dict = build_ret_data(OP_SUCCESS, str(req_data))
        return render_json(return_dict)
    except Exception as e:
        print(e)
        return_dict = build_ret_data(THROW_EXP, str(e))
        return render_json(return_dict)
```

- 用户注册类继承自 DRF 的 APIView 类，其核心方法就是 post() 方法。由于 APIView 类继承自 Django 的 view 类，所以 view 中的 request 请求在 APIView 中一样可以获取到。如代码中的写法，前端通过 POST 方法传过来的参数，后端可以通过 request.data 这个数据结构来获得。
- user.objects.create_user()，这个新增用户的 ORM 语法在第 6 章已介绍过。
- swagger_auto_schema 与 swagger 集成的语法本节接下来讲述。
- return render_json 自定义返回值，本节接下来讲述。

APIView 是 DRF 框架概念体系中最基本的类视图，也是其所有视图的基类，继承自 Django 的 view 父类。DRF APIView 与 Django View 的不同之处在于：

- 传入到视图方法中的是 REST framework 的 Request 对象，而不是 Django 的 HttpRequeset 对象。
- 视图方法可以返回 REST framework 的 Response 对象，视图会为响应数据设置 render 符合前端要求的格式。
- 任何 APIException 异常都会被捕获到且处理成合适的响应信息。
- 在进行 dispatch() 分发前，会对请求进行身份认证、权限检查和流量控制。

DRF APIView 支持定义的属性：

- authentication_classes 列表或元祖，身份认证类。
- permissoin_classes 列表或元祖，权限检查类。
- throttle_classes 列表或元祖，流量控制类。

在 APIView 中仍以常规的类视图定义方法来实现 get()、post() 或者其他请求方式的方法。

2. bifangback 项目的 swagger 文档依赖库

swagger 是后台开发中很好用的交互式文档，Django 原本的 Django Swagger 已经停止维护了，现在一般用 drf_yasg 这个包来实现文档，它里面支持 swagger

和 redoc，redoc 是静态的，作为导出文档的话不错，不过一般用 swagger，因为可以在文档里面调试，非常方便。

在 bifangback 项目中使用 drf_yasg 这个库，来实现 swagger 的文档管理。在项目的 requirements.txt 中已经安装这个依赖库，在 settings.py 的 INSTALLED_APPS 列表中也已存在。此处看看如何简单地配置和使用 drf_yasg 库。本书后面代码的 drf_yasg 部分不再重复。

（1）在 bifangback 项目汇总 urls.py 中，drf_yasg 的部分配置如下：

```python
from rest_framework import permissions
from drf_yasg.views import get_schema_view
from drf_yasg import openapi

schema_view = get_schema_view(
    openapi.Info(
        title=" 毕方 (Bifang) 自动化部署平台 API",
        default_version='v1',
        description=" 毕方 (Bifang) 自动化部署平台，引导你进入 devops 开发的领域。",
        terms_of_service="https://github.com/aguncn/bifang",
        contact=openapi.Contact(email="aguncn@163.com"),
        license=openapi.License(name="BSD License"),
    ),
    public=True,
    permission_classes=(permissions.AllowAny,),
)

urlpatterns += [
    re_path(r'^swagger(?P<format>\.json|\.yaml)$',
        schema_view.without_ui(cache_timeout=0),
        name='schema-json'),
    path('swagger/',
        schema_view.with_ui('swagger', cache_timeout=0),
        name='schema-swagger-ui'),
    path('redoc/',
        schema_view.with_ui('redoc', cache_timeout=0),
        name='schema-redoc'),
]
```

（2）在 API 类中写好注释，因为注释把这些参数注明后，可以被 swagger 自动提取，UserRegisterView() 类开头的三引号注释就是这类写法。当然，这里也可以在 @swagger_auto_schema() 定义更专业的参数。这里省略了，读者可自行练习。

```
@swagger_auto_schema(
    tags=['Users']
)
```

drf_yasg 里有个 AutoSchema，可以自动扫描 ViewSet 和 APIView 这类可以提供接口的地方，和 Spring 里面基于注解的文档定义不同，一般在 Drf 里不需要手动配置每个接口的名称和说明，只要写在 pydoc 里就可以。不过这个 AutoSchema 也不是很准确，所以这里使用 @swagger_auto_schema() 装饰器自定义配置 tags，将注册用户这个 url 放在 Users 这个 tag 组里。

（3）启动服务，打开浏览器访问：http://127.0.0.1:8000/swagger/，可以看到用户注册这个 url 已在 swagger 文档中，如图 7-5 所示。

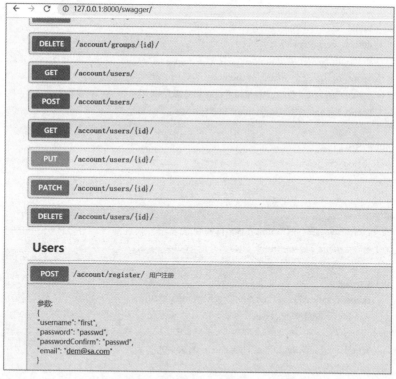

图 7-5 bifangback 用户注册 swagger 文档

3. 统一 bifangback 项目的自定义返回值

统一一个项目的 web 返回值是很有意义的。它可以标准化前端的 API 返回值，统一处理正确和错误的代码或返回值含义，且可以让后端返回更有语义性的值，减少前后端的沟通成本，提高工作效率。

bifangback 项目的自定义返回值并不是本书作者原创，只是看各种源代码的返回值看得多了，大家都会遵守同样的最佳实践。本书后面代码的自定义返回值部分不再重复。

（1）在 bifangback 项目根目录下新建一个 utils 目录，用作 bifangback 项目的公共功能的文件存放。在 utils 目录下新建一个 ret_code.py 文件，内容如下所示：

```python
import json
from django.http import HttpResponse


OP_SUCCESS = 0
THROW_EXP = 1000
OP_DB_FAILED = 1001
CHECK_PARAM_FAILED = 1002
FILE_FORMAT_ERR = 1003
NOT_POST = 1004
NOT_GET = 1005
NOT_PERMISSION = 2000

ERR_CODE = {
    0: ' 操作成功 ',
    1000: " 抛出异常 ",
    1001: " 数据库操作失败 ",
    1002: " 参数检查失败 ",
    1003: " 文件格式有误 ",
    1004: " 非 post 请求 ",
    1005: " 非 get 请求 ",
    2000: " 无权限 "
}


def build_ret_data(ret_code, data=''):
    return {'code': ret_code, 'message': ERR_CODE[ret_code], 'data': data}


def render_json(dictionary={}):
    response = HttpResponse(json.dumps(dictionary), content_type="application/json")
    response['Access-Control-Allow-Origin'] = '*'
    response["Access-Control-Allow-Headers"] = "Origin, X-Requested-With, Content-Type"
```

```
response["Access-Control-Allow-Methods"] = "GET, POST, PUT, OPTIONS"
return response
```

在这个文件里，render_json() 函数自定义了 Django 的 HttpResponse 对象，让它以 json 的格式返给前端内容。当然，Django 本身也有 JsonResponse 对象来实现这一功能。我感觉这样的自定义灵活性更好一些。

build_ret_data() 方法用于自定义需要 json 化的 Python 字典。为了让返回值更有标准语义性，使用 ERR_CODE 字典来聚合所有的返回信息。

（2）在具体的视图中使用统一返回值的过程如下：

```
from utils.ret_code import *

# 成功时的 json 数据
return_dict = build_ret_data(OP_SUCCESS, ' 成功想返回的数据 ')
# 失败时的 json 数据
return_dict = build_ret_data(THROW_EXP, ' 失败原因说明 ')
# 自定义返回
return render_json(return_dict)
```

这里的代码，并列地显示了成功或失败后执行代码（在真正的源代码文件里，要么成功，要么失败，不会同时返回两种结果），其具体效果在本节最后展示。

4. 用户注册路由

在 account 的 urls.py 文件里，新增如下用户注册路由：

```
from .views import UserRegisterView

urlpatterns = [
    ...
    path('register/', UserRegisterView.as_view(), name='register'),
]
```

视图类在路由中调用时，需要使用 as_view() 方法，进行实例化。

5. Postman 测试功能

因为到目前为止，BiFang 系统的前端项目还没有开始。因此，这里选用 Postman 工具，在 Windows 系统上做可视化的 API 测试。

（1）用户注册成功，如图 7-6 所示。

第 7 章 后端用户模块

图 7-6 用户注册成功

（2）用户注册失败，如图 7-7 所示。

图 7-7 用户注册失败

7.4 bifangback 用户 JWT 认证

本章前述内容已实现了用户的注册，接下来应该进行用户的登录操作。在 Django 以前的开发模式中，主要是通过 views 视图将数据传递给 Template 模板，在模板文件中结合 JS 前端代码和视图数据，在服务器端完成网页渲染工作，再将所有数据返回给前端浏览器。在这种开发模式下，用户登录主要使用 session 登录机制。

现在，在大型网站开发或功能复杂、多个使用方调用的场景下，流行"前后端分离"的开发模式。在这种模式下，后端只提供 API 数据，前端调用后端，本身使用 Vue、React、Angular 等框架开发，实现复杂的用户交互和 HTML DOM 双向绑定。在这种开发模式下，主要用 JWT 验证机制。

本书采用的即是前后端分离的开发模式，用户登录验证采用 JWT（Json web token）。

1. JWT 认证简介

（1）JWT 是为了在网络应用环境间传递声明而执行的一种基于 JSON 的开放标准（RFC 7519）。该 token 被设计为紧凑且安全的，特别适用于分布式站点的单点登录(SSO)场景。JWT 的声明一般被用来在身份提供者和服务提供者间传递被认证的用户身份信息，以便于从资源服务器获取资源，也可以增加一些额外的、其他业务逻辑所必需的声明信息，该 token 也可直接被用于认证，也可被加密。

（2）JWT 表现为一段字符串，由三段信息构成，将这三段信息文本用".(点)"链接在一起就构成了 JWT 字符串。例如：

```
eyJ0eXAiOiJKV1QiLCJhbGciOiJIUzI1NiJ9.
eyJ1c2VyX2lkIjoxNywidXNlcm5hbWUiOiJza3kiLCJlZXhwIjoxNjQ5MTU5MzQ5LCJlbWFpbCI6ImFkbWluNAY
WQuY29tIiwib3JpZ19pYXQiOjE2MTc2MjMzNDI9.
FqElj3QyromtLg_aUGJGdFrjrSqT3vBNwau4CYs9-Gw
```

第一部分称为头部（header），第二部分称为载荷（payload），第三部分为签证（signature）。

JWT 的头部承载两部分信息：①声明类型；②声明加密的算法。完整的头部就像下面的 json：

```
{
    'typ': 'JWT',
    'alg': 'SHA256'
}
```

然后将头部进行 base64 加密（该加密是可以对称解密的），构成了第一部分：

eyJ0eXAiOiJKV1QiLCJhbGciOiJIUzI1NiJ9

载荷就是存放有效信息的地方。这些有效信息包含三个部分：①标准中注册的声明；②公共的声明；③私有的声明。将这些信息进行 base64 加密，得到 JWT 的第二部分：

eyJ1c2VyX2lkIjoxNywidXNlcm5hbWUiOiJza3liiwiZXhwIjoxNjQ5MTU5MzQ5LCJlbWFpbCI6ImFkYmNNAYWQuY29tIiwib3JpZ19pYXQiOjE2MTc2MjMzNDl9

JWT 的第三部分是一个签证信息，这个签证信息由三部分组成：① header（base64 后的）；② payload（base64 后的）；③ secret——这个部分需要 base64 加密后的 header 和 base64 加密后的 payload，使用".（点）"连接组成的字符串，然后通过 header 中声明的加密方式进行加盐 secret 组合加密，然后就构成了 JWT 的第三部分。

FqEIj3QyromtLg_aUGJGdFrjrSqT3vBNwau4CYs9-Gw

注意：secret 是保存在服务器端的，JWT 的签发生成也是在服务器端的，secret 就是用来进行 JWT 的签发和验证，所以它就是你服务端的私钥，在任何场景都不应该流露出去。一旦客户端得知这个 secret，那就意味着客户端可以自我签发 JWT。

总结一下 JWT 生成的流程，如图 7-8 所示。

图 7-8　JWT 生成流程

JWT 的优点：
- 天生适用于 SSO 单点登录。
- 服务器端只需要存储秘钥，不需要分别存放用户的 JWT 信息，降低了服务器的存储压力。

- 比 cookie 更加安全可靠。
- 还可以用于提供数据加密传输的功能。

JWT 的缺点：
- 因为 JWT 是存放在客户端的，所以一旦签发以后，服务端是无法控制的或提前回收。
- 增加了客户端的数据传输加密的功能。

2. JWT 视图文件

bifangback 项目的 requirements.txt 文件里已包含了 djangorestframework-jwt 依赖库，这个库主要用来实现 DRF 框架的 JWT 用户认证功能。djangorestframework-jwt 库可以在 settings.py 文件里自定义 JWT_AUTH 这个字典，来实现一些自定义的 JWT 认证配置。bifangback 项目的 settings.py 文件里，关于这个部署的设置如下：

```
# 自定义 obtain_jwt_token 登录参数验证
AUTHENTICATION_BACKENDS = (
    'account.jwt_views.CustomJwtBackend',
)
JWT_AUTH = {
    'JWT_EXPIRATION_DELTA': datetime.timedelta(days=365),
    'JWT_AUTH_HEADER_PREFIX': 'Bearer',
    'JWT_ALLOW_REFRESH': True,
}
```

- AUTHENTICATION_BACKENDS，随后马上实现。
- JWT_AUTH 中的 JWT_EXPIRATION_DELTA 用于定义 JWT 的过期时间，这时演示定义了一年的过期时间，用于企业内网的话建议设置更短时间，也不能为了方便就忘记了安全。
- JWT_AUTH 中的 JWT_AUTH_HEADER_PREFIX 用于定义令牌请求时，请求头的字段值前缀，正常的前缀有 bearer 和 authorization，这里取前者，与前端请求配合。
- JWT_AUTH 中的 JWT_ALLOW_REFRESH 用于允许 JWT 的刷新请求。

由于我们要自定义统一的错误返回及在 payload 里自定义用户数据，所以我们整个自定义了 CustomWebTokenAPIView。

在 account 目录下新建一个 jwt_views.py，内容如下：

```
from datetime import datetime
```

```python
from django.contrib.auth.backends import ModelBackend
from django.db.models import Q
from rest_framework import status
from rest_framework.response import Response
from rest_framework_jwt.settings import api_settings
from rest_framework_jwt.views import JSONWebTokenAPIView
from rest_framework_jwt.views import ObtainJSONWebToken
from rest_framework_jwt.views import RefreshJSONWebToken
from rest_framework_jwt.views import VerifyJSONWebToken
from django.contrib.auth import get_user_model
from drf_yasg.utils import swagger_auto_schema
from django.contrib.auth.models import Group

User = get_user_model()


class CustomJwtBackend(ModelBackend):
    """
    自定义用户验证，定义完之后还需要在 settings 中进行配置
    """
    def authenticate(self, request, username=None, password=None, **kwargs):
        try:
            user = User.objects.get(Q(username=username) | Q(email=username))
            # django 里面的 password 是加密的，前端传过来的 password 是明文，
            # 调用 check_password 就会对明文进行加密，比较两者是否相同
            if user.check_password(password):
                return user
        except Exception as e:
            return None


# 在不改 rest_framework_jwt 源码的情况下，自定义登录后成功和错误的返回，最优雅
def jwt_response_payload_handler(token, user=None, expiration=None):
    """
    自定义 JWT 认证成功返回数据
    """
    # 向前端请求返回用户角色列表，admin 为管理员角色，其他为非管理员角色
    # 如果用户属于多个用户组，只要其中一个为 admin 组，就为管理员角色
    # permissions 仅为 demo 演示，在本项目中未使用
    # is_superuser 为是否能登录 django admin 后台管理界面
    roles = list()
    user_groups = Group.objects.filter(user=user)
    for user_group in user_groups:
```

```python
                roles.append({
                    'id': user_group.name
                })
            data = {
                'token': token,
                'expireAt': expiration,
                'user_id': user.id,
                'user': {
                    'id': user.id,
                    'name': user.username,
                    'email': user.email,
                    'avatar': ''},
                'is_superuser': user.is_superuser,
                'permissions': [{'id': 'demo', 'operation': ['demo']}],
                'roles': roles,
            }
    return {'code': 0, 'message': ' 欢迎回来 ', 'data': data}


def jwt_response_payload_error_handler(serializer, requst=None):
    """
    自定义 JWT 认证错误返回数据
    """
    data = {
        'message': " 用户名或者密码错误 ",
        'status': 400,
        'detail': serializer.errors,
    }
    return {'code': -1, 'data': data}


# JWT 的返回由 JSONWebTokenAPIView 自定义它的调用和返回即可
class CustomWebTokenAPIView(JSONWebTokenAPIView):
    """
    用户 token 验证登录

    参数：
    username
    password
    """

    @swagger_auto_schema(
        tags=['Users']
```

```python
        )
        def post(self, request, *args, **kwargs):
            serializer = self.get_serializer(data=request.data)
            if serializer.is_valid():
                user = serializer.object.get('user') or request.user
                token = serializer.object.get('token')
                expiration = (datetime.utcnow() +
                    api_settings.JWT_EXPIRATION_DELTA)
                response_data = jwt_response_payload_handler(token, user, expiration)
                response = Response(response_data)
                if api_settings.JWT_AUTH_COOKIE:
                    response.set_cookie(api_settings.JWT_AUTH_COOKIE,
                                        token,
                                        expiration=expiration,
                                        httponly=True)
                return response
            error_data = jwt_response_payload_error_handler(serializer, request)
            return Response(error_data, status=status.HTTP_200_OK)


class CustomObtainJSONWebToken(ObtainJSONWebToken, CustomWebTokenAPIView):
    pass


class CustomRefreshJSONWebToken(RefreshJSONWebToken, CustomWebTokenAPIView):
    pass


class CustomVerifyJSONWebToken(VerifyJSONWebToken, CustomWebTokenAPIView):
    pass


obtain_jwt_token = CustomObtainJSONWebToken.as_view()
refresh_jwt_token = CustomRefreshJSONWebToken.as_view()
verity_jwt_token = CustomVerifyJSONWebToken.as_view()
```

CustomJwtBackend 类自定了 JWT 认证时的用户和密码组合认证，这个类被 settings.py 里的 AUTHENTICATION_BACKENDS 引用即生效。

CustomWebTokenAPIView 类里引用了自定义的错误返回函数 jwt_response_payload_error_handler()，也自定义了 jwt_response_payload_handler 函数，用于定制 payload 的内容。其中 'permissions' 这个字段，我看很多开源的管理框架里都有这个内容，也就象征性地填充了一下内容，在 BiFang 系统中这个字段并没有用到。整个 BiFang 系统权限控制系统，为了保持轻量都是在后端进行权限控制的。

网上有其他一些文件，为了自定义 JWT 的错误返回值，采用了更改 djangorestframework-jwt 库源码的形式来实现，我觉得那并不可取，还是要将代码的变化控制在一定范围内。辐射到源码的行为，是绝对不可以的。我在此借鉴了网上的一个文档，使用全自定义 CustomWebTokenAPIView 这个继承类组合 mixin 的方式，最优雅。当然，如果后期 djangorestframework-jwt 库将自定义 jwt_response_payload_error_handler() 函数的功能开发出来，可以更简单地实现。

在稍后的测试中都可以看到这些自定义内容的具体的形式。

3. JWT 路由设置

在汇总 urls.py 文件里增加以下内容，用于用户 JWT 认证的相关路由：

```
from account import jwt_views


urlpatterns = [
    ...
    path('jwt_auth/', jwt_views.obtain_jwt_token),
    path('refresh_jwt_auth/', jwt_views.refresh_jwt_token),
    path('verify_jwt_auth/', jwt_views.verity_jwt_token),
]
```

这里引用的三个变量，都是在 jwt_views.py 文件最后已实例化的路由变量，从而达到路由请求的目的。

4. Postman 测试功能

（1）用户认证成功，如图 7-9 所示。

在用户认证成功后，前端就跳转到用户首页，然后将 token 值缓存到本地，前端每次发起后端请求时，都在 Bearer 请求头里带上这个值，后端就可以准备获取到这个 token，然后进行解析，从而获取其用户值。

同时，用户名密码认证成功后，后端还向前端传递了用户的其他信息，这些信息也可以缓存在本地，使用 vue 的 store 组件来做网页的信息展示和状态同步。这些内容，在本书后面的前端讲解时都会一一涉及。

图 7-9　JWT 用户认证成功

（2）用户认证失败，如图 7-10 所示。

图 7-10　JWT 用户认证失败

7.5　小　　结

本章主要开发完了 bifangback 项目关于用户功能相应的代码。

第 8 章主要开发 bifangback 的项目和应用管理的代码。这些代码的实现将在本节的基础上引入更多的内容，比如分页、过滤等，敬请期待。

第 8 章　后端项目及应用模块

> 一阴一阳之谓道，继之者善也，成之者性也。
>
> ——《易经·系辞上》

本章 GitHub 代码地址：https://github.com/aguncn/bifang-book/tree/main/bifang-ch8。

第 7 章主要讲解了如何实现后端的用户注册和登录的功能。本章主要来实现后端的项目和应用的管理功能，主要包括项目和应用的增删改查，以及基于应用的权限管理。

为了让 bifang 项目适应更大规模的管理，我们设计了基于项目（project）和应用（app）的二级管理模式。设计的出发点是每一个项目，就是一个大的微服务模块，而每一个应用就是一个具体的微服务组件，或是一个部署单元。每一个项目可以包含一个或多个应用，而每一个应用必属于某一个项目。

废话不多说，直接上硬菜！

8.1　实现 bifangback 项目列表 API

毕方（BiFang）项目管理的列表如图 8-1 所示，本节就主要来实现这个网页的后端功能。可以看到，除了基本的项目列表显示，我们还提供了分页、基于名称和时间的查询搜索功能。

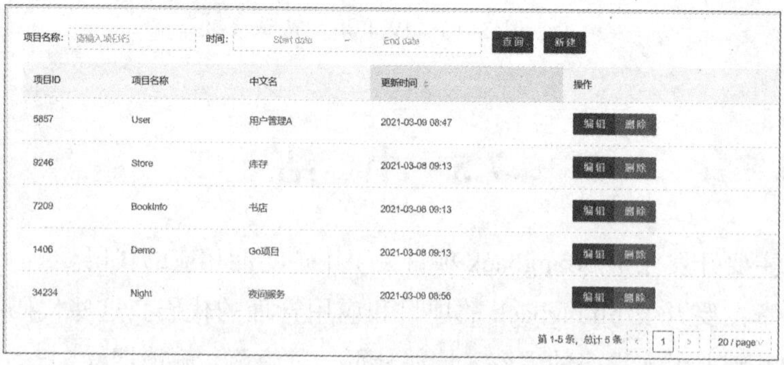

图 8-1　毕方（BiFang）项目管理

1. 新增一个序列化文件

在 bifangback 项目根目录的 project 子目录新增一个 serializers.py 文件，文件内容如下：

```python
from rest_framework import serializers
from cmdb.models import Project


class ProjectSerializer(serializers.ModelSerializer):
    create_user_name = serializers.CharField(source='create_user.username',
                                              read_only=True)

    class Meta:
        model = Project
        fields = '__all__'
```

我们知道前后端常用 json 数据结构交互，在后端我们常想把一个 Python 对象返回给前端。但是 json 序列化是不能序列化对象，所以就有了 DRF 的序列化组件。

在 DRF 中是通过引入 rest_framework 框架中的 serializers 模块来实现序列化和反序列化的功能。一般来说，把 Django 模型对象转换成字典，经过 response 以后变成 json 字符串的过程，称为序列化。而把客户端发送过来的数据，经过 request 以后完成数据校验功能，将其变成字典并把字典转成 Django 模型的过程，称为反序列化。

序列化及反序列化的知识点很多，一条条列举也很枯燥。我们可以通过实践一条条地积累，最后总结出来的知识点才是自己真正理解并掌握的知识。

- ProjectSerializer 类继承自 serializers.ModelSerializer，它可以快速地将一个 model 序列化，而无需一个字段一个字段地定义。同时，它也保留一些自定义的字段的技巧，可以满足很多场合下的模型序列化。
- 类中类 Meta，表示此序列化类的操作模型是 Project 数据表（此表的字段定义见第 6 章），默认支持所有字段的序列化操作。
- create_user_name 行自定义了一个序列化字段。Source 表示其取值是 create_user 的外键关联名称，read_only=True 表示这个字段仅用于给前端显示序列，而不会在 request 请求中包含此字段，也无须对其进行反序列化操作。此效果会在后面章节看到，到时再进行 callback 提示。

2. 新增一个过滤器文件

在 DRF 框架的使用中有一个比较让人头疼的问题，即怎么满足前端那"变态"的数据过滤需求，特别是当前端提供所谓的灵活查询，过滤条件更是五花八门。我们这里提供的是按日期区间和项目名称的过滤。

在 bifangback 项目根目录的 project 子目录中新增一个 filters.py 文件，文件内容如下：

```python
from django_filters import OrderingFilter
from django_filters.rest_framework import FilterSet
from django_filters import filters
from cmdb.models import Project


class ProjectFilter(FilterSet):
    name = filters.CharFilter(field_name='name', lookup_expr='icontains',)
    begin_time = filters.DateTimeFilter(field_name='create_date', lookup_expr='gte',)
    end_time = filters.DateTimeFilter(field_name='create_date', lookup_expr='lte',)
    sort = OrderingFilter(fields=('create_date','update_date'))

    class Meta:
        model = Project
        fields = ['name', 'begin_time', 'end_time']
```

DRF 本身也提供了内置的 Filter 过滤功能，但为了支持更强的功能，我在这里使用了一个第三方包——django_filters。在 7.2 节中已自定义了 DEFAULT_FILTER_BACKENDS，这里对 DRF 的过滤器做进一步的定制。前端的搜索过滤写法，会在本章后面进行测试。

- ProjectFilter 类继承自 django_filters 包中的 FilterSet 类。
- 过滤器类和 Django 中表单类极其类似，写法基本一样，目的是指明过滤的时候使用哪些字段进行过滤，每个字段可以使用哪些运算。运算符的写法基本参照 Django 的 ORM 中查询的写法，比如大于等于，小于等于用 "gte" "lte" 等。
- Class Meta 类中类指明了以 Project 这个 model 模型过滤。Fields 指明了使用哪几个字段参与过滤条件 ['name', 'begin_time', 'end_time']。
- Name 类变量定义使用 name 作为 fields，忽略大小写匹配。
- Begin_time 和 endtime 分别匹配搜索字段中的日期上下限。

3. 新增一个分页文件

如果我们要获取数据库中一张表内所有的数据，但是这张表的数据可能会达到千万条的级别，我们不太可能需要一次把所有的数据都展示出来，因为数据量很大，对服务端的内存压力比较大，而且在网络传输过程中耗时也会比较大，使用分页就可以优化这些问题。DRF 也内置支持不同的分页模式，这里先用 PageNumberPagination 的分页模式。

在 bifangback 项目根目录的 utils 子目录新增一个 pagination.py 文件，文件内容如下：

```python
from rest_framework.pagination import PageNumberPagination


class PNPagination(PageNumberPagination):
    page_size = 15
    max_page_size = 1000
    page_size_query_param = 'pageSize'
    page_query_param = 'currentPage'

    '''
    age_query_param：表示 url 中的页码参数
    page_size_query_param：表示 url 中每页数量参数
    page_size：表示每页的默认显示数量
    max_page_size：表示每页最大显示数量，做限制使用，避免突然大量的查询数据致使数据
                   库崩溃
    '''
```

在 7.2 节中已自定义的 DEFAULT_PAGINATION_CLASS 指向的就是这个文件。这个文件就定义了一个类。如果在 DRF 的类视图中没有单独定义分页类的话，整个 bifang 项目默认都是使用这个分页类。

- PNPagination 类继承自 pagination 包中的 PageNumberPagination 类。
- 类变量的定义已在源代码中有注释，在请求的 url 参数中，需要使用 currentPage 来定义获取的记录，默认每次获取的记录数 pageSize 为 15。

4. 实现项目列表类视图

Django 中基于类的视图对于旧式风格的视图来说是良好的替代品，DRF 提供了一个 APIView 类，它是 Django 的 view 类的子类。APIView 类和一般的 view 类有以下不同：被传入处理方法的请求不会是 Django 的 HttpRequest 类的实例，而是 REST framework 的 Request 类的实例；处理方法可以返回 REST framework 的 Response，而不是 Django 的 HttpRequest；视图会管理内容协议，

给响应设置正确的渲染器；任何 APIException 异常都会被捕获，并且传递给合适的响应；进入的请求将会经过认证，合适的权限和（或）节流检查会在请求被派发到处理方法之前运行。

DRF 中的类视图继承关系如图 8-2 所示。

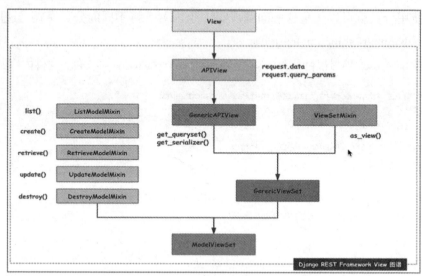

图 8-2　DRF 类视图继承关系

在 bifangback 项目根目录的 project 子目录中的 views.py 文件新增一个 ProjectListView 类，内容如下：

```
from cmdb.models import Project
from .serializers import ProjectSerializer
from rest_framework.generics import ListAPIView
from utils.ret_code import *
from .filters import ProjectFilter


class ProjectListView(ListAPIView):
    queryset = Project.objects.all()
    serializer_class = ProjectSerializer
    filter_class = ProjectFilter

    def get(self, request, *args, **kwargs):
        res = super().get(self, request, *args, **kwargs)
        return_dict = build_ret_data(OP_SUCCESS, res.data)
        return render_json(return_dict)
```

可以看到，在使用 DRF 类视图时，相关的定义都很精简和规范，这也是我们推荐读者好好利用类视图的原因。

- ProjectListView 类继承自 rest_framework.generics 包中的 ListAPIView 类。此视图类主要用于显示模型里的多条列表记录。
- Queryset 变量的 Django ORM 默认显示 Project 模型里的所有记录。
- serializer_class 为此节第 1 小节中定义的序列化类。
- filter_class 为此节第 2 小节中定义的过滤器类。
- 为了能自定义视类图的返回内容，我们使用了 7.3 节定义的 retcode 文件。后面大量的类定义都使用了这个技巧，到时不再重复讲解。

5. 增加路由

定义好项目的类视图后，就可以加入项目的路由了。在 project 目录中的 urls.py 文件内，增加如下内容：

```
from django.urls import path
from . import views

app_name = "project"

urlpatterns = [
    path('list/', views.ProjectListView.as_view(), name='list'),
    path('create/', views.ProjectCreateView.as_view(), name='create'),
    path('detail/<pk>/', views.ProjectRetrieveView.as_view(), name='detail'),
    path('update/<pk>/', views.ProjectUpdateView.as_view(), name='update'),
    path('delete/<pk>/', views.ProjectDestroyView.as_view(), name='delete'),
]
```

为了方便，这里将 project 项目涉及的所有路由定义都在此列明，在之后的 project 讲解章节中，不再说明这些路由的定义环境。如果要分别测试每一个 url，则要先写好类视图，再在 urls.py 里增加对应的路由，不然启动会报错，表示对应的类视图定义还不存在。

<pk> 这样的语法是指前端传递过来 url 时，最后一个 "//" 之间的参数将会作为 pk 变量名传递到类视图中，这也是类视图默认接收的变量。

As_view() 这样的语法是将类视图转化为函数，因为在 urlpatterns 列表中 path 的定义里，第二个参数应该为函数。

6. 测试项目列表 API

（1）完成前面的操作后启动 bifangback 服务。

（2）按 7.4 节定义，启动 postman，使用用户密码通过认证后，获取并记录好 JWT 值。

（3）在 postman 和 authorization 的 Bearer Token 里，输入刚才的 JWT 值，如图 8-3 所示。这个动作是以后所有的后端 API 测试时都需要使用的步骤，到时就不再重复演示这一步了。

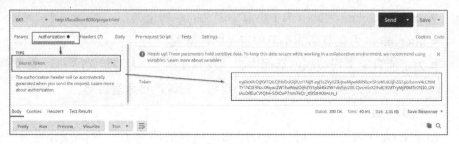

图 8-3　JWT 用于后端 API 认证

（4）在 postman 的 url 中输入：http://localhost:8000/project/list/，得到如图 8-4 所示的数据，表示一切正常。由于我们的模拟项目未到 15 个，所以输出的 next 和 previous 都为 null，表示没有分页。后面如果有超过一页的数据时，读者会看到这里有分页数据，到时再 callback。另外，读者应注意 create_user_name 的输出，这是对应于本节 1 中的 callback，向前端返回一个 model 里没有的字段。

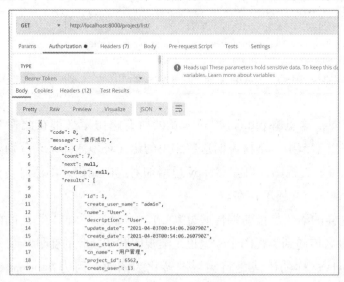

图 8-4　获取所有项目列表

（5）在 postman 的 url 中输入：http://localhost:8000/project/list/?name=Store，

得到如图 8-5 所示数据，表示过滤功能正常。这里指定了 url 的 name 参数，后端就只输出包含 Store 关键字的数据。

图 8-5　项目列表的过滤

8.2　实现 bifangback 新增项目的 API

本节来实现项目新增的 API。前端界面展示如图 8-6 所示，在编写后端 API 时，结合前端的展示效果就能和前端开发同事更有效率地配合。

图 8-6　新增项目

1. 实现新增项目的类视图

在 bifangback 项目根目录的 project 子目录中的 views.py 文件新增一个 ProjectCreateView 类，内容如下：

```python
from rest_framework.generics import CreateAPIView


class ProjectCreateView(CreateAPIView):
    serializer_class = ProjectSerializer

    def post(self, request):
        """
        {
            "name": "projectA",
            "cn_name":"A 项目 ",
            "description": " 这是一个 A 项目 ",
            "project_id": 4325
        }
        """
        req_data = request.data
        data = dict()
        data['project_id'] = req_data['project_id']
        data['name'] = req_data['name']
        data['description'] = req_data['description']
        data['cn_name'] = req_data['cn_name']
        # 从 drf 的 request 中获取用户（对 django 的 request 做了扩展的）
        data['create_user'] = request.user.id
        serializer = ProjectSerializer(data=data)
        if serializer.is_valid() is False:
            return_dict = build_ret_data(THROW_EXP, str(serializer.errors))
            return render_json(return_dict)
        data = serializer.validated_data
        Project.objects.create(**data)
        return_dict = build_ret_data(OP_SUCCESS, serializer.data)
        return render_json(return_dict)
```

- ProjectCreateView 类继承自 CreateAPIView 类，它主要实现了记录的新增功能。我们在这个类中改写了 post 方面，用于自定义接收浏览器的参数，并将这些接收到的参数插入数据库中。
- serializer_class 还是使用上一节定义的序列化类，因为这个项目的 model 里的所有字段都可以获取和更新，所以可以共用同一个序列化类，而无须再自定义一个。但在比较复杂的应用中，针对同一个 model，因为不同的查询、新增和更新需求，是可以定义不同的序列化类的。
- """ 三引号的跨行注释中，定义了从浏览器前端获取到的 json 数据格式，方便在后面写代码时，不会忘记所有的字段参数提取。

- DRF 中的 request.data 返回请求主题的解析内容，与标准的 Django 中的 request.POST 和 request.FILES 类似，并且还具有其他特点：它包括所有解析的内容，文件（file）和 非文件（non-file inputs）；它支持解析 post 以外的 HTTP method，比如 PUT、PATCH；它更加灵活，不仅仅支持表单数据，传入同样的 json 数据一样可以正确解析，并且不用做额外的处理（意思是前端不管提交的是表单数据，还是 json 数据，".data" 都能正确解析）。
- 接下来使用一个 data 字典，承载一个 project 项目的数据。
- 后面，对 data 进行数据有效性验证，如果验证通过，则将从前端获取到的数据插入数据库中，从而完成一个项目数据的插入。如果失败，则使用自定义的错误返回。

2. 测试新增项目 API

在本节前面的步骤操作完成后（这里省略了路由的增加和 JWT 的获取），启动 bifangback 应用。在 Postman 的 url 中输入：http://localhost:8000/project/create/，以 post 方式请求，再设计一个简单的 json 数据，然后发送请求 (send)。默认出现图 8-7 所示的结果，表示编码无误。同时，登录 django admin 管理后台，可以看到新增的项目已入库，如图 8-8 所示。在之后的后端 API 开发中，可以用同样的方式通过 Django admin 管理后台，来验证相关数据是否正确。这是一个小技巧，比直接在数据库里查看这些记录方便多了。

图 8-7　测试新增项目 API

图 8-8　Dajngo admin 后台查看新增项目

8.3　实现 bifangback 查看具体项目的 API

对数据库记录的增、删、改、查，是写一般数据库类程序应用的基本功。前面章节已实现了对 bifang 应用数据库里项目数据表的列表查询，新增了记录的功能。接下来就来实现对项目数据表里具体记录的查看、修改和删除吧。

本节来实现对具体项目的查看功能。

1. 实现查看具体项目的类视图

在 bifangback 项目根目录的 project 子目录中的 views.py 文件新增一个 ProjectRetrieveView 类，内容如下：

```
from rest_framework.generics import RetrieveAPIView

class ProjectRetrieveView(RetrieveAPIView):
    queryset = Project.objects.all()
    serializer_class = ProjectSerializer

    def get(self, request, *args, **kwargs):
```

```
        res = super().get(self, request, *args, **kwargs)
        return_dict = build_ret_data(OP_SUCCESS, res.data)
        return render_json(return_dict)
```

- ProjectRetrieveView 类继承自 RetrieveAPIView 类，它主要实现了对数据库单个记录的查看。
- 如果不需要自定义 get 返回值，其实实现一个查看具体的 API，只需要两行代码（queryset 及 serializer_class）。

如果向上追溯 RetrieveAPIView 的 Mixin 类，会看到如下代码：

```
f class GenericAPIView(views.APIView):
    """
    Base class for all other generic views.
    """
    queryset = None
    serializer_class = None

    lookup_field = 'pk'
    lookup_url_kwarg = None

    filter_backends = api_settings.DEFAULT_FILTER_BACKENDS
    pagination_class = api_settings.DEFAULT_PAGINATION_CLASS

    def get_queryset(self):
        ......
```

- 可以看到，如果存在有指定具体记录的 key，我们需要传入一个 key 值，而此 key 值的默认参数名称为 pk。这个值刚好就对应于 8.1 节中第 5 小节所说的查看具体项目的路由定义：path('detail/<pk>/', views.ProjectRetrieveView.as_view(), name='detail')。此 pk 值如何传入将在接下来的测试中会看到。在之后编辑某一记录时，也存在这样的情况，这里讲解清楚了，到时不再说明。

2. 测试查看具体项目信息的 API

在本节前面的步骤操作完成之后（这里省略了路由的增加和 JWT 的获取），启动 bifangback 应用。在 Postman 的 url 中输入：http://localhost:8000/project/detail/8/，以 get 方式请求，然后发送请求（send）。如果出现如图 8-8 所示的结果，表示编码无误。上面 url 中最后的 8 就是对应于 1 中的 pk 值，表示想要获取 id 为 8 的项目的具体信息（在这本书测试时，这个 id 为 8 的记录，即对应

8.2 节新增的那条记录）。

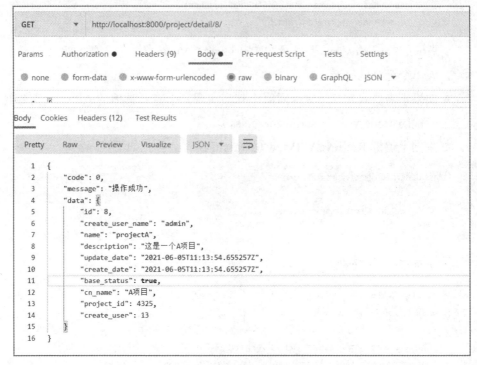

图 8-9 测试查看具体项目的 API

8.4 实现 bifangback 修改具体项目的 API

本节实现对具体项目信息的修改更新功能。

1. 实现修改项目的类视图

在 bifangback 项目根目录的 project 子目录中的 views.py 文件新增一个 ProjectUpdateView 类，内容如下：

```
from class ProjectUpdateView(UpdateAPIView):
    """
    url 获取 pk,修改时指定序列化类和 query_set
    """
    serializer_class = ProjectSerializer
    queryset = Project.objects.all()

    # 前端使用 patch 方法到达这里
```

```python
def patch(self, request, *args, **kwargs):
    req_data = request.data
    pid = req_data['id']
    name = req_data['name']
    cn_name = req_data['cn_name']
    description = req_data['description']
    project_id = req_data['project_id']
    # 这样更新可以把那些 update_date 字段自动更新，而使用 filter().update() 则是不会
    try:
        _p = Project.objects.get(id=pid)
        _p.name = name
        _p.cn_name = cn_name
        _p.description = description
        _p.project_id = project_id
        _p.save()
        return_dict = build_ret_data(OP_SUCCESS, str(req_data))
        return render_json(return_dict)
    except Exception as e:
        print(e)
        return_dict = build_ret_data(THROW_EXP, str(e))
        return render_json(return_dict)
```

- ProjectUpdateView 类继承自 UpdateAPIView 类，主要实现了对数据库单个记录的修改。
- 改写了 UpdateAPIView 类中的 patch() 方法，用于更新一个具体的项目信息记录。
- 和新增项目信息记录同样的思路，先获取所有前端传递过来的参数。
- 然后通过 try...except 语句将这条项目信息的更新记录保存到数据库中，返回操作成功的信息。如果失败则抛出对应的错误。

2. 测试查看具体项目信息的 API

在本节前面的步骤操作完成后（这里省略了路由的增加和 JWT 的获取），启动 bifangback 应用。在 Postman 的 url 中输入：http://localhost:8000/project/update/8/，以 patch 方式请求。同时，按格式要求传递一个 json 格式的数据，然后发送请求（send）。如果出现如图 8-10 所示结果，表示编码无误。

图 8-10 更新具体项目记录的 API

8.5 实现 bifangback 删除具体项目的 API

本节实现对具体项目信息的删除功能。

1. 实现删除项目的类视图

在 bifangback 项目根目录的 project 子目录中的 views.py 文件新增一个 ProjectDestroyView 类，内容如下：

```
from class ProjectDestroyView(DestroyAPIView):
    queryset = Project.objects.all()

    def destroy(self, request, *args, **kwargs):
        try:
            res = super().destroy(self, request, *args, **kwargs)
            return_dict = build_ret_data(OP_SUCCESS, str(res))
            return render_json(return_dict)
        except Exception as e:
            print(e)
            return_dict = build_ret_data(THROW_EXP, str(e))
            return render_json(return_dict)
```

- ProjectDestroyView 类继承自 DestroyAPIView 类，它主要实现了对数据库

单个记录的修改。
- Queryset 类变量定义了在哪个模型表中删除记录。
- 改写 destroy() 方法，主要是自定义返回信息，调用了 super().destory() 方法来实现真正的内置删除方法。

2. 测试删除具体项目信息的 API

本节前面的步骤操作完成后（这里省略了路由的增加和 JWT 的获取），启动 bifangback 应用。在 Postman 的 url 中输入：http://localhost:8000/project/delete/8/，以 delete 方式请求，然后发送请求（send）。如果出现如图 8-11 所示结果，表示编码无误。如果登录 Django admin 管理后台会发现相应的记录已被删除。

图 8-11 删除具体项目记录的 API

8.6 实现 bifangback 应用增删查改的 API

本章前述部分已实现了项目的增删查改的 API。因为在项目中不涉及数据库外键，所以比较简单，但其框架代码已很好地演示了基本功能的实现。相信读者认真学完之后，应该可以了解使用 DRF 写后端 API 的基本套路。

所以，在实现应用 app 的增删查改功能（包括之后类似的功能）时，为不浪费纸张和时间，不再重复讲解这些内容，而只着重讲解代码中特定的实现。在讲解 app 的增删查改功能时，我会统一列出文件的所有代码，并讲解其主要的功能和需要关注的技术点。

1. 新增一个序列化文件

在 bifangback 项目根目录的 app 子目录中新增一个 serializers.py 文件，文件内容如下：

```python
from rest_framework import serializers
from cmdb.models import App


class AppSerializer(serializers.ModelSerializer):
    project_name = serializers.CharField(source='project.name', default="None")
    git_name = serializers.CharField(source='git.name', default="None")

    class Meta:
        model = App
        fields = '__all__'
```

在序列化时，为了方便浏览器显示，会同时返回 project_name 和 git_name 的外键名称，而不只是外键 id。

2. 新增一个过滤器文件

在 bifangback 项目根目录的 app 子目录中新增一个 filters.py 文件，文件内容如下：

```python
from django_filters import OrderingFilter
from django_filters.rest_framework import FilterSet
from django_filters import filters
from cmdb.models import App


class AppFilter(FilterSet):
    name = filters.CharFilter(field_name='name', lookup_expr='icontains',)
    begin_time = filters.DateTimeFilter(field_name='create_date', lookup_expr='gte',)
    end_time = filters.DateTimeFilter(field_name='create_date', lookup_expr='lte',)
    sort = OrderingFilter(fields=('create_date',))

    class Meta:
        model = App
        fields = ['name', 'begin_time', 'end_time']
```

和项目 project 里的过滤器一样，只定义了按名称、起止创建时间进行过滤的方式。如果这样的过滤器多了，可以考虑合成一个文件。但为了灵活配置，各自独立的文件也不算冗余。如何应用就让时间来沉淀吧。

3. 实现应用 app 增删查改的类视图

修改 bifangback 项目根目录的 app 子目录中的 views.py 文件，将增删查改

的类视图一并写出,内容如下:

```python
from from cmdb.models import App
from cmdb.models import Project
from cmdb.models import GitTb
from .serializers import AppSerializer
from rest_framework.generics import ListAPIView
from rest_framework.generics import CreateAPIView
from rest_framework.generics import RetrieveAPIView
from rest_framework.generics import UpdateAPIView
from rest_framework.generics import DestroyAPIView
from utils.ret_code import *
from .filters import AppFilter
from utils.permission import is_project_admin


class AppListView(ListAPIView):
    queryset = App.objects.all().order_by('-id')
    serializer_class = AppSerializer
    filter_class = AppFilter

    def get(self, request, *args, **kwargs):
        res = super().get(self, request, *args, **kwargs)
        return_dict = build_ret_data(OP_SUCCESS, res.data)
        return render_json(return_dict)


class AppCreateView(CreateAPIView):
    serializer_class = AppSerializer

    def post(self, request):
        """
        {
            "name": "appA",
            "cn_name":"A 应用 ",
            "description": " 这是一个 A 应用 ",
            "app_id": 4325,
            "git_app_id": 2,
            "git_id": 1,
            "project_id": 6,
            "git_trigger_token": "i98sdfasdf935345",
            "build_script": "script/build.sh",
            "deploy_script": "script/bifang.sh",
            "zip_package_name": "go-demo.zip",
```

```python
            "service_port": 8080,
}
"""
req_data = request.data
print(req_data, "#####################")
user = request.user
project_id = req_data['project_id']
"""
# 前端开发完成后开启权限测试
if not is_project_admin(project_id, user):
    return_dict = build_ret_data(THROW_EXP, '你不是项目创建者，不能在此项目下创建 app 应用！')
    return render_json(return_dict)
"""


data = dict()
data['name'] = req_data['name']
data['description'] = req_data['description']
data['cn_name'] = req_data['cn_name']
data['app_id'] = req_data['app_id']
data['git_app_id'] = req_data['git_app_id']
# 外键关联
data['git'] = req_data['git_id']
data['project'] = project_id
# 普通字段
data['git_trigger_token'] = req_data['git_trigger_token']
data['build_script'] = req_data['build_script']
data['deploy_script'] = req_data['deploy_script']
data['zip_package_name'] = req_data['zip_package_name']
data['service_port'] = req_data['service_port']
# 在此版本中默认使用 root 账号启动应用
# 此功能留待今后扩展，因为传入这个参数也没有传入 saltypie 去执行
data['service_username'] = 'root'
data['service_group'] = 'root'
# 从 drf 的 request 中获取用户（对 django 的 request 做了扩展的）
data['create_user'] = user.id
serializer = AppSerializer(data=data)
if serializer.is_valid() is False:
    return_dict = build_ret_data(THROW_EXP, str(serializer.errors))
    return render_json(return_dict)
data = serializer.validated_data
App.objects.create(**data)
return_dict = build_ret_data(OP_SUCCESS, serializer.data)
```

```python
        return render_json(return_dict)


class AppRetrieveView(RetrieveAPIView):
    queryset = App.objects.all()
    serializer_class = AppSerializer

    def get(self, request, *args, **kwargs):
        res = super().get(self, request, *args, **kwargs)
        return_dict = build_ret_data(OP_SUCCESS, res.data)
        return render_json(return_dict)


class AppUpdateView(UpdateAPIView):
    """
    url 获取 pk, 修改时指定序列化类和 query_set
    """
    serializer_class = AppSerializer
    queryset = App.objects.all()

    # 前端使用 patch 方法，到达这里
    def patch(self, request, *args, **kwargs):
        req_data = request.data
        pid = req_data['id']
        name = req_data['name']
        cn_name = req_data['cn_name']
        description = req_data['description']
        app_id = req_data['app_id']
        # 这样更新，可以自动更新那些 update_date 字段，而使用 filter().update() 则是不会
        try:
            _a = App.objects.get(id=pid)
            _a.name = name
            _a.cn_name = cn_name
            _a.description = description
            _a.app_id = app_id
            _a.save()
            return_dict = build_ret_data(OP_SUCCESS, str(req_data))
            return render_json(return_dict)
        except Exception as e:
            print(e)
            return_dict = build_ret_data(THROW_EXP, str(e))
            return render_json(return_dict)
```

```python
class AppDestroyView(DestroyAPIView):
    queryset = App.objects.all()

    def destroy(self, request, *args, **kwargs):
        try:
            res = super().destroy(self, request, *args, **kwargs)
            return_dict = build_ret_data(OP_SUCCESS, str(res))
            return render_json(return_dict)
        except Exception as e:
            print(e)
            return_dict = build_ret_data(THROW_EXP, str(e))
            return render_json(return_dict)
```

与 project 的 views.py 文件对比基本一致，只是因为有外键存在，所以会有相关 id 的获取。还有几个细节点要注意，git_trigger_token、build_script、deploy_script、zip_package_name 和 service_port 这几个和 app 应用绑定的变量，需要结合本书第 3 章和第 4 章的编译和部署脚本一起来学习。这一过程相当于把手工操作 Web 化了。

4. 增加路由

定义好项目的类视图后就可以加入项目的路由了。在 app 目录中的 urls.py 文件内增加如下内容：

```python
from django.urls import path
from . import views

app_name = "app"

urlpatterns = [
    path('list/', views.AppListView.as_view(), name='list'),
    path('create/', views.AppCreateView.as_view(), name='create'),
    path('detail/<pk>/', views.AppRetrieveView.as_view(), name='detail'),
    path('update/<pk>/', views.AppUpdateView.as_view(), name='update'),
    path('delete/<pk>/', views.AppDestroyView.as_view(), name='delete'),
]
```

和 project 的 urls.py 文件几乎一模一样，无须过多解释，在 git 代码里也都有这些内容。但考虑有些读者喜欢在地铁、公交，或是河边、阴凉处和草丛中看书，或是不喜欢看电子产品，本书再列举少量代码也是有必要的。希望不喜欢的读者可以理解并谅解。

5. 测试应用列表 API

这里只挑一个应用列表进行测试，url 为 http://localhost:8000/app/list/，显示效果如图 8-12 所示。其他 API 测试不再占用公共资源，请读者自行进行。

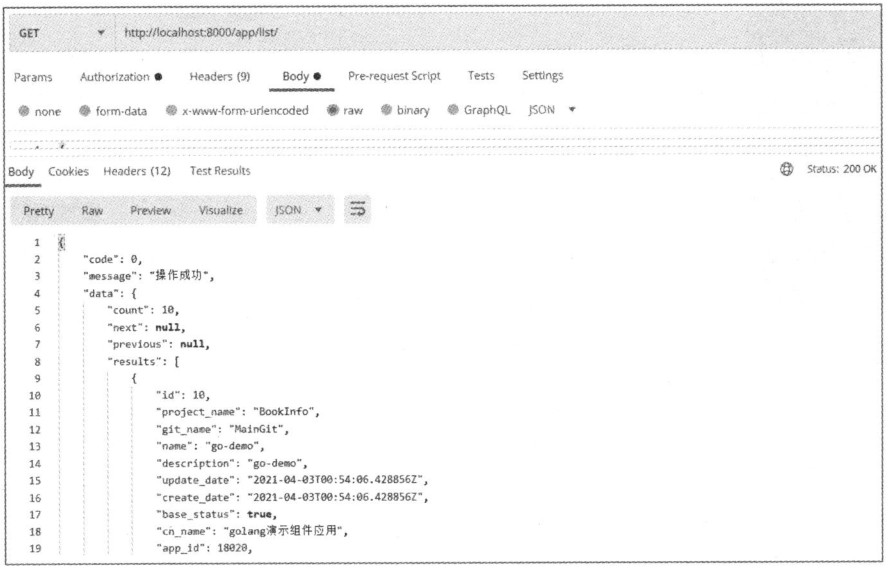

图 8-12 获取所有应用列表的 API

8.7 实现 bifangback 基于应用的权限管理

本来本章讲到此处就可以结束了，但还有一个功能，我认为放在这里讲比较合适，那就是关于权限的设计。

BiFang 项目设计的权限模型是基于应用级别的。每个 BiFang 用户可以将自己项目的权限授予其他用户。权限分为三类：新建发布单权限（和构建同权）、环境流转权限和部署权限。

前端操作授权及操作界面如图 8-13 和图 8-14 所示。通过这个界面可以在应用级别对权限进行展示、增加和删除。

在真正进行权限卡点时，植入一个类似权限管理的 hook 钩子函数，通过比对数据表中的权限来阻止或放行用户的相应操作。这些内容以及 hook 钩子函数会在本节讲解，而如何调用这些钩子函数则留待后面对应的代码中讲解。

图 8-13 基于应用的权限管理

图 8-14 基于应用的权限编辑

1. 实现展示权限列表的类视图

读者翻回本书 6.9 节,那里讲了权限数据库的设计。一个应用的具体权限主要与 Action 权限、应用组件和操作用户三者关联,用于确定一个用户对一个应用的 Action（Create 对应新建发布单及构建权限、Env 对应环境流转权限、Deploy 对应部署权限）的唯一权限。

与本章前面章节的套路一样,先建立序列化和过滤器文件,再建立类视图和路由。为了内容紧凑,这里一并讲解。

（1）在 bifangback 项目根目录的 permission 子目录中新增一个 serializers.py

文件，内容如下：

```python
from rest_framework import serializers
from cmdb.models import Permission


class PermissionSerializer(serializers.ModelSerializer):
    app_name = serializers.CharField(source='app.name', read_only=True)
    action_name = serializers.CharField(source='action.name', read_only=True)
    create_username = serializers.CharField(source='create_user.username', read_only=True)
    pm_username = serializers.CharField(source='pm_user.username', read_only=True)

    class Meta:
        model = Permission
        fields = '__all__'
```

PermissionSerializer 类除了将 Permission 数据表所有字段序列化外，还显示 app_name、action_name 和 pm_username 以及 create_username 的真实名称（而不只是默认序列化的 id 序号）。

（2）在 bifangback 项目根目录的 permission 子目录中新增一个 filters.py 文件，内容如下：

```python
from django_filters import OrderingFilter
from django_filters.rest_framework import FilterSet
from django_filters import filters
from cmdb.models import Permission


class PermissionFilter(FilterSet):
    app = filters.CharFilter(field_name='app__name', lookup_expr='icontains',)
    action = filters.CharFilter(field_name='action__name', lookup_expr='icontains', )
    begin_time = filters.DateTimeFilter(field_name='create_date', lookup_expr='gte',)
    end_time = filters.DateTimeFilter(field_name='create_date', lookup_expr='lte',)
    sort = OrderingFilter(fields=('create_date',))

    class Meta:
        model = Permission
        fields = ['app', 'action', 'begin_time', 'end_time']
```

PermissionFilter 类支持按 app、action 名称进行过滤，也支持将起止时间进行过滤，默认以创建时间进行降序排列，便于在开始显示最新的数据，符合一般查看习惯。

（3）在 bifangback 项目根目录的 permission 子目录中，将默认的 views.py 的内容更新如下：

```python
from django.contrib.auth import get_user_model
from rest_framework.generics import ListAPIView
from rest_framework.generics import CreateAPIView
from rest_framework.generics import DestroyAPIView
from cmdb.models import App
from cmdb.models import Action
from cmdb.models import Permission
from .serializers import PermissionSerializer
from .filters import PermissionFilter
from utils.ret_code import *

User = get_user_model()


class PermissionListView(ListAPIView):
    queryset = Permission.objects.all()
    serializer_class = PermissionSerializer
    filter_class = PermissionFilter

    def get(self, request, *args, **kwargs):
        res = super().get(self, request, *args, **kwargs)
        return_dict = build_ret_data(OP_SUCCESS, res.data)
        return render_json(return_dict)


class PermissionCreateView(CreateAPIView):
    serializer_class = PermissionSerializer

    def post(self, request):
        """
        {
            "action": "Create",
            "user_id": 15,
            "app_name": "Ratings",
        }
        """
        req_data = request.data
        data = dict()
        user = User.objects.get(id=req_data['user_id'])
        app = App.objects.get(name=req_data['app_name'])
```

```python
        action = Action.objects.get(name=req_data['action'])
        permission_name = '{}-{}-{}'.format(app.name, action.name, user.username)
        data['name'] = permission_name
        data['description'] = permission_name
        data['app'] = app.id
        data['action'] = action.id
        data['pm_user'] = user.id
        # 从 drf 的 request 中获取用户（对 django 的 request 做了扩展的）
        data['create_user'] = request.user.id
        serializer = PermissionSerializer(data=data)
        if serializer.is_valid() is False:
            return_dict = build_ret_data(THROW_EXP, str(serializer.errors))
            return render_json(return_dict)
        data = serializer.validated_data
        Permission.objects.create(**data)
        return_dict = build_ret_data(OP_SUCCESS, serializer.data)
        return render_json(return_dict)


class PermissionDestroyView(DestroyAPIView):
    queryset = Permission.objects.all()

    def destroy(self, request, *args, **kwargs):
        try:
            res = super().destroy(self, request, *args, **kwargs)
            return_dict = build_ret_data(OP_SUCCESS, str(res))
            return render_json(return_dict)
        except Exception as e:
            print(e)
            return_dict = build_ret_data(THROW_EXP, str(e))
            return render_json(return_dict)
```

- PermissionListView 类用于显示权限列表，支持过滤功能。
- PermissionCreateView 类用于用户权限的创建，以及防止 name 冲突；permission_name 命名将 app、acton 和 username 结合起来，既解决了冲突，又使命名具有一定语义性。
- PermissionDestroyView 类用于用户权限的删除。

（4）在 bifangback 项目根目录的 permission 子目录中，将 urls.py 的内容更新如下：

```python
from django.urls import path
```

```
from . import views

app_name = "permission"

urlpatterns = [
    path('list/', views.PermissionListView.as_view(), name='list'),
    path('create/', views.PermissionCreateView.as_view(), name='create'),
    path('delete/<pk>/', views.PermissionDestroyView.as_view(), name='delete'),
]
```

在进行权限删除时，需要指定对应的 pk 值。此 pk 的意义前面已讲过。

2. 测试权限列表的 API

在本节前面的步骤操作完成之后（这里省略了路由的增加和 JWT 的获取），启动 bifangback 应用。在 Postman 的 url 中输入 http://localhost:8000/permission/list/，以 get 方式请求，然后发送请求（send）。如果出现如图 8-15 所示的结果，则表示编码无误。至于权限的增加和删除，读者按项目 API 的方式自行测试即可。

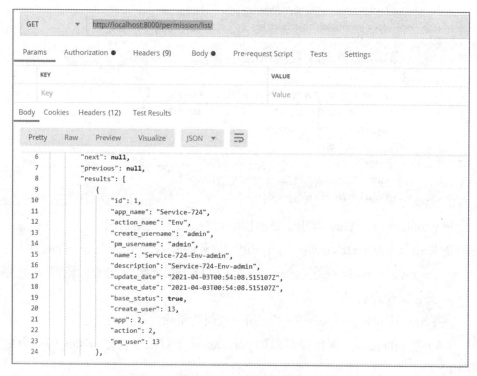

图 8-15　测试权限列表 API

3. 权限验证函数

做好权限数据表后，如何验证权限呢？我们的思路是，在每一个权限卡点时，传入用户名、Action 权限名和 app 应用名给一个权限验证函数作为参数，如果返回 true，则表示有权限；如果返回 false，则表示无权限。

在 bifangback 项目根目录的 utils 子目录中新增一个 permission.py 文件，内容如下：

```python
from from django.contrib.auth.models import Group
from django.core.exceptions import ObjectDoesNotExist
from cmdb.models import Project
from cmdb.models import App
from cmdb.models import Permission


# 判断是否为管理员组
def is_admin_group(user):
    try:
        user_groups = Group.objects.filter(user=user)
        print("user groups, ", user_groups)
    except ObjectDoesNotExist as e:
        print(e)
        return False
    for user_group in user_groups:
        if user_group.name == 'admin':
            return True
    return False


# 判断是否为 Project 管理员
def is_project_admin(project_id, user):
    project = Project.objects.get(id=project_id)
    if user == project.create_user or is_admin_group(user):
        return True
    return False


# 判断是否为 APP 管理员
def is_app_admin(app_id, user):
    app = App.objects.get(id=app_id)
    if user == app.create_user or is_admin_group(user):
        return True
    return False
```

```python
# 获取 APP 管理员
def get_app_admin(app_id):
    return App.objects.get(id=app_id).create_user


# 获取 APP 的各个权限的相关成员
def get_app_user(app_id, action_id):
    filter_dict = dict()
    filter_dict['app__id'] = app_id
    filter_dict['action__id'] = action_id
    permission_set = Permission.objects.get(**filter_dict)
    user_set = permission_set.pm_user.all()
    return user_set


# 判断是否具有 APP 的相关环境的相关权限
def is_right(app_id, action_id, user):
    # 管理员直接有相关权限
    if is_app_admin(app_id, user):
        return True
    filter_dict = dict()
    filter_dict['app__id'] = app_id
    filter_dict['action__id'] = action_id
    try:
        permission_set = Permission.objects.filter(**filter_dict)
        for permission in permission_set:
            if user == permission.pm_user:
                return True
        return False
    except Permission.DoesNotExist as e:
        print(e)
        return False
```

is_right 为主函数，用于权限判断，返回 true 或 false。其实现思路：如果用户为超级管理员、项目管理员或 app 管理员，则直接返回 true。如果是其他用户，则从 Permission 数据表中查询有相关权限的用户，如果用户在其中则返回 true，如果不在则返回 false。

8.8 小　结

本章主要开发了 bifangback 项目和应用管理，以及权限管理相应的代码。因为这些代码有一定的开发普适性，所以在开始项目的增删查改的代码实现和验证方面都讲得很细致。

当读者熟悉了这一流程的代码套路后，在随后的应用增删查改方面就加快了讲解进度，希望读者能接受我这种写书风格。在讲解完项目和应用之后，我顺便也讲解了应用权限的管理，算是一个相似的功能实现。

本章最后，留给读者一个自我学习的机会。其实，bifangback 项目的服务器模块的代码和项目及应用的代码大体是一致的，就不在本书以后的章节浪费纸张来讲解了，读者看一下 server 目录下的代码就可以学会。postman 测试效果如图 8-16 所示。注意看服务器列表的输出，由于我们的模拟超过 15 台服务器，所以 API 输出会产生分页链接。这也是 8.1 节中分页展示的 callback。

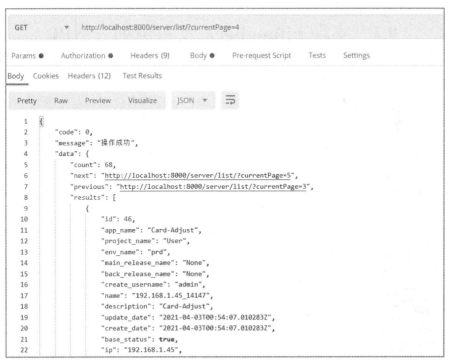

图 8-16　测试服务器列表 API

第 9 章 发布单及环境流转

若无必要，勿增实体。

——威廉·奥卡姆

本章 GitHub 代码地址：https://github.com/aguncn/bifang-book/tree/main/bifang-ch9。

第 8 章实现了后端项目和应用的管理，权限 API 及服务器管理功能。本章主要实现发布单和环境功能，主要包括发布单的创建、软件包的构建、发布单在各个环境之间的流转。本章就来看看如何实现这些后端 API。

9.1 实现 bifangback 发布单列表 API

BiFang 发布单的列表如图 9-1 所示，本节主要实现这个网页的后端功能。同第 8 章一样，除了基本的发布单列表显示，还提供了分页、基于项目、组件应用和发布单号的查询搜索功能。

图 9-1 发布单列表

1. 新增一个序列化文件

在 bifangback 项目根目录的 release 子目录中新增一个 serializers.py 文件，文件内容如下：

```python
from rest_framework import serializers
from cmdb.models import Release


class ReleaseSerializer(serializers.ModelSerializer):
    project_name = serializers.CharField(source='app.project.name', default="None")
    app_name = serializers.CharField(source='app.name', default="None")
    service_port = serializers.CharField(source='app.service_port', default="None")
    env_name = serializers.CharField(source='env.name', default="None")
    create_user_name = serializers.CharField(source='create_user.username')
    release_status_name = serializers.CharField(source='release_status.name', default="None")
    git_url = serializers.CharField(source='app.git.git_url', default="None")
    git_app_id = serializers.CharField(source='app.git_app_id', default="None")

    class Meta:
        model = Release
        fields = '__all__'
```

- Release 数据表中有很多的外键字段依赖。这个外键字段默认序列化输出时是以关联 id 的形式输出的。为了能在前端页面展示其具体的名称，通常会获取这些 id 关联的具体名称。在 ReleaseSerializer 类中，project_name 和 app_name 等这些新增的序列化字段就起到了这样的效果。在其 source 的属性定义中，我们使用"字段.外键名称（app.name）"的形式获取这些值。它甚至支持不断级联取值的语法（app.git.git_url）。
- 读者可以结合后面的 API 输出进行进一步的把握和学习这种序列化的语法。

2. 新增一个过滤器文件

在 bifangback 项目根目录的 release 子目录中新增一个 filters.py 文件，文件内容如下：

```python
from django.db.models import Q
from django_filters import OrderingFilter
from django_filters.rest_framework import FilterSet
from django_filters import filters
from cmdb.models import Release
```

```python
class ReleaseFilter(FilterSet):
    name = filters.CharFilter(field_name='name', lookup_expr='icontains',)
    begin_time = filters.DateTimeFilter(field_name='create_date', lookup_expr='gte',)
    end_time = filters.DateTimeFilter(field_name='create_date', lookup_expr='lte',)
    sort = OrderingFilter(fields=('create_date','update_date'))
    # 自定义过滤方法
    release_status = filters.CharFilter(field_name='release_status', method='release_status_filter')
    project_name = filters.CharFilter(field_name='project_name', method='project_name_filter')
    app_name = filters.CharFilter(field_name='app_name', method='app_name_filter')

    def release_status_filter(self, queryset, field_name, value):
        # 前端传过来的 deploy_status 为字符串,将其转换为列表就可以过滤需要的发布单了
        filter_list = value.split(',')
        # 外建基于列表过滤的语法
        return queryset.filter(release_status__name__in=filter_list)

    def project_name_filter(self, queryset, field_name, value):
        # 外建基于列表过滤的语法
        return queryset.filter(app__project__name__exact=value)

    def app_name_filter(self, queryset, field_name, value):
        # 外建基于列表过滤的语法
        return queryset.filter(app__name__exact=value)

    class Meta:
        model = Release
        fields = ['name', 'project_name', 'app_name', 'begin_time', 'end_time', 'release_status']
```

- 这个过滤器与前面章节的过滤器写法有些不一样。新的代码知识在于自定义了三个过滤方法,并将这些过滤方法与其类变量关联。其间的不同是因为我们不是基于 release 本身的字段过滤,而是基于 release 数据表的外键里的其他字段进行过滤。如 release 数据表本身有一个 app 的外键字段,那就需要使用这样的过滤方法来实现。不然,就只能基于 app 的 id 值来进行过滤搜索,那样就没有实际的使用价值了。

3. 实现发布单列表类视图

在 bifangback 项目根目录的 release 子目录中的 views.py 文件中新增一个 ReleaseListView 类,内容如下:

```python
class ReleaseListView(ListAPIView):
```

```
queryset = Release.objects.all().order_by('-create_date')
serializer_class = ReleaseSerializer
filter_class = ReleaseFilter

def get(self, request, *args, **kwargs):
    res = super().get(self, request, *args, **kwargs)
    return_dict = build_ret_data(OP_SUCCESS, res.data)
    return render_json(return_dict)
```

这个代码没什么新的知识点，只是定义好 ReleaseListView 类的 queryset、serializer_class 和 filter_class 变量即可。

4. 增加路由

定义好项目的类视图之后，就可以加入项目的路由了。在 release 的目录中的 urls.py 文件内增加如下内容：

```
from django.urls import path
from . import views
from . import build_views

urlpatterns = [
    path('list/', views.ReleaseListView.as_view(), name='list'),
    path('create/', views.ReleaseCreateView.as_view(), name='create'),
    path('detail/<pk>/', views.ReleaseRetrieveView.as_view(), name='detail'),
    path('build/', build_views.ReleaseBuildView.as_view(), name='build'),
    path('build_status/', build_views.ReleaseBuildStatusView.as_view(), name='build_status'),
    path('statistics/', views.ReleaseStatisticsView.as_view(), name='statistics')
]
```

为了方便，这里将 release 发布单涉及的所有路由定义都在此列明了，之后的 release 讲解章节不再说明这些路由的定义环境。如果要分别测试每一个 url，则要先写好类视图之后，再在 urls.py 里增加对应的路由，不然启动会报错，表示对应的类视图定义还不存在。

5. 测试发布单列表 API

在本节前面的步骤操作完成后（这里省略了路由的增加和 JWT 的获取），启动 bifangback 应用。在 Postman 的 url 中输入 http://localhost:8000/release/list/?app_name=go-demo，以 get 方式请求，然后发送请求（send）。输出的主要数据结构如图 9-2 和图 9-3 所示。

读者需要留心，在输出的 json 里，除了有 app、env 这样的纯 id 数值外，也输出了 app_name、env_name 这样有真正名称的字段。而这些额外的非数据表里原有的字段名称就是在 9.1 节中自定义序列化字段的输出。另外，在请求的 url 里，我们刻意带上了 app_name=go-demo 这样的具名参数，输出了所有 app 名称为 go-demo 的发布单列表，这是在 9.1 节中定义的过滤器的作用。

DRF 有了这两个功能，前端显示和过滤这些日常开发中的常用功能也就能很容易地通过套路化的方式实现啦。

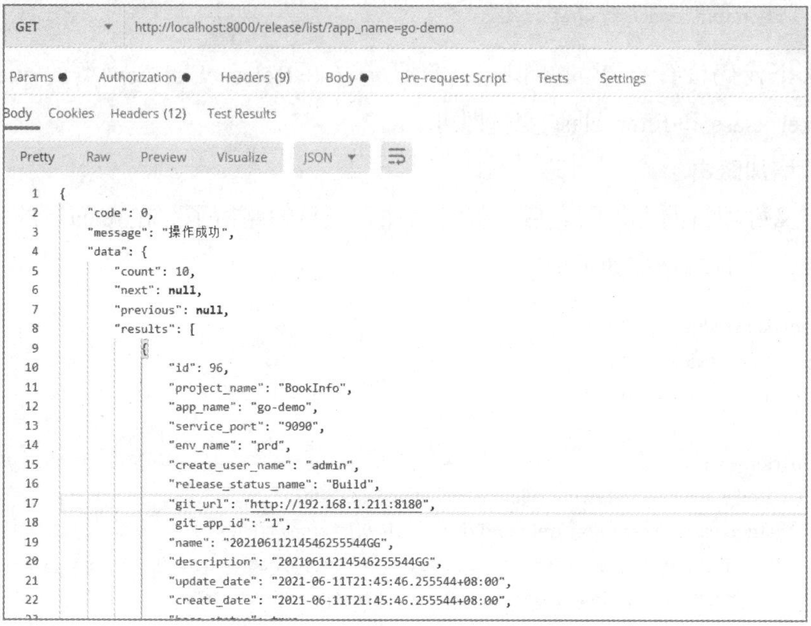

图 9-2 发布单列表 API 输出列表


```
{
    "id": 96,
    "project_name": "BookInfo",
    "app_name": "go-demo",
    "service_port": "9090",
    "env_name": "prd",
    "create_user_name": "admin",
    "release_status_name": "Build",
    "git_url": "http://192.168.1.211:8180",
    "git_app_id": "1",
    "name": "20210611214546255544GG",
    "description": "20210611214546255544GG",
    "update_date": "2021-06-11T21:45:46.255544+08:00",
    "create_date": "2021-06-11T21:45:46.255544+08:00",
    "base_status": true,
    "deploy_no": 0,
    "git_branch": "master",
    "pipeline_id": 0,
    "pipeline_url": "http://www.demo.com",
    "deploy_script_url": "http://1.2.3.4/a/b/bifang.sh",
    "zip_package_url": "http://1.2.3.4/a/b/go-demo.zip",
    "create_user": 13,
    "app": 10,
    "env": 2,
    "release_status": 4
}
```


图 9-3 发布单列表 API 输出单个 json

9.2 实现 bifangback 新增发布单的 API

本节来实现发布单的新增 API。前端界面展示如图 9-4 所示，在编写后端 API 时，结合前端的展示效果就能和前端开发同事更有效率地配合。

图 9-4 新建发布单

1. 新增一个序列化文件

在 bifangback 项目根目录的 release 子目录的 serializers.py 文件中新增一个 ReleaseCreateSerializer 类，内容如下：

```
class ReleaseCreateSerializer(serializers.ModelSerializer):
    class Meta:
        model = Release
        # fields = '__all__'
        fields = ['name', 'description', 'git_branch', 'app', 'release_status', 'create_user']
```

- 当同一个数据表用于不同的用途时，可以制订不同的序列化类，来控制在序列化过程中输出和写入的不同字段。与序列化发布单列表时的字段不一样，ReleaseCreateSerializer 类只序列化 name 和 description 等 6 个字段。
- Fields 为 '__all__' 时，取所有的字段，而指定列表时则只序列化指定的列表中的项。

2. 实现新增发布单的类视图

在 bifangback 项目根目录的 release 子目录中的 views.py 文件中新增一个 ReleaseCreateView 类，内容如下：

```python
class ReleaseCreateView(CreateAPIView):
    serializer_class = ReleaseCreateSerializer

    def post(self, request):
        """
        {
            "description": " 这是一个测试发布单 ",
            "app_id": 1,
            "git_branch": "master"
        }
        """
        req_data = request.data
        user = request.user
        app_id = req_data['app_id']

        # 前端开发完成后开启权限测试
        action = Action.objects.get(name='Create')
        if not is_right(app_id, action.id, user):
            return_dict = build_ret_data(NOT_PERMISSION, ' 你无此应用的新建及建构发布单权限！ ')
            return render_json(return_dict)

        data = dict()
        random_letter = ''.join(random.sample(string.ascii_letters, 2))
        name = datetime.datetime.now().strftime("%Y%m%d%H%M%S%f") + random_letter.upper()
        release_status_name = 'Create'
        release_status = ReleaseStatus.objects.get(name=release_status_name)
        data['name'] = name
        data['description'] = req_data['description']
        data['git_branch'] = req_data['git_branch']
        data['app'] = app_id
        data['release_status'] = release_status.id

        # 从 drf 的 request 中获取用户（对 Django 的 request 做了扩展）
        data['create_user'] = user.id
        serializer = ReleaseCreateSerializer(data=data)
        if serializer.is_valid() is False:
            return_dict = build_ret_data(THROW_EXP, str(serializer.errors))
            return render_json(return_dict)
        data = serializer.validated_data
```

```
release = Release.objects.create(**data)
write_release_history(release_name=release.name,
                      env_name=None,
                      release_status_name=release_status_name,
                      deploy_type=None,
                      log='Create',
                      user_id=user.id)
return_dict = build_ret_data(OP_SUCCESS, serializer.data)
return render_json(return_dict)
```

- 总体代码思路和 9.1 节创建项目的代码一致，但也有一些不同点。
- serializer_class 指定的是 9.1 节的序列化类。
- 判断是否有新建发布单的权限时，将 Action 关键字（此处为 'Create'）的实例、用户 id 和应用 id 这三者传入第 8 章写好的 is_right() 函数中，根据返回 true 或 false 就可以判断是否有相应的权限。在之后的环境流转和部署权限中，都是以同样的思路来实现的，只是 Action 关键字（后两者分别是 'Env' 和 'Deploy'）不一样。
- 发布单的名称是以时间戳加两位随机大写字母生成，读者看 name 变量定义即知。
- 我们同时维护了一个发布单的操作历史记录表，此表的内容通过 write_release_history() 函数实现数据的插入，其参数以清楚记录发布单的操作过程为宜。其代码位于 utils/write_history.py 文件中，函数定义如下：

```
#更新发布单历史，这样可以串联起来发布单的操作历史
#但操作服务器历史不在此之列，在下一个函数
def write_release_history(release_name=None, env_name=None,
                          release_status_name=None, deploy_type=None,
                          deploy_no=None, log=None, user_id=None):
    name = uuid.uuid1()
    release_status = ReleaseStatus.objects.get(name=release_status_name)
    release = Release.objects.get(name=release_name)
    create_user = None
    if user_id is not None:
        create_user = User.objects.get(id=user_id)
    env = None
    if env_name is not None:
        env = Env.objects.get(name=env_name)
    ReleaseHistory.objects.create(name=name,
                                  release=release,
```

```
                            env=env,
                            release_status=release_status,
                            deploy_type=deploy_type,
                            log_no=deploy_no,
                            log=log,
                            create_user=create_user)
```

如果有些阶段不存在对应值的话，就将其字段置为 None。

3. 测试新增发布单 API

在本节前面的步骤操作完成后（这里省略了路由的增加和 JWT 的获取），启动 bifangback 应用。在 Postman 的 url 中输入 http://localhost:8000/release/create/，以 post 方式请求，再设计一个简单的发布单 json 数据，然后发送请求（send）。默认出现如图 9-5 所示的结果，表示编码无误。同时登录 Django Admin 管理后台，可以看到新增的发布单已入库。

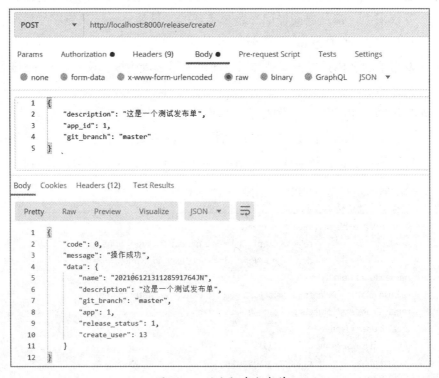

图 9-5　测试新建发布单 API

9.3 实现 bifangback 软件构建的 API

本节的功能基于第 3 章的 GitLab 的 CI/CD 功能来实现软件构建，其基本流程是 bifangback 将 GitLab 所需要的参数传递进去，通过 trigger token 触发软件构建过程。然后，GitLab 启动 docker 容器进行真正的软件构建。在软件构建过程中，将 sonarcube 的报告发送给 sonarcobe 服务器。在软件构建完成后，将软件包上传到 file-server 文件服务器上，从而完成整个软件构建的过程。其中 bifangback 操作 GitLab 的过程是通过 python-gitlab 这个第三方 Python 软件包实现的。

对于这些具体的手工操作流程，还请读者再把第 3 章的内容复习一次，做到心中有数，在看下面的代码时才不会迷路。软件构建的前端界面如图 9-6 所示。

图 9-6　发布单构建软件包

1. 实现 bifangback 触发软件构建的类视图

由于序列化类的写法和实现都大同小异，相信读者已经掌握，此后不再列这个文件的内容。在 bifangback 项目根目录的 release 子目录中的 build_views.py 文件中新增一个 ReleaseBuildView 类，内容如下：

```python
参数：
app_name
release_name
git_branch
"""
def post(self, request):
    # 序列化前端数据，并判断是否有效
    serializer = ReleaseBuildSerializer(data=request.data)
    if serializer.is_valid():
        ser_data = serializer.validated_data
        user = request.user
        app_name = ser_data['app_name']
        release_name = ser_data['release_name']
        git_branch = ser_data['git_branch']
        app = App.objects.get(name=app_name)
        zip_package_name = app.zip_package_name
        git_url = app.git.git_url
        git_access_token = app.git.git_token
        git_trigger_token = app.git.git_trigger_token
        project_id = app.git_app_id
        build_script = app.build_script
        deploy_script = app.deploy_script
        file_up_server = settings.FILE_UP_SERVER

        # 前端开发完成后开启权限测试
        action = Action.objects.get(name='Create')
        if not is_right(app.id, action.id, user):
            return_dict = build_ret_data(NOT_PERMISSION, '你无此应用的新建及建构发布单权限！')
            return render_json(return_dict)

        # 先触发编译，但由于编译时间较长，为防连接过期，需异步一下，先返回 id，再
          获取编译状态
        try:
            pipeline = gitlab_trigger(git_url, git_access_token,
                                     project_id, app_name, release_name,
                                     git_branch, git_trigger_token,
                                     build_script, deploy_script,
                                     zip_package_name, file_up_server)
        except Exception as e:
            print(e)
            return_dict = build_ret_data(THROW_EXP, 'gitlab 触发错误，请确认 gitlab 连接、
                运行及配置正确')
            return render_json(return_dict)
```

```
                # 在编译前更新一下发布单的状态，待写编译历史库记录
                release_status_name = 'Building'
                release_status = ReleaseStatus.objects.get(name=release_status_name)
                release = Release.objects.filter(name=release_name).update(pipeline_id=pipeline.id,
                                pipeline_url=pipeline.web_url,
                                release_status=release_status)
                write_release_history(release_name=release_name,
                                env_name=None,
                                release_status_name=release_status_name,
                                deploy_type=None,
                                log='Building',
                                user_id=user.id)

                return_dict = build_ret_data(OP_SUCCESS, 'gitlab ci pipeline id: {}'.format(pipeline.id))
                return render_json(return_dict)
            else:
                return_dict = build_ret_data(THROW_EXP, ' 序列化条件不满足 ')
                return render_json(return_dict)
```

- 由于 ReleaseBuildView 类的自定义程度比较高，所以我们让它继承自 APIView 类，这样 post 方法和返回都需要自己实现。
- 从前端获取到需要的参数后，其他 GitLab CI/CD 需要的参数就可以在数据库里检索得到。
- 在所有参数准备完成后，就调用 gitlab_trigger() 函数来触发构建过程。
- 根据 gitlab_trigger() 的返回结果来决定本次构建是否成功还是失败（成功或失败不是最终软件包的生成，而是触发构建这一动作，真正的软件包的生成判断是在后面的构建状态中判断的）。
- 在这一过程中将相关的结果通过 write_release_history() 函数写入发布单状态数据表（ReleaseHistory）中，便于后期展示发布单时间线。

gitlab_trigger() 函数就是第 3 章 python-gitlab 库测试代码的封装，内容如下：

```
def gitlab_trigger(git_url, git_access_token,
                project_id, app_name, release,
                git_branch, git_trigger_token,
                build_script, deploy_script,
                zip_package_name, file_up_server):
    git_url = git_url
    git_access_token = git_access_token
```

```
gl = gitlab.Gitlab(git_url, private_token=git_access_token)

project = gl.projects.get(project_id)

pipeline = project.trigger_pipeline(git_branch,
                                    git_trigger_token,
                                    variables={"RELEASE": release,
                                               'APP_NAME': app_name,
                                               'BUILD_SCRIPT': build_script,
                                               'DEPLOY_SCRIPT': deploy_script,
                                               'ZIP_PACKAGE_NAME': zip_package_name,
                                               'FILE_UP_SERVER': file_up_server})

return pipeline
```

这个脚本和 3.4 节的脚本相差不大。只是 3.4 节中的内容为了测试输出而直接调用 print()，这里将 pipeline 变量返回给调用方，调用方根据结果来判断。

2. 实现 bifangback 获取软件构建状态的类视图

9.2 节中触发了 GitLab 的 CI/CD 过程，但这个构建过程有没有意外，有没有正常生成软件包，有没有将 sonarcube 的代码质量报告成功发送，有没有将生成的软件包上传到 file-server 中？对这些结果的判断就需要一个新的 API 来获取。

在实现前端时，先调用上一节的 API，然后再调用本节的 API。

在 bifangback 项目根目录的 release 子目录中的 build_views.py 文件中新增一个 ReleaseBuildStatusView 类，内容如下：

```
class ReleaseBuildStatusView(APIView):
    """
    获取编译软件状态

    参数：
    app_name
    release_name
    """
    def post(self, request):
        # 序列化前端数据，并判断是否有效
        # 在获取状态时，前端可以不传 pipeline id 过来，因为这个 id 可以通过数据库获取
        serializer = ReleaseBuildStatusSerializer(data=request.data)
        if serializer.is_valid():
            ser_data = serializer.validated_data
            user = request.user
```

```python
app_name = ser_data['app_name']
release_name = ser_data['release_name']
app = App.objects.get(name=app_name)
zip_package_name = app.zip_package_name
deploy_script = app.deploy_script
# 部署脚本在上传时只取了最后的文件名,目录名被忽略,这里也要做转换
deploy_script = os.path.basename(deploy_script)
git_url = app.git.git_url
git_access_token = app.git.git_token
project_id = app.git_app_id
release = Release.objects.get(name=release_name)
pipeline_id = release.pipeline_id


pipeline = pipeline_status(git_url, git_access_token, project_id, pipeline_id)
if pipeline.finished_at is None:
    print(pipeline.status)
    return_dict = build_ret_data(OP_SUCCESS, 'ing')
    return render_json(return_dict)
elif pipeline.status != 'success':
    print(pipeline)
    print(pipeline.id)
    print(pipeline.status)
    print(pipeline.ref)
    print(pipeline.web_url)
    print(pipeline.duration)
    release_status_name = 'BuildFailed'
    release_status = ReleaseStatus.objects.get(name=release_status_name)
    release = Release.objects.filter(name=release_name).update(release_status=release_status)
    write_release_history(release_name=release_name,
                          env_name=None,
                          release_status_name=release_status_name,
                          deploy_type=None,
                          log='Building',
                          user_id=user.id)
    return_dict = build_ret_data(OP_SUCCESS, 'error')
    return render_json(return_dict)
else:
    print(pipeline.status)
    file_down_server = settings.FILE_DOWN_SERVER
    deploy_script_url = '{}/{}/{}/{}'.format(file_down_server, app_name, release_name,
                        deploy_script)
    zip_package_url = '{}/{}/{}/{}'.format(file_down_server, app_name, release_name,
                      zip_package_name)
```

```
                    release_status_name = 'Build'
                    release_status = ReleaseStatus.objects.get(name=release_status_name)
                    Release.objects.filter(name=release_name).update(release_status=release_status,
                                        deploy_script_url=deploy_script_url,
                                        zip_package_url=zip_package_url)
                    write_release_history(release_name=release_name,
                                        env_name=None,
                                        release_status_name=release_status_name,
                                        deploy_type=None,
                                        log='Build',
                                        user_id=user.id)

                    return_dict = build_ret_data(OP_SUCCESS, 'success')
                    return render_json(return_dict)
            else:
                return_dict = build_ret_data(THROW_EXP, '序列化条件不满足 ')
                return render_json(return_dict)
```

- 由于 ReleaseBuildStatusView 类的自定义程度比较高，所以让它继承自 APIView 类，这样 post 方法和返回都需要自己实现。
- 本小节的代码实现逻辑和 9.2 节一样，都是先从前端获取必要参数，然后再从数据库中获取到所有参数，之后调用 pipeline_status() 函数进行判断。
- 根据 pipeline 的 finished 和 status 这两个状态值，就可以判断 CI/CD 过程是构建当中、构建失败、还是构建成功。如果构建成功，组建好此次构建的 deploy_script_url 和 zip_package_url，回到写发布单数据表 Release 中。
- 在这一过程当中，将相关的结果通过 write_release_history() 函数写入发布单状态数据表（ReleaseHistory）当中，便于后期展示发布单时间线。

pipeline_status () 函数也在 utils/gitlab.py 文件当中，内容如下：

```python
def pipeline_status(git_url, git_access_token, project_id, pipeline_id):
    git_url = git_url
    git_access_token = git_access_token
    gl = gitlab.Gitlab(git_url, private_token=git_access_token)

    project = gl.projects.get(project_id)
    pipeline = project.pipelines.get(pipeline_id)
    time.sleep(1)
    pipeline.refresh()
    return pipeline
```

函数实现很单纯，通过 url 和 token 两个参数认证 GitLab 且通过后，将应用的 id 实例获取，再通过实例获取 pipeline 对象。此次 GitLab CI/CD 的状态值就封装在这个 pipeline 变量当中。调用 pipeline.refresh() 是为了获取刷新后的值，而不是上一次缓存的值。

3. 测试发布单软件构建的 API

本节前面的步骤操作完成后（这里省略了路由的增加和 JWT 的获取），启动 bifangback 应用。同时，为了测试 GitLab 的 CI/CD 功能和 sonarcube 的报告，也要启动这两个服务。

（1）在 Postman 的 url 中输入 http://localhost:8000/release/build_status/，以 post 方式请求，然后发送请求（send）。如果出现如图 9-7 所示的结果则表示编码无误。同时，登录 GitLab 可以看到 CI/CD 已开始运行，如图 9-8 所示。

图 9-7　发布单构建 API

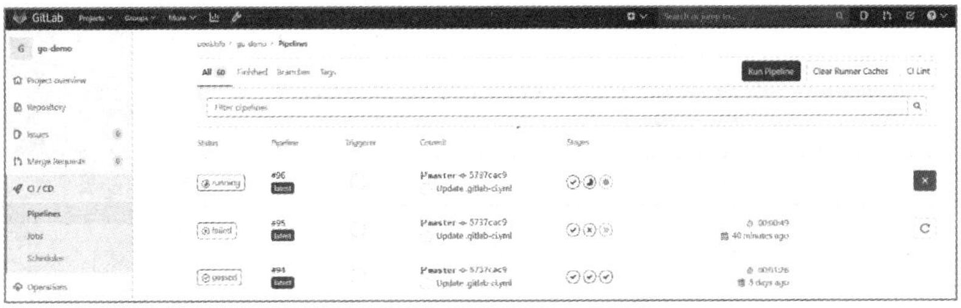

图 9-8　GitLab 中的 CI/CD 状态

（2）软件构建完成，在 Postman 的 url 中输入 http://localhost:8000/release/build_status/，以 post 方式请求，然后发送请求（send）。如果出现如图 9-9 所示的结果则表示编码无误。同时，登录 sonarqube 可以看到代码质量报告已生成，如图 9-10 所示。file-server 文件服务器已上传了相应的部署脚本和软件包，如图 9-11 所示。

图 9-9　发布单构建状态 API

图 9-10　sonarqube 代码质量报告

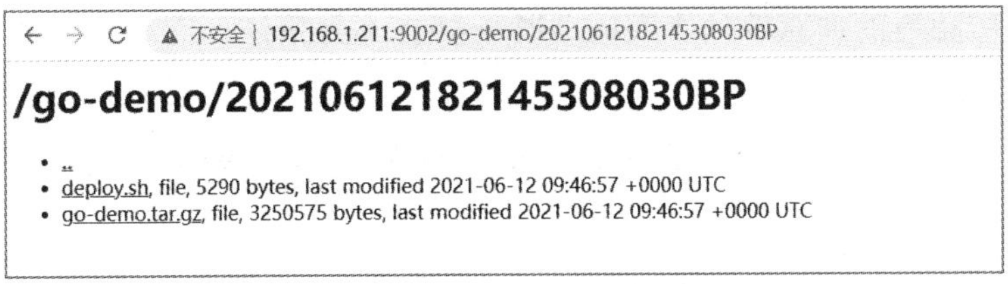

第 9 章 发布单及环境流转

图 9-11 已上传到 file-server 的部署脚本和软件包

9.4 实现 bifangback 发布单环境流转的 API

本节实现发布单的环境流转功能。设计这个功能的初心是因为在现实的企业场景中，一个软件是否能部署在一个环境中，可能取决于测试同事的准入许可。

想象一下，如果一个测试同事正在对 UAT 环境进行一些手工测试，如果这时未经许可地部署，那测试结果的准确性就得不到保证了。所以在这里加了一个发布单环境流转的功能。当然，读者可以根据自己公司的实际情况重新设计公司的发布单功能。发布单的环境流转前端界面如图 9-12 所示。

图 9-12 发布单环境流转

1. 实现环境流转的类视图

在环境流转这个功能中一共提供了两个 API，一是获取所有环境信息的列表，这个功能与前面获取列表类的 API 一样，不再赘述；二是实现环境流转的 API，本节将重点介绍。

在 bifangback 项目根目录的 release 子目录中的 views.py 文件中新增一个 EnvExchangeView 类，内容如下：

```python
class EnvExchangeView(APIView):
    """
    环境流转

    参数：
    env_id
    release_name
    app_id
    """
    def post(self, request):
        # 序列化前端数据，并判断是否有效
        serializer = EnvExchangeSerializer(data=request.data)
        if serializer.is_valid():
            ser_data = serializer.validated_data
            user = request.user
            release_name = ser_data['release_name']
            env_id = ser_data['env_id']
            env = Env.objects.get(id=env_id)
            app_id = ser_data['app_id']
            # 前端开发完成后开启权限测试
            action = Action.objects.get(name='Env')
            if not is_right(app_id, action.id, user):
                return_dict = build_ret_data(NOT_PERMISSION, '你无权在此应用下新建发布单！')
                return render_json(return_dict)

            release_status_name = 'Ready'
            release_status = ReleaseStatus.objects.get(name=release_status_name)
            release = Release.objects.filter(name=release_name).update(env=env,
                                    release_status=release_status)
            write_release_history(release_name=release_name,
                                  env_name=env.name,
                                  release_status_name=release_status_name,
                                  deploy_type=None,
                                  log='target env is {}.'.format(env.name),
                                  user_id=user.id)

            return_dict = build_ret_data(OP_SUCCESS, 'env exchange is ok.')
```

```
                return render_json(return_dict)
            else:
                return_dict = build_ret_data(THROW_EXP, '序列化条件不满足')
                return render_json(return_dict)
```

- 由于自定义程度高，EnvExchangeView 类也继承自 APIView 类。
- 环境流转权限的 Action 关键字是 Env。
- Release.objects.filter().update() 这样的 Django ORM 语法，主要用于更新指定的批量数据。由于发布单号的名称唯一性，所以这里也只更新了一条数据。
- 通过 write_release_history() 记录用户操作环境流转的信息，用于之后的时间线显示。

2. 测试查看具体项目信息的 API

本节前面的步骤操作完成后（这里省略了路由的增加和 JWT 的获取），启动 bifangback 应用。在 Postman 的 url 中输入 http://localhost:8000/env/exchange/，以 post 方式请求。同时，按格式要求传递一个 json 格式的数据，然后发送请求（send）。如果出现如图 9-13 所示的结果，表示编码无误。登录 Django Admin 管理后台会发现发布单的环境字段已置为相应的环境，如图 9-14 所示。

图 9-13　发布单环境流转 API

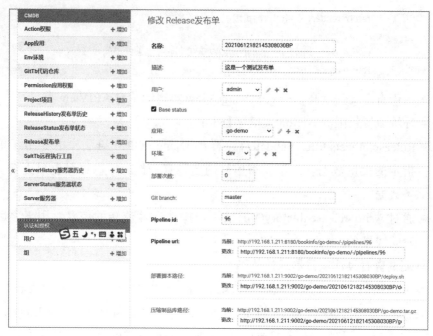

图 9-14 DjangoAdmin 后台验证发布单环境流转

9.5 小 结

本章主要开发了 bifangback 发布单的创建、软件包的构建以及发布单环境的流转，这是 bifangback 项目两大重点功能之一。另一个重点功能——软件的部署，将在第 10 章讲解，敬请期待。

本章使用了之前没有使用过的 DRF 中的 APIView 类。对照本书 8.1 节的内容可知，APIView 是 DRF 概念体系中最基本类视图，也是所有视图的基类，继承自 Django 的 View 父类。

APIView 与 View 的不同之处如下：

- 传入视图方法中的是 REST framework 的 Request 对象，而不是 Django 的 HttpRequeset 对象。
- 视图方法可以返回 REST framework 的 Response 对象，视图会为响应数据设置 render 符合前端要求的格式。
- 任何 APIException 异常都会被捕获到，并且处理成合适的响应信息。

- 在进行 dispatch() 分发前，会对请求进行身份认证、权限检查、流量控制。

APIView 支持定义的属性：
- authentication_classes　　列表或元祖，身份认证类。
- permissoin_classes　　　　列表或元祖，权限检查类。
- throttle_classes　　　　　 列表或元祖，流量控制类。

在 APIView 中仍以常规的类视图定义方法来实现 get()、post() 或者其他请求。

第 10 章 自动化部署

痛苦就是被迫离开原地。

——康德

本章 GitHub 代码地址：https://github.com/aguncn/bifang-book/tree/main/bifang-ch10。

前述章节已实现了自动化部署之前的准备工作和流程把握。比如，发布单的创建、环境的流转、项目和应用的管理、服务器的录入等工作。本章就可以来实现本书最重要的技术点——后端的自动化部署功能。

当然，本章代码的主要实现思路，也就是对本书第 4 章内容的网页化包装。所以如果读者已忘了这部分内容，强烈建议先复习并测试好本书第 4 章的内容。

10.1 实现 bifangback 发布单部署列表 API

BiFang 发布单部署列表如图 10-1 所示，如果读者对照图 9-1 的发布单图片，会发现两者在很大程度上是相似的，只是最右侧的字段显示和按钮功能不同。

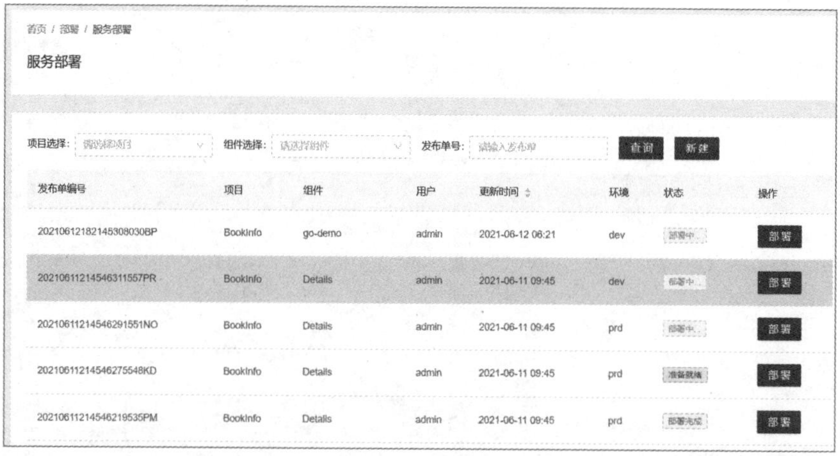

图 10-1 发布单部署列表

其实，后端在这种情况下，并不用专门新写一份代码来满足这种前端需求，而只是重用发布单列表功能，但使用不同的序列化类和过滤器功能，就可以实现这样的前端页面了。

这也是前后端分离开发网站的优势，在实现很多类似的前端显示时，可以重用后端的某一 API 接口。

10.2　实现 bifangback 部署服务器列表 API

BiFang 部署服务器列表如图 10-2 所示，同 10.1 节一样并未使用新的后端 API，而是复用了服务器列表的后端 API，但过滤到了指定环境和指定应用的所有服务器列表。和服务器列表相关的 API，读者可以对照 bifangback 项目根目录的 server 子目录的代码，这里不再重复讲解。

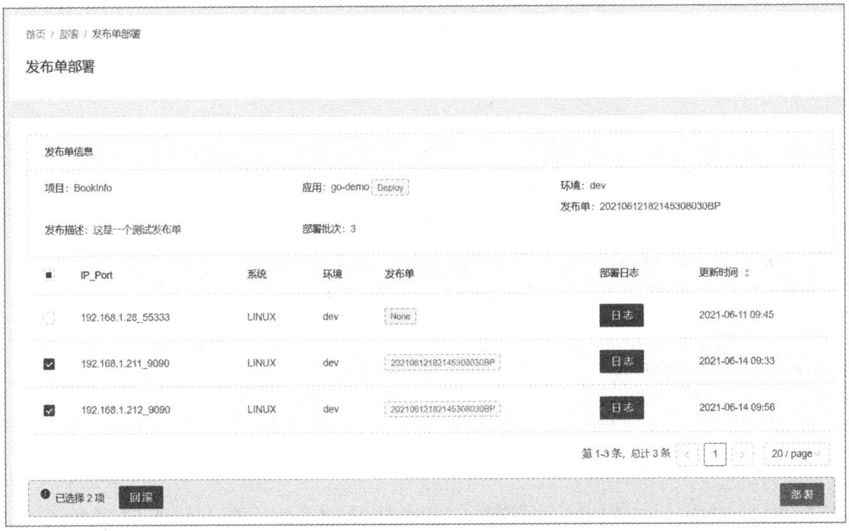

图 10-2　部署服务器列表

10.3　实现 bifangback 自动化部署 API

当用户在图 10-2 所示的界面上选择了要部署的服务器，再单击回滚或部署时就会开始真正的部署。在跳出的界面中会显示部署正在进行中。如果部署完成，则会显示部署的服务器、在部署过程中执行的命令和部署的结果，如图 10-3 和图 10-4 所示。这些部署过程的后端 API 会在本章讲解。当部署完成后，查看服务器的部署日志将在第 11 章讲解。

图 10-3 部署过程页面

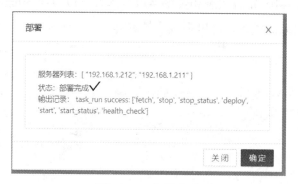

图 10-4 部署成功页面

1. deploy() 函数实现

bifangback 实现自动化部署功能主要由 4 个函数构成：deploy()、task_run()、cmd_run() 和 salt_cmd()，这 4 个函数依次调用，通过参数传递以实现真正的软件部署功能。当然，如果读者掌握了这些技巧之后，可以实现更多更复杂的功能，比如将软件构建和软件部署以可视化拖拽节点的方式来自定义流水线的关键节点，满足更多自定义的设计。

在 bifangback 项目根目录的 deploy 子目录中新增一个 deploy_views.py 文件，新增一个 deploy() 函数，内容如下：

```
def deploy(request):
    if request.method == 'POST':
        """
        {
```

```python
    "app_name": "go-demo",
    "env_name":"dev",
    "user_id":111,
    "service_port":9090,
    "release_name": "20210117132941344089GA",
    "deploy_type": "deploy",
    "deploy_no": 1,
    "op_type": "deploy",
    "target_list": "192.168.1.211,192.168.1.212"
}
"""
req_data = json.loads(request.body.decode('utf-8'))
req_data['target_list'] = req_data['target_list'].split(',')
# 由于这个视图函数不是由 DRF 直接提供的（APIVIEW），而是自己手写的，所以用户
  和用户 ID 的参数，也要自己手动获取
token = request.META.get('HTTP_AUTHORIZATION').split(' ')[1]  # Bearer [token 值 ]
token_user = jwt_decode_handler(token)
user_id = token_user['user_id']
user = User.objects.get(id=user_id)
# 序列化前端数据，并判断是否有效
serializer = DeploySerializer(data=req_data)
if serializer.is_valid():
    ser_data = serializer.validated_data
    app_name = ser_data['app_name']
    service_port = ser_data['service_port']
    release_name = ser_data['release_name']
    env_name = ser_data['env_name']
    # op_type 用于定义是部署应用，还是服务器启停，deploy_type 用于定义操作指令
    op_type = ser_data['op_type']
    deploy_type = ser_data['deploy_type']
    target_list = ser_data['target_list']
    # 部署批次是在当前发布单的部署批次上加 1，然后，将这个新的批次号分别传到
      部署历史和服务器历史中保存
    deploy_no = int(ser_data['deploy_no']) + 1

    # 前端开发完成后开启权限测试
    app = App.objects.get(name=app_name)
    action = Action.objects.get(name='Deploy')
    if not is_right(app.id, action.id, user):
        return_dict = build_ret_data(NOT_PERMISSION, ' 你没有此发布单的部署权限！ ')
        return render_json(return_dict)

    # op_type 的 deploy 用来部署发布，maintenance 用来启停服务
```

```python
            if deploy_type == 'deploy' and op_type == 'deploy':
                action_list = ['fetch', 'stop', 'stop_status', 'deploy', 'start', 'start_status', 'health_
                    check']
                # 更新发布单历史及状态
                write_release_history(release_name, env_name, 'Ongoing', deploy_type, deploy_no,
                    'Ongoing', user_id)
                update_release_status(release_name, app_name, env_name, deploy_no, 'Ongoing')
            # 回滚，只更新服务器操作历史及服务器主备发布单
            elif deploy_type == 'rollback' and op_type == 'deploy':
                action_list = ['stop', 'stop_status', 'rollback', 'start', 'start_status', 'health_check']
            # 之后的代码判断逻辑，可以用来处理单纯的服务器应用启停，而不需要部署发
              布单
            elif deploy_type == 'stop' and op_type == 'maintenance':
                action_list = ['stop', 'stop_status']
            elif deploy_type == 'start' and op_type == 'maintenance':
                action_list = ['start', 'start_status']
            elif deploy_type == 'restart' and op_type == 'maintenance':
                action_list = ['stop', 'stop_status', 'start', 'start_status']
            else:
                return_dict = build_ret_data(THROW_EXP, ' 异常流程参数 ')
                return render_json(return_dict)

            # 以 True 或 False 的返回值来判断任务的成功或失败
            # 写入成功或失败的记录放在真正执行的函数中
            (ret_bool, ret_msg) = task_run(action_list, env_name,
                                    app_name, service_port, release_name,
                                    target_list, user_id,
                                    op_type, deploy_type, deploy_no)
            if ret_bool:
                return_dict = build_ret_data(OP_SUCCESS, ret_msg)
                return render_json(return_dict)
            else:
                return_dict = build_ret_data(THROW_EXP, ret_msg)
                return render_json(return_dict)
    else:
        return_dict = build_ret_data(THROW_EXP, ' 序列化条件不满足 ')
        return render_json(return_dict)
```

- Deploy() 函数的主要作用是接收到前端传递过来的参数后，将参数进行解析，然后根据参数来决定调用哪些命令。代码中本身带有大量注释，如果读者仔细阅读，应该可以跟上 bifangback 的设计思路。

- 前端传过来的 target_list 为字符串形式，需要调用 Python 的 split() 函数将

其转换为列表。
- 由于 deploy() 是标准的 Django 函数视图，而不是经过封闭的 DRF 的视图，所以 token 之类的认证都要自己来写实现。
- op_type 分为 deploy 和 maintenance，分别用于软件的部署回滚和服务器的启停（本项目未实现服务器的启停功能，读者可自实现）。
- deploy_type 类型分为 deploy 和 rollback，分别用于软件的部署及回滚。
- 根据不同的 op_type 和 deploy_type，组合不同的命令序列。
- 将不同的命令序列和参数传递给 task_run() 函数，并使用了多结果返回。
- 最后，根据返回的结果值进行部署成功还是失败的判断。整个函数链都以 True 或 False 作为第一个返回参数做判断。所以，任何一个步骤出错都会使整个部署过程即时中止，防止错误进一步蔓延。

2. task_run() 函数实现

同样在 deploy_views.py 文件，task_run() 函数内容如下：

```
def task_run(action_list, env_name,
             app_name, service_port, release_name,
             target_list, user_id,
             op_type, deploy_type, deploy_no):

    try:
        for action in action_list:
            print('action: ', action)

            (ret_bool, ret_msg) = cmd_run(env_name, app_name,
                                          service_port, release_name,
                                          target_list, action,
                                          user_id, op_type, deploy_type, deploy_no)
            # 如果其中任务一个步骤执行出错，则先记录失败，再返回 False
            if not ret_bool:
                # print('data_false: ', data)
                # 有真正部署，出错时才需要更新发布单历史，其他情况只更新服务器发布历史（暂不考虑回滚失败）
                if deploy_type == 'deploy':
                    # 更新服务器状态
                    update_server_status(target_list, service_port, deploy_no, "Failed")
                    # 写入部署历史
                    write_release_history(release_name, env_name, 'Failed', deploy_type, deploy_
```

```python
                            no, 'Failed', user_id)
                    # 更新发布单状态
                    update_release_status(release_name, app_name, env_name, deploy_no, 'Failed')

                    return False, "{}: {}".format(action, ret_msg)
            # print('data_true: ', data)
        # 这里 for 循环完成后，表示任务已成功执行，可以集中更新数据库记录
        # 只有部署和回滚，才需要更新发布单历史和服务器主备发布单（回滚时，发布单参数
          并没有用上）
        if op_type == 'deploy':
            # 成功更新相关数据表，先更新服务器状态，然后在更新发布单状态时先判断服务
              器状态，有先后顺序
            update_server_status(target_list, service_port, deploy_no, "Success")
            update_server_release(target_list, service_port, release_name, deploy_type)
            write_release_history(release_name, env_name, 'Success', deploy_type, deploy_no,
                'Success', user_id)
            # 当某一次部署完成后，还需要判断是不是所有服务器都已部署完成，所以传入的
              不是 Success，而是 Check
            # 然后将 Check 参数传到 update_release_status 函数中判断
            update_release_status(release_name, app_name, env_name, deploy_no, 'Check')

        print("finish: ", action_list, env_name, app_name, release_name, target_list)
        return True, "task_run success: {}".format(str(action_list))
    except Exception as e:
        print(e)
        return False, "task_run error: {}".format(str(e))
```

- 在 task_run() 函数中，将每一条命令进行循环执行。
- 在每一条命令循环中都调用 cmd_run() 执行命令。同时，根据 cmd_run() 命令的返回结果进行命令执行成功或失败的判断。
- 如果命令执行失败，则同时更新发布单和服务器的状态，并将失败信息写入发布单的部署历史。
- 如果所有命令序列执行成功，则将部署成功的信息，同时到更新发布单的状态和服务器的部署状态，并将成功信息写入发布单的部署历史。
- 最后，需判断这个发布单是不是已在指定环境上部署到了所有的机器上，以决定发布单状态是"部署中"还是"部署完成"。
- 这里没有刻意区分部署和回滚，只是将回滚这个 action 记录到了服务器的状态上。并更新了服务器上的主备发布单。读者如果使用 bifangback，需要在这个环境中加强一下适合自己企业的设计。

- 在上面涉及的几个记录状态的函数中，write_release_history() 在第 9 章已介绍过。下面分列一下 update_release_status ()、update_server_status() 和 update_server_release() 的代码。它们均位于 utils 目录下的 write_history.py 文件中。这些函数都已写好主要的注释，没有必要再讲解。其内容如下：

```python
import uuid
from django.contrib.auth import get_user_model
from cmdb.models import ReleaseStatus
from cmdb.models import ServerStatus
from cmdb.models import Release
from cmdb.models import App
from cmdb.models import Env
from cmdb.models import Server
from cmdb.history_models import ReleaseHistory
from cmdb.history_models import ServerHistory

User = get_user_model()


# 更新发布单状态
def update_release_status(release_name, app_name, env_name, deploy_no, release_status_name):
    if release_status_name != 'Check':
        # 如果是进行中或失败，直接写入
        release_status = ReleaseStatus.objects.get(name=release_status_name)
        release = Release.objects.filter(name=release_name).update(deploy_no=deploy_no, release_status=release_status)
    else:
        # 如果只是完成其中一次的成功部署，Check 时则要判断同一个应用、同一个环境的所有服务器的状态
        # 判断条件有两个：服务器上的主发布单与部署一致，所有服务器的部署状态为成功
        release = Release.objects.get(name=release_name)
        server_status = ServerStatus.objects.get(name='Success')
        release_status = ReleaseStatus.objects.get(name='Success')
        app = App.objects.get(name=app_name)
        env = Env.objects.get(name=env_name)
        servers = Server.objects.filter(app=app, env=env)
        for server in servers:
            if server.main_release != release or server.server_status != server_status:
                print("not meet")
                return
        print("all meet")
        release = Release.objects.filter(name=release_name).update(deploy_no=deploy_no,
```

```python
                                                    release_status=release_status)


# 更新服务器状态
def update_server_status(target_list, service_port, deploy_no, server_status_name):
    for ip in target_list:
        server_name = '{}_{}'.format(ip, service_port)
        server_status = ServerStatus.objects.get(name=server_status_name)
        server = Server.objects.filter(name=server_name).update(deploy_no=deploy_no, server_
            status=server_status)


# 更新服务器的主备发布单
def update_server_release(target_list, service_port, release_name, deploy_type):

    for ip in target_list:
        server = Server.objects.get(name='{}_{}'.format(ip, service_port))
        # 真正部署时,将服务器的主发布单放到备用发布单,用新发布单填充主发布单
        if deploy_type == 'deploy':
            release = Release.objects.get(name=release_name)
            # 如果存在部署,且不是重复部署
            if server.main_release and release != server.main_release:
                back_release = server.main_release
                server.back_release = back_release
            server.main_release = release
            server.save()

        # 如果是回滚,则将主备发布单都设置为备用发布单即可,因为只支持一次回滚。多次
          回滚,不如重新发布
        else:
            release = server.back_release
            server.main_release = release
            server.save()


# 更新服务器操作历史,可用 ajax 展示具体部署过程,也可以查看一个具体服务器的操作历史
def write_server_history(ip=None, service_port=None, release_name=None,
                         env_name=None, op_type=None,
                         action_type=None, deploy_no=None, log=None,
                         user_id=None):
    name = uuid.uuid1()
    server = Server.objects.get(name='{}_{}'.format(ip, service_port))
    release = None
```

```
if release_name is not None:
    release = Release.objects.get(name=release_name)
create_user = None
if user_id is not None:
    create_user = User.objects.get(id=user_id)
env = None
if env_name is not None:
    env = Env.objects.get(name=env_name)
ServerHistory.objects.create(name=name,
                             release=release,
                             env=env,
                             server=server,
                             op_type=op_type,
                             action_type=action_type,
                             log_no=deploy_no,
                             log=log,
                             create_user=create_user)
```

3. cmd_run() 函数实现

同样在 deploy_views.py 文件，cmd_run() 函数内容如下：

```
def cmd_run(env_name, app_name, service_port,
            release_name, target_list,
            action, user_id,
            op_type, deploy_type, deploy_no):
    env = Env.objects.get(name=env_name)
    app = App.objects.get(name=app_name)
    release = Release.objects.get(name=release_name)

    salt_url = env.salt.salt_url
    salt_user = env.salt.salt_user
    salt_pwd = env.salt.salt_pwd
    eauth = env.salt.eauth

    deploy_script_url = release.deploy_script_url
    zip_package_name = app.zip_package_name
    zip_package_url = release.zip_package_url
    # 使用 saltypie 来获取返回值，不必自己写 HTTP 请求，更容易解析结果
    ret = salt_cmd(salt_url, salt_user, salt_pwd, eauth,
                   target_list, deploy_script_url,
                   app_name, release, env,
                   zip_package_name, zip_package_url,
                   service_port, action)
```

```python
        time.sleep(1)
        # salt 找不到服务器，则返回列表里字典中有一个必为空，这一步要提前判断
        if any(not item for item in ret):
            return False, " 找不到 salt minion 客户端： " + str(ret)
        for server in ret:
            for ip, detail in server.items():
                print('ip: ', ip)
                print('retcode: ', detail['retcode'])
                print('stdout: ', detail['stdout'])
                print('stderr: ', detail['stderr'])
                print('pid: ', detail['pid'])
                # 记录服务器操作历史
                write_server_history(ip, service_port, release_name, env_name, op_type, action, deploy_no,
                                     detail['stdout'] + detail['stderr'], user_id)

                # 部署脚本的每一个步骤，成功时必须返回 success 关键字
                if 'success' not in detail['stdout']:
                    return False, " 脚本执行完成，但没有 success 关键字： " + detail['stdout']
                if len(detail['stderr']) > 0 and detail['retcode'] != 0:
                    return False, " 脚本执行完成，有 stderr 错误： " + detail['stderr']

        return True, " 脚本执行完成： " + detail['stdout']
```

- cmd_run() 主要实现了对 salt_cmd() 的调用，以及对 salt_cmd() 返回结果的精确判断。
- 调用 salt_cmd() 的参数比较多，需要提前准备好。deploy_script_url 这个参数是在生成软件包时，同时也生成的 deploy.sh 脚本的 url 地址。
- any(not item for item in ret) 的写法用于快速地判断 ret 里存在一个列表项是否存在值。如果找不到 salt 服务器，就会存在这样的情况。如果还有其他情况，也可以放在这里来判断，这里只是列举其中一种情况。
- 如果读者不熟悉 ret 的格式，可以在 salt_cmd() 里输出来，细细分析。在循环每一个 server 时，先将部署细节更新到服务器中。如果发现有错，即时返回，中止部署。
- 其中的几个 print() 输出，用于确认之前分析的 ret 格式是否正确。因为这一部署成功与否的判断，对于 bifangback 来说极其重要，所以不断地重复一些关键点，对于保证系统的稳定性至关重要。一个正常的 print() 输出，如图 10-5 所示。

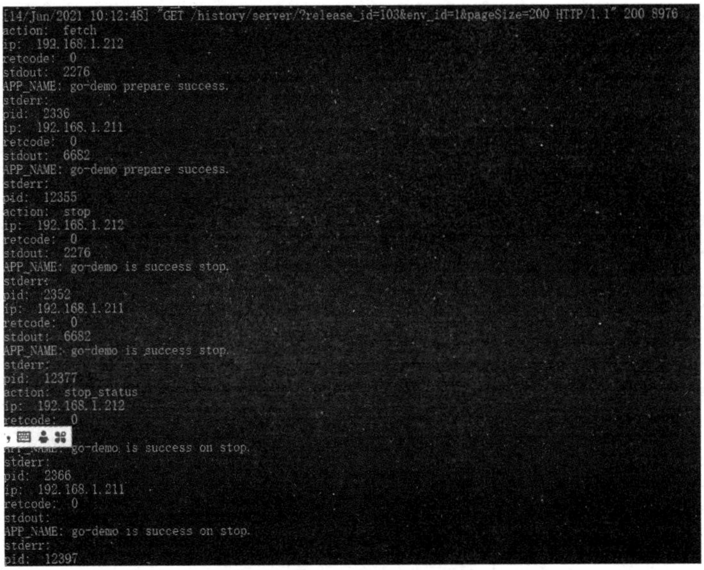

图 10-5　部署过程中的输出台输出

4. salt_cmd() 函数实现

在 utils 目录下新建一个 saltstack.py 文件用于定义 salt_cmd() 函数，内容如下：

```
from saltypie import Salt


def salt_cmd(salt_url, salt_user, salt_pwd, eauth,
             target_list, script_url,
             app, release, env,
             zip_package_name, zip_package_url,
             service_port, action):

    salt = Salt(
        url=salt_url,
        username=salt_user,
        passwd=salt_pwd,
        trust_host=True,
        eauth=eauth
    )

    arg_list = [script_url, '{} {} {} {} {} {}'.format(app, release, env,
                                        zip_package_name, zip_package_url,
                                        service_port, action)]

    # print(arg_list)
```

```python
    exe_return = salt.execute(
        client=Salt.CLIENT_LOCAL,
        target=target_list,
        tgt_type='list',
        fun='cmd.script',
        args=arg_list,
    )
    # print(exe_return)
    return exe_return['return']
```

- salt_cmd() 的函数和本书 4.4 节的内容一致，是对 saltypie 这个第三方库的二次封装。
- arg_list 列表的第一项为 deploy.sh 的 url 地址（支持 http:// 协议，也可以使用 salt:// 协议）。
- arg_list 列表的第二项为传递给 deploy.sh 脚本的 7 个参数。通过这种方式，就可以将部署脚本和参数很方便地联结在一起了。
- salt.execute() 的参数里指定 tgt_list 为 list，表示可以同时在多个服务器上执行命令。Func 为 cmd.script 表示要执行脚本。

5. 测试自动化部署 API

在本节前面的步骤操作完成后（这里省略了路由的增加和 JWT 的获取），启动 bifangback 应用。在 Postman 的 url 中输入 http://localhost:8000/deploy/deploy/，以 post 方式请求，构建一个合适的 json 请求数据，然后发送请求（send）。如果输出如图 10-6 所示，并且可以通过 url 正常访问已部署的应用，如图 10-7 所示，则表示自动化部署的代码基本上没有问题了。如果读者把在本书第 2 章的 demo 项目用来部署，应该也可以成功实现自动化部署。

图 10-6 测试部署 API

图 10-7　测试部署成功后的应用

10.4　小　结

本章主要开发了 bifangback 最核心的功能——自动化部署功能，因为它涉及多个服务间的调用，所以读者在实践本章内容时，需要将相关的服务都启动起来。

另外，本书的截图和服务器是以我的机器的 IP 为示例的，读者在自己的测试环境下实践时，需要进行相应的替换。

不知不觉中快进入后端开发的尾声了。第 11 章是后端开发的最后一章，主要讲解三个方面的内容：显示发布单操作的时间线、服务器日志的时间线和 bifangback 简单的 dashboard。

第 11 章　后端数据展示

> 对于不可言说之物，我们唯有保持沉默。
>
> ——维特根斯坦

本章 GitHub 代码地址：https://github.com/aguncn/bifang-book/tree/main/bifang-ch11。

前述章节已实现了自动化部署的大多数后端 API，本章将讲解几个收尾阶段的 API——显示发布单操作的时间线、服务器日志的时间线、bifangback 简单的 dashboard。

11.1　实现 bifangback 发布单历史 API

BiFang 发布单部署历史如图 11-1 所示，这个 API 的功能可以让我们了解这个发布单的历史操作时间线，为 BiFang 后期进一步地完善度量做好数据统计。同时，也可以对发布问题的时间进行简单的追溯。

图 11-1　发布单历史

右边显示的时间线，即为一个发布单的历史操作记录。这显示是经过前端 Vue 框架处理后的结果，而后端 API 提供的每一条记录的 json 数据格式如下：

```
{
    base_status: true
    create_date: "2021-06-14T10:13:43.161366+08:00"
    create_user: "admin"
    deploy_type: "deploy"
    description: null
    env: "dev"
    id: 13
    log: "Ongoing"
    log_no: 4
    name: "24630779-ccb6-11eb-bd53-1cbfc0db9238"
    release: "20210612182145308030BP"
    release_status: 6
    release_status_name: "Ongoing"
    update_date: "2021-06-14T10:13:43.161366+08:00"
}
```

本节就来看看，如何通过后端输出以上格式的列表。由于在发布单的创建、流转和部署过程中，都已通过 write_release_history() 函数把相关的数据记录插入了数据库表。所以本节的内容主要是从数据库中把记录读出来，很简单。

1. 发布单历史的序列化文件

在 history 子目录中新增一个 serializers.py 文件和一个 ReleaseHistorySerializer 类，内容如下：

```
class ReleaseHistorySerializer(serializers.ModelSerializer):
    env = serializers.CharField(source='env.name', default=None)
    release = serializers.CharField(source='release.name', default=None)
    create_user = serializers.CharField(source='create_user.username', default=None)
    release_status_name = serializers.CharField(source='release_status.name', default=None)

    class Meta:
        model = ReleaseHistory
        fields = '__all__'
```

- 主要操作的数据表是 ReleaseHistory。
- env、release、create_user、release_status_name 都是 ReleaseHistory 表的外键，这里希望输出这些字段的外键真实名称，方便前端显示真实的名称，而不是一个个关联的 id。

2. 发布单历史的过滤器文件

在 history 子目录中新增一个 filters.py 文件和一个 ReleaseHistoryFilter 类，内容如下：

```python
class ReleaseHistoryFilter(FilterSet):
    release = filters.CharFilter(field_name='release__name', lookup_expr='icontains',)
    begin_time = filters.DateTimeFilter(field_name='create_date', lookup_expr='gte',)
    end_time = filters.DateTimeFilter(field_name='create_date', lookup_expr='lte',)
    sort = OrderingFilter(fields=('create_date',))

    class Meta:
        model = ReleaseHistory
        fields = ['release_id', 'begin_time', 'end_time']
```

- 主要操作的数据表是 ReleaseHistory。
- 最主要的过滤项是 release，其他几个用于将来扩展吧。

3. 发布单历史的类视图

在 history 子目录的 views.py 文件中新增一个 ReleaseHistoryListView 类，内容如下：

```python
class ReleaseHistoryListView(ListAPIView):
    queryset = ReleaseHistory.objects.all()
    serializer_class = ReleaseHistorySerializer
    filter_class = ReleaseHistoryFilter

    def get(self, request, *args, **kwargs):
        res = super().get(self, request, *args, **kwargs)
        return_dict = build_ret_data(OP_SUCCESS, res.data)
        return render_json(return_dict)
```

这是 DRF 中非常标准的一个 ListAPIView，使用上面定义的序列化类和过滤器类即可。由于整个 bifangback 都做了自定义的请求返回体，所以在类里复写了 get() 方法。

4. 发布单历史的路由

在 history 子目录的 urls.py 文件中定义好 history 这个 app 的全部路由，内容如下：

```python
from django.urls import path
from . import views
```

```
app_name = "history"

urlpatterns = [
    path('release/', views.ReleaseHistoryListView.as_view(), name='release'),
    path('server/', views.ServerHistoryListView.as_view(), name='server'),
]
```

为了简省，将下一节要讲的服务器部署历史的路由也在这里展示了，过会不再重复。

5. 测试发布单历史 API

本地测试访问：http://localhost:8000/history/release/?release_id=103，出现如图 11-2 所示的输出则表示此 API 成功开发完成。

图 11-2　测试发布单历史 API

11.2　实现 bifangback 服务器部署历史 API

毕方（BiFang）服务器部署历史如图 11-3 所示，这个 API 的功能可以让我们了解在这个服务器上的部署历史。如果有部署错误，也可以通过这个界面找到一些出错的细节并发现修正的可能性。

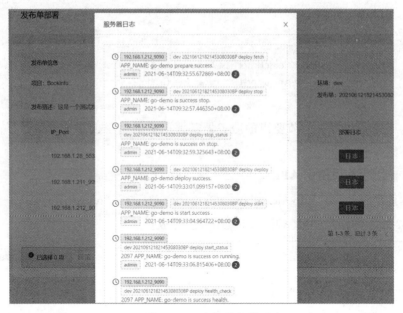

图 11-3 服务器部署历史

在发布单部署过程中,我们都已通过 write_server_history () 函数把相关的数据记录插入了数据库表。所以本节的内容和 11.1 节一样,主要是从数据表中把记录读出来。

1. 服务器部署历史的序列化文件

在 history 子目录的 serializers.py 文件中新增一个 ServerHistoryFilter 类,内容如下:

```
class ServerHistoryFilter(FilterSet):
    release_name = filters.CharFilter(field_name='release__name', lookup_expr='icontains',)
    env_name = filters.CharFilter(field_name='env__name', lookup_expr='icontains', )
    begin_time = filters.DateTimeFilter(field_name='create_date', lookup_expr='gte',)
    end_time = filters.DateTimeFilter(field_name='create_date', lookup_expr='lte',)
    sort = OrderingFilter(fields=('create_date',))

    class Meta:
        model = ServerHistory
        fields = ['release_name', 'env_name', 'release_id', 'env_id', 'log_no', 'begin_time', 'end_time']
```

- 主要操作的数据表是 ServerHistory。
- env_name、release_name 都是 ServerHistory 表的外键,这里希望输出这些字段的外键真实名称,方便前端显示真实的名称,而不是一个个关联的 id。

2. 服务器部署历史的过滤器文件

在 history 的 filters.py 文件中新增一个 ServerHistorySerializer 类，内容如下：

```python
class ServerHistorySerializer(serializers.ModelSerializer):
    server = serializers.CharField(source='server.name', default=None)
    env = serializers.CharField(source='env.name', default=None)
    release = serializers.CharField(source='release.name', default=None)
    create_user = serializers.CharField(source='create_user.username')

    class Meta:
        model = ServerHistory
        fields = '__all__'
```

- 主要操作的数据表是 ServerHistory。
- 最主要的过滤项是 server、env、release。

3. 服务器部署历史的类视图

在 history 子目录的 views.py 文件中新增一个 ServerHistoryListView 类，内容如下：

```python
class ServerHistoryListView(ListAPIView):
    queryset = ServerHistory.objects.all()
    serializer_class = ServerHistorySerializer
    filter_class = ServerHistoryFilter

    def get(self, request, *args, **kwargs):
        res = super().get(self, request, *args, **kwargs)
        return_dict = build_ret_data(OP_SUCCESS, res.data)
        return render_json(return_dict)
```

这是 DRF 中非常标准的 ListAPIView，使用上面定义的序列化类和过滤器类即可。由于整个 bifangback 都做了自定义的请求返回体，所以在类里复写了 get() 方法。

4. 测试服务器部署历史 API

本地测试访问：http://localhost:8000/history/server/?release_id=103&env_id=1&pageSize=200，出现如图 11-4 所示的输出则表示此 API 成功开发完成。

图 11-4　测试服务器部署历史 API.png

11.3　实现 bifangback 简单的 dashboard

每个 bifangback 都应该有自己的 dashboard，BiFang 项目也不例外。由于本项目是出于 demo 性质，所以在 dashboard 方面设计得比较简略。但麻雀虽小，五脏俱全，所以在 dashbaord 里还输出了数字、柱状图和环状图，分别来反映 BiFang 项目的项目、应用、发布单数量、发布失败的统计及发布单在各应用之间的排行，如图 11-5 所示。

本节就来讲解如何通过 DRF 后端来输出前端渲染所需要的这些数据。

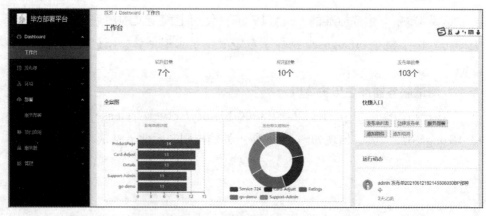

图 11-5　BiFang 项目的 dashboard

1. Dashboard 的函数视图

在 stats 子目录的 views.py 文件中新增三个 Django 的函数视图，用于直接将数据返回前端、all_count()、release_top5()、release_failed_top5()，内容如下：

```python
from django.db.models import F
from django.db.models import Count
from cmdb.models import Project
from cmdb.models import App
from cmdb.models import Release
from cmdb.models import ReleaseStatus
from utils.ret_code import *


# 输出项目、组件、发布单总数
def all_count(request):
    project_count = Project.objects.count()
    app_count = App.objects.count()
    release_count = Release.objects.count()
    data = {
        'project': project_count,
        'app': app_count,
        'release': release_count
    }
    return_dict = build_ret_data(OP_SUCCESS, data)
    return render_json(return_dict)


# 输出各组件发布单总量 top5
def release_top5(request):
    queryset = Release.objects.values('app_id') \
                    .annotate(release_count=Count('name'), app_name=F('app__name')) \
                    .order_by('-release_count')[:5]
    data = {'app_name': [],
            'release_count': []}
    for item in queryset:
        data['app_name'].append(item['app_name'])
        data['release_count'].append(item['release_count'])
    print(data)
    return_dict = build_ret_data(OP_SUCCESS, data)
    return render_json(return_dict)


# 输出各组件发布单错误数量 top5
```

```python
def release_failed_top5(request):
    release_status = ReleaseStatus.objects.get(name='Failed')
    queryset = Release.objects.filter(release_status=release_status).values('app_id') \
                    .annotate(release_count=Count('name'), app_name=F('app__name')) \
                    .order_by('-release_count')[:5]
    data = {'app_name': [],
            'release_count': []}
    for item in queryset:
        data['app_name'].append(item['app_name'])
        data['release_count'].append(item['release_count'])
    print(data)
    return_dict = build_ret_data(OP_SUCCESS, data)
    return render_json(return_dict)
```

- 如注释，all_count() 用于输出项目、组件、发布单总数，以字典形式返回。
- release_top5() 用于输出各组件发布单总量 top5，这里使用了 Django ORM 中的 annotate 聚合函数来返回对应的数据，其返回值形式在随后的测试 API 中展示。读者在学习这类语法时，对照结果输出来调整参数，是最快速的。
- release_failed_top5 () 用于输出各组件发布单错误数量 top5，这个数值能引起运维同事对相关组件的部署重视，这里使用了 Django ORM 中的 F 查询，用来取出 app 的每一个外键真正值。

2. 发布单历史的路由

在 stats 子目录的 urls.py 文件中定义好 stats 这个 app 的全部路由，内容如下：

```python
from django.urls import path
from . import views

app_name = "stats"

urlpatterns = [
    path('all_count/', views.all_count, name='all_count'),
    path('release_top5/', views.release_top5, name='release_top5'),
    path('release_failed_top5/', views.release_failed_top5, name='release_failed_top5'),
]
```

3. 测试发布单历史 API

项目数量访问：http://127.0.0.1:8000/stats/all_count/，如图 11-6 所示。

第 11 章 后端数据展示

图 11-6 项目数量 API

发布单排行访问：http：//127.0.0.1:8000/stats/release_top5/，如图 11-7 所示。

图 11-7 发布单排行 API

发布单失败排行访问：http://127.0.0.1:8000/stats/release_failed_top5/，如图 11-8 所示。

图 11-8　发布单失败排行 API

11.4　小　结

本章实现了后端数量的一些 dashboard 输出展示，当然，对于真正的企业应用系统，这些数据是远远不够的。有兴趣的读者可以多设计一些这方面的指标，用于真正推进公司的 devops 进程。

由于 BiFang 项目是一个前后端分离的开发项目，所以从第 12 章开始进入前端网页的开发。使用的语言也不再是 Python，而是以 JS 语言为主。

如果本章之前的内容还没有消化完，测试结束后建议再多看几次、多测试几次、多修改几次。

准备好了，我们就出发。

第 12 章　前端项目选型与搭建

> 业精于勤，荒于嬉，行成于思，毁于随。
>
> ——韩愈

本章 GitHub 代码地址：https://github.com/aguncn/bifang-book/tree/main/bifang-ch12。

前文主要讲解了 BiFang 自动化部署系统服务端相关技术的使用以及相关模块的设计，就像已经有了骨骼和灵魂，必须辅以血肉肌肤（即 GUI 界面），才能成为一个完整的人。

本章将介绍前端项目的搭建、项目框架的选型以及相关技术的介绍，基础项目的搭建，让读者能有一个全局的概览。在 BiFang 项目中以 Vue.js 为前端开发语言，ant-design-vue 为前端 UI 框架，辅以 Less、ES6、Vuex、Vue-router 等技术来实现整个 Web 项目。

12.1　项目开发语言选型

前端 Web 项目当前最火的三个框架为 Angular、React 以及 Vue，这三兄弟都是以数据驱动视图达到页面的渲染，即 MVVM（Model-View-ViewModel）。

老大哥 Angular 由大名鼎鼎的谷歌公司率先推出，当时震动业界，在以 jquery 和 zepto 为主流框架的 Web 项目中开辟了一条新的航道，但由于过度设计，在其推出时就带了很多的根本性问题，不得不在 2016 年重新推出了版本 2.0，那时大势已去，被老二老三远远甩开。

二哥 React 是由国际知名的 Facebook 推出，一经推出立马被各大公司列为 Web 项目首选框架，被评价为 JS 界在意识层面向前迈出的一大步，它以一种实用简洁的方式向人们展示了真正的函数式编程，应用数量直线上升，后续推出的 React native 更是如虎添翼，一次开发运行在 H5、Andriod、iOS 三端的框架是吸引各公司使用它的最核心的缘由。

三弟 Vue 是由当时不知名的前谷歌公司员工 Evan You 个人开发的，在借鉴了 Angular 和 React 的核心优势后，最轻量级的 MVVM 框架诞生，优雅的接口设计、不依赖于 JSX 的 html/js/css 分明的组件设计、国人的爱国情怀都把它推向一个高峰。

1. 底层原理

先来简单说一下三兄弟的底层实现和工作原理。

Angular 将会遍历 DOM 模板，来生成相应的 NG 指令，所有的指令都负责针对 view(即 ng-model) 来设置数据绑定。$scope 中的数据更新，使用脏值检查方式实现，在双向绑定中，scope 下的对象一直处于 digest loop 中，loop 通过遍历这些对象来发现他们是否改变，如果改变就会调用相应的处理方法来实现双向绑定。每次变化，所有 watcher 都要重新计算。

React 推崇的是函数式编程和单向数据流：给定原始界面(或数据)，施加一个变化，就能推导出界面或者数据的更新。React 的渲染建立在 Virtual DOM 上，当状态发生变化时，React 重新渲染 Virtual DOM，比较计算之后给真实 DOM 打补丁。

Vue 借鉴 Angular 的数据监听方式，通过 Object.defineProperty 对对象添加 getter 和 setter 方法，对象变化独立触发。在视图渲染上，Vue 借鉴了 React 的 Virtual DOM，通过 diff 算法实现真实 DOM 节点的最小渲染。

由于 Vue 一开始是尤雨溪抽取 Angular 核心模块来仿写的，所以 Angular 和 Vue 都是通过类似模板的语法，描述界面状态与数据的绑定关系，然后通过内部转换，把这个结构建立起来，当界面发生变化的时候，按照配置规则去更新相应的数据，然后，再根据配置好的规则去从数据更新界面状态。

2. 优缺点

Angular 是一个大而全的前端框架，包含服务、模板、数据双向绑定、模块化、路由、过滤器和依赖注入等所有功能，自带了极其丰富的 Angular 指令。同时 ng 模块化，每个模块的代码独立拥有自己的作用域，model、controller 等能很容易地写出可复用的代码，对于团队敏捷开发特别重要。

缺点：Angular 包含许多概念，对于初学者上手难度较高。同时由于 Angular 在设计之初就包含了开发所需的所有模块，体量较大，不如其他两个兄弟灵活。

React 提出了 Virtual Dom 这种新颖的思路，这种思路还衍生出了 React Native，有可能会统一 Web/Native 开发，同时，Virtual Dom 技术在数据变更时，先更新 Virtual DOM，然后和实际 DOM 比较，最后再更新实际的 DOM，操作 DOM 次数少，速度更快。React 支持服务端渲染，利于搜索引擎优化。

缺点：第三方组件不如 Angular 多，依赖社区的贡献，社区组件代码质量参差不齐。JSX 的学习成本也是初学者面临的一个问题。

Vue.js 是构建数据驱动的 Web 界面的渐进式框架，目标是通过尽可能简单的 API 实现响应的数据绑定和组合的视图组件。核心是一个响应的数据绑定系统，集各家之所长，实现了自身的 Virtual Dom 技术，同时支持服务端渲染。组件化模板、行为、样式一目了然，模板使用 html 语法，初学者可以轻松入门。

缺点：国际认可度不如 Angular 和 React，且不如 React 和 Angular 成熟。在移动端 Native 化的探索项目 weex 不如 React Native 成熟。

每一个框架都有它自身的差异化设计，有一句至理名言：合适的才是最好的。

- 如果你想要平缓的学习曲线以及轻量级框架：Vue。
- 如果你喜欢真正干净的代码：Vue。
- 如果你的项目团队很大：React 或 Angular。
- 如果你构建一个跨平台应用程序：React。

综上所述，我们需要的是一个小而美，且适合大家上手的前端 Web 项目框架，此处优先选择 Vue。

12.2 ant-design-vue

Ant Design 作为一门设计语言面世，经历过多年的迭代和积累，它对 UI 的设计思想已经成为一套事实标准，受到众多前端开发者及企业的追捧和喜爱，也是 React 开发者手中的神兵利器。希望 ant-design-vue 能够让 Vue 开发者也享受到 Ant Design 的优秀设计体验。

ant-design-vue 是 Ant Design 的 Vue 实现，组件的风格与 Ant Design 保持同步，组件的 html 结构和 css 样式也保持一致，真正做到了样式零修改，组件 API 也尽量保持了一致。

12.3 css、less 与 scss/sass

less、scss/sass 都是动态样式语言，less 比 css 多出很多功能（如变量、继承、运算、函数），scss/sass 在 less 基础上多了嵌套、运算和 Mixin 等。

1. 变量作用域

less、scss 的变量符不一样，less 是 @，scss 是 $，变量作用域也不一样。

```
@color: #00c;    --- 蓝色
#header {
    @color: #c00;              --- 红色
    border: 1px solid @color;  --- 红色边框
}
#footer {
    border: 1px solid @color;  --- 蓝色边框
}
```

Less- 作用域编译后：

```
#header{border:1px solid #cc0000;}
#footer{border:1px solid #0000cc;}
```

scss- 作用域：

```
$color: #00c;    --- 蓝色
#header {
    $color: #c00;              --- 红色
    border: 1px solid $color;  --- 红色边框
}
#footer {
    border: 1px solid $color;  --- 蓝色边框
}
```

Sass- 作用域编译后：

```
#header{border:1px solid #c00}
#footer{border:1px solid #c00}
```

2. 条件循环

less 不支持条件语句，scss 语句支持 if-else{}、for{} 循环语句

```
@if lightness($color) > 30% {
```

```
    #header {
        $color: #c00;              --- 红色
        border: 1px solid $color;  --- 红色边框
    }
} @else {
    #header {
        $color: #c00;;             --- 红色
        border: 1px solid $color;  --- 红色边框
    }
}
@for $i from 1 to 10 {             --- 循环
    .border-#{$i} {
        border: #{$i}px solid red;
    }
}
```

3. 引入外部样式文件

less 引用外部文件和 css 中的 @import 没什么差异。scss 引用的外部文件命名必须以 "_" 开头，文件名如果以下划线（_）开头的话，sass 会认为该文件是一个引用文件，不会将其编译为 css 文件。源代码：

```
@import "_test1.scss";
@import "_test2.scss";
@import "_test3.scss";
编译后：
h1 {
    font-size: 17px;
}
h2 {
    font-size: 17px;
}
h3 {
    font-size: 17px;
}
```

总结：sass/scss 或 less 都可以看作一种基于 css 之上的高级语言，其目的是使得 css 开发更灵活以及更强大。sass 的功能比 less 强大，基本可以说是一种真正的编程语言，less 则相对清晰明了、易于上手，对编译环境要求比较宽松，在实际开发中更倾向于选择 less。

12.4 使用 Vue CLI 搭建项目

Vue CLI 是一个基于 Vue.js 进行快速开发的完整系统,致力于将 Vue 生态中的工具基础标准化。它确保了各种构建工具能够基于智能的默认配置即可平稳衔接,这样你可以专注在撰写应用上,而不必花好几天去纠结配置的问题。与此同时,它也为每个工具提供了调整配置的灵活性,无须 eject。

以上是 Vue CLI 的官方介绍,通过 Vue CLI 命令可以根据所需配置项初始化 Vue 项目,而且项目所需要的依赖也已下载完毕,可以直接通过命令运行项目。

1. 安装依赖

Node.js 是运行在服务端的 JavaScript,是使用 Vue.js 所需要的依赖。BiFang 项目使用的 Vue CLI 版本为 4.5.10,需要 Node.js v8.9 或更高版本(推荐 v10 以上),这里使用 12.10.0 版本。

打开命令提示符窗口或 shell 窗口,输入如下命令:

```
node -v

v12.10.0   // 已安装
node: command not found // 未安装
```

如果你的本地已安装 Node.js,会展示具体的 node 版本信息,如果本地未安装或版本太低,需重新安装 Node.js。打开 https://nodejs.org/dist/v12.10.0/,下载对应环境的安装包。

(1)在 Windows 上安装 Node.js。

下载 node-v12.10.0-x64.msi 或 node-v12.10.0-x86.msi,按照提示安装检测 PATH 环境变量是否配置了 Node.js,单击开始→运行→输入 "cmd" →输入命令 "path",若已包含 Node.js 安装目录路径,进入下一步;若未包含,需通过:我的电脑→设置→高级设置→环境变量→找到 Path,把 Node.js 安装目录路径加入环境变量。

(2)在 Linux 上安装 Node.js。

Node 官网已把 Linux 下载版本更改为已编译好的版本,可以直接下载解压后使用。下载 node-v12.10.0-linux-x64.tar.xz 到对应目录,使用 tar 命令解压:

```
tar xf node-v12.10.0-linux-x64.tar.xz
```

进入解压目录:

```
cd node-v12.10.0-linux-x64/
```

执行 node 命令，查看版本 v12.10.0：

```
./bin/node -v
```

解压文件的 bin 目录，底下包含了 node、npm 等命令，可以使用 ln 命令来设置软连接：

```
ln -s /usr/software/nodejs/bin/npm /usr/local/bin/
ln -s /usr/software/nodejs/bin/node /usr/local/bin/
```

（3）在 Mac OS 上安装 Node.js。

读者可以通过以 brew 命令在 Mac OS 上来安装 node：

```
brew install node@12.10
```

2. 安装 Vue CLI

使用如下命令安装 Vue CLI：

```
npm install -g @vue/cli
```

安装之后就可以在命令行中访问 vue 命令，通过如下命令查看版本信息：

```
vue --version
```

如需升级全局的 Vue CLI 包，需运行：

```
npm update -g @vue/cli
```

3. 创建项目

运行以下命令来创建一个新项目：

```
vue create bifang-demo
```

- 选择预设

出现如下选项：

```
? Please pick a preset: (Use arrow keys)
> myPreset ([Vue 2] less, babel, router, vuex, eslint)
    Default ([Vue 2] babel, eslint)
    Default (Vue 3 Preview) ([Vue 3] babel, eslint)
```

```
Manually select features
```

如果之前使用过 Vue CLI 创建过项目并保存过选择过的配置项，会展示在此菜单中，如第一项名为 myPreset。若之前未设置过则不会出现第一行，第一次设置选择 Manually select features。

- 手动选择特性：

```
? Please pick a preset: Manually select features
? Check the features needed for your project:
 >◉ Choose Vue version
   ◉ Babel
   ○ TypeScript
   ○ Progressive Web App (PWA) Support
   ◉ Router
   ◉ Vuex
   ◉ CSS Pre-processors
   ◉ Linter / Formatter
   ○ Unit Testing
   ○ E2E Testing
```

在此步中选择"选择 Vue 版本"、Babel、Router、Vuex、CSS Pre-processors、Linter/Formatter 选项。选择 Vue 版本，BiFang 项目使用 Vue2.x 版本。

TypeScript 表示项目中使用 TypeScript 语法，会自动加入相关编译语法依赖，BiFang 项目暂不使用 TS。

Progressive Web App 表示项目中使用 PWA，即谷歌推出的一项浏览器存储技术，一般用于移动端 H5 项目。支持 PWA 技术的移动设备，用户在访问 H5 页面时，会自动缓存页面相关资源文件，在二次打开时可大幅提升页面的加载速度。

Unit Testing 和 E2E Testing 用于前端自动化。Unit 测试用于开发者写好逻辑来测试逻辑是否正确，E2E 测试用于在黑盒环境下，测试需求是否正确实现。

- 配置 Vue router：

```
Use history mode for router? (Requires proper server setup for index fallback
in production) (Y/n)
```

配置 Vue router 是否使用 history 模式，默认使用的是 hash 模式。hash 模式打开用户模块，使用的链接如下：

```
http://yoursite.com/#/user?id=1
```

history 模式打开用户模块，使用的链接如下：

```
http://yoursite.com/user/id
```

history 模式要用好，还需要后台配置支持，这里选择 hash 模式。
- 配置样式处理器：

```
Pick a CSS pre-processor (PostCSS, Autoprefixer and CSS Modules are supported
by default):
    Sass/SCSS (with dart-sass)
    Sass/SCSS (with node-sass)
>   Less
    Stylus
```

BiFang 项目使用 less 作为样式处理器，less 则相对清晰明了，易于上手，对编译环境要求比较宽松。
- 配置代码检测模式：

```
Pick a linter / formatter config:
    ESLint with error prevention only
    ESLint + Airbnb config
    ESLint + Standard config
>   ESLint + Prettier
```

JavaScript 是一个动态的弱类型语言，在开发中比较容易出错。因为没有编译程序，为了寻找 JavaScript 代码错误，通常需要在执行过程中不断调试。ESLint 可以让程序员在编码的过程中发现问题而不是在执行的过程中。

Prettier 插件是 VS Code 常用的插件，ESLint 配合 Prettier 更方便使用。

```
? Pick additional lint features:
 >Lint on save
    Lint and fix on commit
```

选择在保存文件时检查代码。
- 配置 Babel、ESLint 存放：

```
? Where do you prefer placing config for Babel, ESLint, etc.? (Use arrow keys)
> In dedicated config files
In package.json
```

选择单独文件存放 Babel 和 ESLint 配置。
- 保存本次项目配置：

Save this as a preset for future projects? (y/N)

保存本次项目配置，可用于下次创建 Vue 项目，不再需要重新选择配置项目。

配置完成，耐心等待依赖下载完成。

4. 运行项目

项目搭建完毕，使用如下命令进入项目目录并运行项目：

```
cd bifang-demo
npm run serve
```

待项目编译完毕后，打开浏览器输入地址，如图 12-1 所示：

http://localhost:8080/

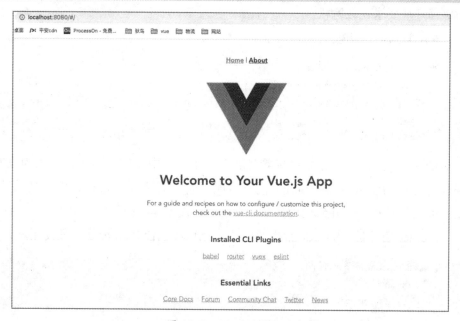

图 12-1　VueCli 项目运行图

项目启动成功，Vue 初始化项目已搭建完毕，接下来结合 ant-design-vue UI 框架构建简单的前端 UI 界面。

5. 项目目录结构

使用 Vue CLI 创建的项目包含如下目录，接下来介绍主要目录和文件的具体作用：

```
> node_modules
    > public
    > src
        > asset
        > components
        > router
        > store
        > view
    > babel.config.js
    > package.json
```

- node_modules：当前项目的所有依赖项，通过 npm install 安装的项目依赖。
- public：存放入口文件以及浏览器图标。
- asset：资源文件存放目录，包含图片、gif 等。
- components：组件存放目录，存放 Vue 项目中所有组件、主要工作目录。
- router：路由配置目录，新增页面需要在此配置路由。
- store：使用 Vuex 实现的全局数据共享，Vuex 参考了 react redux。
- view：MVVM 中的 view 层，存放路由所对应的展示层、主要工作目录。
- babel.config.js：babel 编译配置文件，Vue CLI3 默认使用 cli-plugin-babel，默认配置使用 Babel7+babel-loader，可以配置使用其他的 Babel preset 或 plugin。
- package.json：项目依赖插件，可以使用 npm install xx --save 来更新列表。

介绍完创建项目的默认文件列表，MVVM 作为 MVC 的变体，文件目录还缺少 service 层来处理接口请求，缺少 mock 层做前后端分离架构中假数据的配置，所以需要在 src 目录新增 service 目录和 mock 目录。

12.5 项目引入 Antd 组件库

基础项目使用 Vue CLI 搭建完毕，需要引入 ant-design-vue 组件库才能使用其中的 UI 组件。

- 安装依赖：

```
npm install --save ant-design-vue
```

- 完整使用组件：

打开入口文件 main.js，新增如下代码：

```
import Antd from 'ant-design-vue'
import 'ant-design-vue/dist/antd.css'
Vue.config.productionTip = false;
Vue.use(Antd);
```

以上代码便完成了 Antd 的引入。需要注意的是，样式文件需要单独引入。

12.6 小 结

本章我带领读者熟悉了前端 Web 项目的框架选型，介绍了 Angular、React 以及 Vue 前端框架的历史、底层实现原理和各框架的优缺点。没有最好的框架，只有最适合项目的框架，相信读者在学习了解后，能在后续的项目中根据具体问题具体分析，能选择适合自身项目的框架。

接下来讲解了 less、sass/scss，通过 demo 可以了解它们与 css 的区别，相信没有了解过相关知识点的读者能大有收货。

最后介绍了如何安装并使用 Vue CLI 创建一个前端 Web 项目，也在文中介绍了前端知识点，有想对这部分知识点有更深层次了解的读者，可以查阅相关方面的资料。

针对前端框架的选型，第 13 章将用 Angular、React 和 Vue 分别实现待完成列表 demo，也就是大名鼎鼎的 todolist，来看看三个框架的实现区别。

未来和未知，都是人生旅途中的星辰大海。保持激情，积累元气，再出发！

第 13 章　前端框架 ToDoList 实现

Talk is cheap; show me the code.

——Linus Torvalds

本章 GitHub 代码地址：https://github.com/aguncn/bifang-book/tree/main/bifang-ch13。

第 12 章简单介绍了前端框架选型中基本的三大框架 Angular、React 和 Vue 的底层原理实现和优缺点，让读者对框架有了大概的认识，但读者可能依旧对三者的实现区别不甚了解。Linux 创始人 Linus 曾说："Talk is cheap; show me the code."为了让读者对三个框架有更深层次的认识，本章通过大名鼎鼎的 ToDoList 实现来详细讲解三者的实现区别。

13.1　ToDoList 介绍

ToDoList 是一款非常优秀的任务管理软件，用户可以用它方便地组织和安排计划，界面设计优秀，初级用户也能够快速上手。因为众所周知且功能交互优秀，ToDoList 项目也成了业界最火的一个 demo 项目，各路大神在学习新的 UI 框架时总是拿它当练手项目，以此来熟悉与其他框架的区别。它有专门的网站，地址为 www.todolist.cn，在浏览器输入地址，可以看到如图 13-1 所示界面。

图 13-1　ToDoList 截图

1. 功能介绍

ToDoList 可分为三部分：头部（header）、列表（list）和底部（footer）。

（1）头部包含 ToDoList 标签说明和一个添加任务的输入框，在输入框中输入任务，单击回车按钮，任务自动添加进入待完成列表。

（2）列表包含正在进行中的任务列表和已完成的任务列表，通过头部输入框添加的任务，默认进入到正在进行中的任务列表中，此列表中的任务交互有以下两点：

- 单击任务前方的 checkbox 框，表示任务已完成，任务进入已完成列表中。
- 单击任务右侧的按钮，任务被删除。任务栏右侧展示目前进行中的任务数量。

已完成列表中的任务交互也有两点：

- 单击任务前方的 checkbox 框，表示任务依旧未完成，任务进入正在进行的任务列表中。
- 单击任务右侧的按钮，任务被删除。任务栏右侧展示目前已完成的任务数量。

（3）底部包含版权声明和一个清除按钮，单击清除按钮会删除列表中的所有任务。插入任务后的截图如图 13-2 所示。

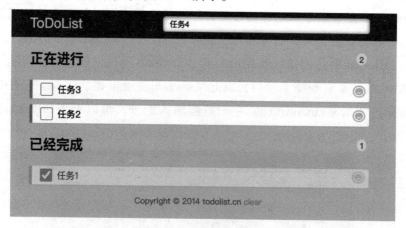

图 13-2　ToDoList 添加任务

2. 模块组件设计

接下来设计 ToDoList 模块，如上文介绍，页面的头部、列表和底部是如下三个基础模块组件，同时在主入口存储任务的列表数据，相关模块的数据更新通知主入口做对应的操作。

任务存储的数据结构如下：

```
list: [{
    id: 0, // 任务 ID
    name: ' 任务 1', // 任务名称
    isDone: true // 任务完成状态
}, {
    id: 1,
    name: ' 任务 2',
    isDone: false
}]
```

头部：监听 input 输入框的 enter 事件，把输入框中的数据同步到任务存储层。

列表：获取存储层存储的任务列表数据，拆分已完成任务和正在进行中的任务；监听 checkbox 的单击事件，把任务相关状态同步到任务存储层；监听删除按钮的单击事件，把需要删除任务的 id 同步到任务存储层。

底部：监听 clear 按钮的单击事件，并通知任务存储层把所有的任务清除。

现在已经拆分好了 ToDoList 相关模块和数据结构，下面看看使用 Angular、React 以及 Vue 如何实现 ToDoList 项目。

13.2　Angular 实现 ToDoList

Angular 是一个用于构建移动和桌面 Web 应用程序的开发平台，用于创建高效、复杂、精致的单页面应用，通过新的属性和表达式扩展了 HTML，实现一套框架、多种平台，移动端和桌面端，使用 Typescript/JavaScript 和其他语言。

AngularJS 通过为开发者呈现一个更高层次的抽象来简化应用的开发。如同其他的抽象技术一样，这也会损失一部分灵活性。换句话说，并不是所有的应用都适合用 AngularJS 来做。AngularJS 主要考虑的是构建 CRUD 应用。幸运的是，至少 90% 的 Web 应用都是 CRUD 应用。但是要了解什么适合用 AngularJS 构建，就得了解什么不适合用 AngularJS 构建。

如游戏、图形界面编辑器，这种 DOM 操作很频繁也很复杂的应用，和 CRUD 应用就有很大的不同，它们不适合用 AngularJS 来构建。像这种情况用一些更轻量、简单的技术，如 jQuery 可能会更好。

1. 安装 Angular Cli

安装 Angular Cli，前置需要 Node 环境，前面章节已介绍如何安装 Node 环境，本次不再赘述，运行一下命令全局安装 Angular Cli：

```
npm install -g @angular/cli
```

2. 创建项目

Angular 运行以下命令来创建一个新项目：

```
ng new angular-todolist
```

3. 创建头部组件

在 Angular 中可以直接通过 ng 命令创建组件，使用命令创建 header 组件如下：

```
ng g component components/header
```

在 src/components 目录下自动创建 header 目录，目录包含以下 4 个文件：

- header.component.css：编写样式。
- header.component.html：编写 html dom 元素。
- header.component.ts：编写组件实现。
- header.component.spec.ts：组件的单元测试文件。

其中 header.component.html 部分代码如下，label 标签展示 AngularToDoList 名称，input 输入框绑定数据中声明的 todo 变量，通过 keydown 函数监听回车事件：

```html
<header>
    <section>
        <form action="javascript:postaction()" id="form">
            <label for="title">AngularToDoList</label>
            <input type="text" [(ngModel)]='todo' (keydown)='addTodo($event)'
                placeholder=" 添加 ToDo" [ngModelOptions]="{standalone: true}" />
        </form>
    </section>
</header>
```

header.component.ts 脚本部分代码如下,声明变量 todo 接收 input 输入框输入的值,回车事件执行 addTodo 函数,使用 EventEmitter 通知父组件在任务列表中添加输入的任务:

```typescript
import { Component, OnInit, Output,EventEmitter } from '@angular/core';
@Component({
    selector: 'app-header',
    templateUrl: './header.component.html',
    styleUrls: ['./header.component.less']
})
export class HeaderComponent implements OnInit {

    constructor() { }
    public todo: any = '';

    @Output()
    add: EventEmitter<Object> = new EventEmitter()

    addTodo(e) {
        let todoObj = {
            todo: this.todo,
            isDone: false
        }

        if (e.keyCode == 13) { // 表示回车按钮
            this.add.emit(todoObj)
            this.todo=""
        }
    }

    ngOnInit(): void {
    }

}
```

header.component.less 样式部分代码如下,参考 www.todolist.cn 网站中的样式,设置 header 组件所需要的样式:

```less
header {
    height:50px;
    background:#333;
    background:rgba(47,47,47,0.98);
```

```
    }
    section{
        margin:0 auto;
    }
    label{
        float:left;
        width:100px;
        line-height:50px;
        color:#DDD;
        font-size:24px;
        cursor:pointer;
        font-family: "Helvetica Neue",Helvetica,Arial,sans-serif;
    }
    header input{
        float:right;
        width:60%;
        height:24px;
        margin-top:12px;
        text-indent:10px;
        border-radius:5px;
        box-shadow: 0 1px 0 rgba(255,255,255,0.24), 0 1px 6px rgba(0,0,0,0.45) inset;
        border:none;
    }
    input:focus{
        outline-width:0
    }
```

4. 创建列表组件

Angular 中可以直接通过 ng 命令创建组件，使用命令创建 list 列表组件如下：

```
ng g component components/list
```

在 src/components 目录下自动创建 list 目录，目录中包含以下 4 个文件：

- list.component.css：编写样式。
- list.component.html：编写 html dom 元素。
- list.component.ts：编写组件实现。
- list.component.spec.ts：组件的单元测试文件。

其中 list.component.html 部分代码如下，列表分为两部分：正在进行的任务以及已完成的任务，单击任务中的 checkbox，执行 changeTodo 函数，传入任务对应的 id 和状态，通过 EventEmitter 方法通知父组件更新任务列表中对应任务的状态。单击右侧减号按钮，执行 deleteTodo，通知父组件删除对应的任务：

```html
<section>
    <h2> 正在进行 </h2>
    <ol class="demo-box">
        <ng-container *ngFor="let item of todoList let key = index">
            <li *ngIf="!item.isDone">
                <input type="checkbox" (click)="changeTodo(key,true)">
                <p>{{item.todo}}</p>
                <a (click)="deleteTodo(key)">-</a>
            </li>
        </ng-container>
    </ol>
    <h2> 已经完成 </h2>
    <ul>
        <ng-container *ngFor="let item of todoList let key = index">
            <li *ngIf="item.isDone" >
                <input type="checkbox" (click)="changeTodo(key,false)" checked='checked'>
                <p>{{item.todo}}</p>
                <a (click)="deleteTodo(key)">-</a>
            </li>
        </ng-container>
    </ul>
</section>
```

list.component.ts 脚本部分代码如下，通过 @Input todoList 变量接收父组件传入的任务列表，遍历任务列表通过 isDone 来判断任务列表中哪些任务是正在进行的任务，哪些任务是已完成的任务：

```typescript
import { Component, OnInit, Input, Output, EventEmitter } from '@angular/core';

@Component({
    selector: 'app-list',
    templateUrl: './list.component.html',
    styleUrls: ['./list.component.less']
})
export class ListComponent implements OnInit {

    @Input() todoList

    @Output()
    delete: EventEmitter<Number> = new EventEmitter()
```

```
@Output()
change: EventEmitter<Object> = new EventEmitter()

constructor() { }

ngOnInit(): void {
}

/**
 * 删除数据
 */
deleteTodo(index) {
    this.delete.emit(index)
}
/**
 * 转换状态
 */
changeTodo(index, done) {
    this.change.emit({index:index,done:done})
}
}
```

list.component.css 样式部分代码如下，参考 www.todolist.cn 网站中的样式，设置 list 组件所需要的样式：

```
h2{ position:relative; }
span{
    position:absolute;top:2px;right:5px;display:inline-block;
    padding:0 5px;height:20px;border-radius:20px;
    background:#E6E6FA;line-height:22px;text-align:center;
    color:#666;font-size:14px;
}
ol,ul{ padding:0;list-style:none; }
li input{
    position:absolute;top:2px;left:10px;width:22px;
    height:22px;cursor:pointer;
}
p{ margin: 0; }
li p input{
    top:3px;left:40px;width:70%;height:20px;line-height:14px;
    text-indent:5px;font-size:14px;
}
```

```
li{
    height:32px;line-height:32px;background: #fff;
    position:relative;margin-bottom: 10px;padding:0 45px;
    border-radius:3px;border-left: 5px solid #629A9C;
    box-shadow: 0 1px 2px rgba(0,0,0,0.07);
}
ol li{cursor:move;}
ul li{border-left: 5px solid #999;opacity: 0.5;}
li a{
    position:absolute;top:2px;right:5px;display:inline-block;width:14px;
    height:12px;border-radius:14px;border:6px double #FFF;
    background:#CCC;line-height:14px;text-align:center;color:#FFF;
    font-weight:bold;font-size:14px;cursor:pointer;
}
```

5. 创建底部组件

在 Angular 中可以直接通过 ng 命令创建组件，使用命令创建 footer 底部列表组件如下：

```
ng g component components/footer
```

在 src/components 目录下自动创建 footer 目录，目录中包含以下 4 个文件：

- footer.component.css：编写样式。
- footer.component.html：编写 html dom 元素。
- footer.component.ts：编写组件实现。
- footer.component.spec.ts：组件的单元测试文件。

其中 footer.component.html 部分代码如下，监听清空按钮的单击事件，单击触发声明的 clearData 函数：

```
<footer>
    Copyright © 2014 todolist.cn <a (click)="clear()">clear</a>
</footer>
```

footer.component.ts 脚本部分代码如下，声明 clearData 函数，内部调用 EventEmitter 方法通知父组件清除任务列表：

```
import { Component, OnInit, Output,EventEmitter } from '@angular/core';

@Component({
    selector: 'app-footer',
    templateUrl: './footer.component.html',
```

```
    styleUrls: ['./footer.component.less']
})
export class FooterComponent implements OnInit {

    constructor() { }

    @Output()
    clear: EventEmitter<Object> = new EventEmitter()

    /**
     * 清除所有数据
     */
    clearData() {
        this.clear.emit()
    }

    ngOnInit(): void {
    }
}
```

footer.component.css 样式部分代码如下，参考 www.todolist.cn 网站中的样式，设置 footer 组件所需要的样式：

```
<style>
    footer{
        color:#666;
        font-size:14px;
        text-align:center;
    }
    footer a{
        color:#666;
        text-decoration:none;
        color:#999;
    }
</style>
```

6. 创建主入口组件

通过上面的介绍，发现主入口组件 app.component 是最重要的组件，它不仅负责存储任务，同时监听三个子组件发送的事件，同步更新任务列表；引用头部组（header.component）、列表组件（list.component）和底部组件（footer.component）并传递任务列表到列表组件。

默认生成的 app.component.html 添加代码如下，引用的三个子组件根据顺序挂载到 id 为 app 的容器中；监听 Header 组件的 add 事件，监听 List 组件的 change 和 delete 事件，监听 Footer 组件的 clear 事件；把任务列表 tasklist 传递给 List 组件：

```html
<div id="app">
    <app-header (add)="addLisener($event)"></app-header>
    <app-list
        [todoList]="todoList"
        (delete)="deleteTodoLisener($event)"
        (change)="changeTodoLisener($event)"></app-list>
    <app-footer (clear)="clearData()"></app-footer>
</div>
```

app.component.ts 脚本部分代码如下，声明监听子组件的处理函数：addLisener（添加任务）、changeLisener（处理任务的状态）、deleteLisener（删除对应的任务）和 clearData（清空任务列表）：

```ts
import { Component } from '@angular/core';
@Component({
    selector: 'app-root',
    templateUrl: './app.component.html',
    styleUrls: ['./app.component.less']
})
export class AppComponent {
    title = 'angular-todolist';
    public todo: any = '';
    public todoList = []; // 要做的数组

    /**
     * 删除数据
     */
    deleteLisener(index) {
            this.todoList.splice(index, 1);
    }
    /**
     * 转换状态
     */
    changeLisener({index, done}) {
        if (done) {
            var tempTodo = this.todoList[index]
            tempTodo.isDone = true
```

```
        } else {
            var tempDone = this.todoList[index]
            tempDone.isDone = false
        }
    }
    /**
     * 清除所有数据
     */
    clearData() {
        localStorage.clear();
        this.todoList = [];
    }
    /**
     * 添加数据
     * @param todoObj
     */
    addLisener(todoObj){
        this.todoList.push(todoObj);
        console.log("outerLisener",todoObj)
    }
}
```

主入口组件的 app.component.css 样式为空,但要设置全局样式。在 src 目录下添加 style.css,内容如下,并在 angular.json 中通过 styles 配置引入:

```
section{margin:0 auto;}
@media screen and (max-device-width: 620px)
{
    section{width:96%;padding:0 2%;}
}
@media screen and (min-width: 620px)
{
    section{width:600px;padding:0 10px;}
}
```

7. 运行项目

通过启动命令运行项目:

```
ng serve --open
```

运行后的页面截图如图 13-3 所示。

第 13 章 前端框架 ToDoList 实现

图 13-3　Angular ToDoList 实现

13.3　React 实现 ToDoList

React 是用于构建用户界面的 JavaScript 库，由于其设计思想极其独特，属于革命性创新，性能出众，代码逻辑却非常简单，因此，越来越多的人开始关注并使用它，认为它可能是将来 Web 开发的主流工具。

这个项目本身也越滚越大，从最早的 UI 引擎变成了一整套前后端通吃的 Web App 解决方案。衍生的 React Native 项目，目标更宏伟，希望用写 Web App 的方式去写 Native App。如果这个方式能实现，整个互联网行业都会被颠覆，因为同一组人只需要写一次 UI，就能同时在服务器、浏览器和手机上渲染。

React 框架的特点如下：

- 声明式设计：React 采用声明范式，可以轻松描述应用。
- 高效：React 通过对 DOM 的模拟，最大限度地减少与 DOM 的交互。
- 灵活：React 可以与已知的库或框架很好地配合。
- JSX：JSX 是 JavaScript 语法的扩展。React 开发不一定使用 JSX，但建议使用它。
- 组件：通过 React 构建组件，使得代码更加容易被复用，能够很好地应用在大项目的开发中。
- 单向响应的数据流：React 实现了单向响应的数据流，从而减少了重复代码，这也是它为什么比传统数据绑定更简单的原因。

接下来使用 React 脚手架创建应用并实现 ToDoList 项目。

1. 安装 React 脚手架

运行以下命令来安装 create-react-app 脚手架，通过它无须配置就能快速构建 React 开发环境：

```
npm install -g create-react-app
```

2. 创建项目

运行以下命令来创建一个新项目：

```
create-react-app react-todolist
```

3. 创建头部组件

在 src/components 目录下创建 header 目录，该目录存放 Header 组件以及相关 css 文件，在 header 目录添加 Header.js 文件即为头部组件，代码如下：

```
import React from 'react';
import { v4 as uuid } from 'uuid';
import "./style.css"

class Header extends React.Component {
    state = {
        todo:""
    }

    changeTodo(e){
        this.setState({
            todo:e.target.value
        })
    }

    addTodo (event) {
        event.preventDefault()
        if (!this.state.todo.trim()) {
            alert(" 任务不能为空 ")
            return
        }

        let id = uuid();
        this.props.addLisener({id,text:this.state.todo,complete:false});
        this.setState({
            todo:""
        })
    }

    render () {
```

```
            return (
                <header>
                    <section>
                        <form id="form">
                            <label>ReactToDoList</label>
                            <input
                                type="text"
                                ref='text'
                                name="title"
                                placeholder=" 添加 ToDo"
                                required="required"
                                value={this.state.todo}
                                onChange={this.changeTodo.bind(this)}
                                onBlur={this.addTodo.bind(this)} />
                        </form>
                    </section>
                </header>
            )
        }
    }

export default Header;
```

React 组件与 Angular、Vue 组件代码格式有明显的差异，React 组件采用 jsx 语法糖，它认为 html 也是 javascript 的一部分，可以通过 JS 生成对应的 DOM 元素，所以我们看到组件 render 函数是返回的一串 DOM 结构，父组件的事件和属性通过 this.props 方式可以访问。

Header 组件中声明变量 todo 接收 input 输入框输入的值，鼠标移出 onBlur 事件执行 addTodo 函数，使用 this.props.addLisener 通知父组件在任务列表中添加输入的任务。

在 header 目录添加 style.css 文件，样式文件代码如下，参考 www.todolist.cn 网站中的样式，设置 header 组件所需要的样式：

```
header {
    height:50px;
    background:#333;
    background:rgba(47,47,47,0.98);
}
section{
    margin:0 auto;
}
label{
```

```css
        float:left;
        width:100px;
        line-height:50px;
        color:#DDD;
        font-size:24px;
        cursor:pointer;
        font-family: "Helvetica Neue",Helvetica,Arial,sans-serif;
}
header input{
        float:right;
        width:60%;
        height:24px;
        margin-top:12px;
        text-indent:10px;
        border-radius:5px;
        box-shadow: 0 1px 0 rgba(255,255,255,0.24), 0 1px 6px rgba(0,0,0,0.45) inset;
        border:none
}
input:focus{
        outline-width:0
}
```

4. 创建列表组件

在 src/components 目录下创建 list 目录，该目录存放 List 组件以及相关的 css 文件，在 list 目录添加 List.js 文件即为列表组件，代码如下：

```
import React from 'react'
import Todo from './todo'
import "./style.css"

class List extends React.Component {
    OnDeleteItem (id) {
        this.props.deleteLisener(id)
    }


    onChangeItem (id) {
        this.props.changeLisener(id);
    }
    render () {
        let doneCount=0,todoCount=0
        const doneList = this.props.data.map(({ id, text, complete }, index) => {
            if (complete) {
                doneCount++
```

```jsx
                    return <Todo
                                    key={index}
                                    id={id}
                                    text={text}
                                    complete={complete}
                                    changeItem={this.onChangeItem.bind(this)}
                                    deleteItem={this.OnDeleteItem.bind(this)}
                                />
                }
                else{
                    return ""
                }

            })

            const todoList = this.props.data.map(({ id, text, complete }, index) => {
                console.log("list render",id,text,complete)
                if (!complete) {
                    todoCount++
                    return <Todo
                                    key={index}
                                    id={id}
                                    text={text}
                                    complete={complete}
                                    changeItem={this.onChangeItem.bind(this)}
                                    deleteItem={this.OnDeleteItem.bind(this)}
                                />
                }
                else{
                    return ""
                }

            })

            return (
                <section>
                    <h2>正在进行 <span id="todocount">{todoCount}</span></h2>
                    <ol id="todolist" className="demo-box">
                        {todoList}
                    </ol>
                    <h2>已经完成 <span id="donecount">{doneCount}</span></h2>
                    <ul id="donelist">
                        {doneList}
```

```
            </ul>
        </section>
        )
    }
}

export default List;
```

列表分为两部分：正在进行的任务以及已完成的任务，单击任务中的 checkbox，执行 changeTodo 函数，传入任务对应的 id 和状态，通过 $emit 方法通知父组件更新任务列表中对应任务的状态。单击右侧减号按钮，执行 deleteTodo，通知父组件删除对应的任务。

使用 React，再把任务单独抽做一个 Todo 组件，在 list 目录下新增 Todo.js 文件即为 Todo 组件，内容如下：

```
import React from 'react'
class Todo extends React.Component {
    changeState () {
        this.props.changeItem(this.props.id);
    }

    handleDelete () {
        this.props.deleteItem(this.props.id);
    }
    render () {
        return (
            <li draggable="true">
                <input
                    type="checkbox"
                    checked={this.props.complete?'checked':''}
                    onChange={this.changeState.bind(this)} />
                <p>{this.props.text}</p>
                <a href="javascript:void(0)" onClick={this.handleDelete.bind(this)}>-</a>
            </li>
        )
    }
}

export default Todo;
```

在 list 目录添加 style.css 文件，样式文件代码如下，参考 www.todolist.cn 网站中的样式，设置 List 组件所需要的样式：

```
h2{ position:relative; }
span{
    position:absolute;top:2px;right:5px;display:inline-block;
    padding:0 5px;height:20px;border-radius:20px;
    background:#E6E6FA;line-height:22px;text-align:center;
    color:#666;font-size:14px;
}
ol,ul{ padding:0;list-style:none; }
li input{
    position:absolute;top:2px;left:10px;width:22px;
    height:22px;cursor:pointer;
}
p{ margin: 0; }
li p input{
    top:3px;left:40px;width:70%;height:20px;line-height:14px;
    text-indent:5px;font-size:14px;
}
li{
    height:32px;line-height:32px;background: #fff;
    position:relative;margin-bottom: 10px;padding:0 45px;
    border-radius:3px;border-left: 5px solid #629A9C;
    box-shadow: 0 1px 2px rgba(0,0,0,0.07);
}
ol li{cursor:move;}
ul li{border-left: 5px solid #999;opacity: 0.5;}
li a{
    position:absolute;top:2px;right:5px;display:inline-block;width:14px;
    height:12px;border-radius:14px;border:6px double #FFF;
    background:#CCC;line-height:14px;text-align:center;color:#FFF;
    font-weight:bold;font-size:14px;cursor:pointer;
}
```

5. 创建底部组件

在 src/components 目录下创建 footer 目录，该目录存放 Footer 组件以及相关的 css 文件，在 footer 目录添加 Footer.js 文件即为底部组件，代码如下：

```
import React from 'react'
import "./style.css"

class Footer extends React.Component {
    clear () {
        this.props.Clear();
    }
```

```
        render () {
            return (
                <footer>
                    Copyright © bifang
                    <a href="javascript:void(0)" onClick={this.clear.bind(this)}>clear</a>
                </footer>
            )
        }
    }
export default Footer;
```

单击 clear 按钮调用 clear 函数，内部通过 this.props 调用父组件传入的 Clear 方法清除所有任务列表：

```
<script>
export default {
    name:"Footer",
    data(){
        return {
        }
    },
    methods:{
        clear(){
            this.$emit("clear")
        }
    }
}
</script>
```

在 footer 目录添加 style.css 文件，样式文件代码如下，参考 www.todolist.cn 网站中的样式，设置 Footer 组件所需要的样式：

```
footer{
        color:#666;
        font-size:14px;
        text-align:center;
}
footer a{
        color:#666;
        text-decoration:none;
        color:#999;
}
```

6. 创建主入口组件

通过上面的介绍，发现主入口组件 App.js 是最重要的组件，它不仅负责存储任务，同时监听三个子组件发送的事件，同步更新任务列表；引用头部组件（Header.js）、列表组件（List.js）和底部组件（Footer.js）并传递任务列表到列表组件。

主入口组件 App.js 代码如下，引用的三个子组件根据顺序挂载到 id 为 app 的容器中；监听 Header 组件的 onAddLisener 事件，监听 List 组件的 onChangeLisener 和 onDeleteLisener 事件，监听 Footer 组件的 Clear 事件；把任务列表 tasklist 传递给 List 组件：

```
import React from 'react'
import List from './components/list/list.js'
import Header from './components/header/header.js'
import Footer from './components/footer/footer.js'

class App extends React.Component {
    state = {
        data: []
    }

    ClearData () {
        this.setState({data:[]});
    }

    onAddLisener (newItem) {
        let newdata = this.state.data.concat(newItem);
        this.setState({data : newdata});
    }

    onChangeLisener (id) {
        let newdata = this.state.data.map((item,index) => {
            if(item.id === id) {
                item.complete = !item.complete;
            }
            return item;
        })
        this.setState({data : newdata});
    }

    onDeleteLisener (id) {
```

```
            let delItem = this.state.data.findIndex(item=>item.id===id)
            this.state.data.splice(delItem,1)
            let newdata =this.state.data
            console.log(newdata)
            this.setState({ data: newdata })
        }

        render () {
            return (
                <div className='ui comments'>
                    <Header
                        addLisener={this.onAddLisener.bind(this)} />
                    <List
                            data={this.state.data}
                            changeLisener={this.onChangeLisener.bind(this)}
                            deleteLisener={this.onDeleteLisener.bind(this)}
                    />
                    <Footer
                        Clear={this.ClearData.bind(this)}
                    />
                </div>
            )
        }
}

export default App;
```

主入口组件的 style 样式为空，但我们要设置全局样式。在 src 目录下默认生成的 index.css 添加内容如下，并在 index.js 中通过 import 方式引入：

```
section{margin:0 auto;}
@media screen and (max-device-width: 620px)
{
    section{width:96%;padding:0 2%;}
}
@media screen and (min-width: 620px)
{
    section{width:600px;padding:0 10px;}
}
```

7. 运行项目

通过启动命令运行项目：

```
ng start
```

运行后的页面截图如图 13-4 所示。

图 ch13-4　React ToDoList 实现

13.4　Vue 实现 ToDoList

使用 Vue 框架实现 ToDoList，使用 Vue Cli 创建基础项目，在第 12 章中详细讲述了如何使用 Vue Cli 创建项目，在此不再赘述，如读者有不清楚的地方可返回第 12 章仔细研读。

1. 创建项目

运行以下命令来创建一个新项目：

```
vue create vue-todolist
```

2. 创建头部组件

在 src/components 目录下创建 Header.vue 文件，Header.vue 文件包含三部分：template（模板）、script（脚本）和 style（样式）。

其中 template 部分代码如下，label 标签展示 VueToDoList 名称，input 输入框绑定 data 数据中声明的 todo 变量，通过 keyup.enter 函数监听回车事件：

```
<template>
    <header>
        <section>
            <form id="form">
                <label for="title">VueToDoList</label>
                <input type="text" v-model='todo' @keyup.enter='addTodo' placeholder=" 添加任务 " />
            </form>
        </section>
    </header>
```

</template>

script 脚本部分代码如下，声明变量 todo 接收 input 输入框输入的值，回车事件执行 addTodo 函数，使用 $emit 方法通知父组件在任务列表中添加输入的任务：

```
export default {
    data() {
        return {
            todo: ""
        }
    },
    methods: {
        addTodo() {
            if (this.todo === "") {
                alert(" 请输入待办事项后再进行添加 ");
                return;
            }
            this.$emit("add", this.todo);
            this.todo = "";
        }
    }
}
</script>
```

style 样式部分代码如下，参考 www.todolist.cn 网站中的样式，设置 header 组件所需要的样式，在 Header.vue 中添加以下内容：

```
<style>
    header {
        height:50px;
        background:#333;
        background:rgba(47,47,47,0.98);
    }
    section{
        margin:0 auto;
    }
    label{
        float:left;
        width:100px;
        line-height:50px;
        color:#DDD;
        font-size:24px;
```

```css
        cursor:pointer;
        font-family: "Helvetica Neue",Helvetica,Arial,sans-serif;
    }
    header input{
        float:right;
        width:60%;
        height:24px;
        margin-top:12px;
        text-indent:10px;
        border-radius:5px;
        box-shadow: 0 1px 0 rgba(255,255,255,0.24), 0 1px 6px rgba(0,0,0,0.45) inset;
        border:none
    }
    input:focus{
        outline-width:0
    }
</style>
```

3. 创建列表组件

在 src/components 目录下创建 List.vue 文件，List.vue 文件内容包含三部分：template（模板）、script（脚本）和 style（样式）。

template 部分代码如下，列表分为两部分：正在进行的任务以及已完成的任务，单击任务中的 checkbox，执行 changeTodo 函数，传入任务对应的 id 和状态，通过 $emit 方法通知父组件更新任务列表中对应任务的状态。单击右侧减号按钮，执行 deleteTodo，通知父组件删除对应的任务：

```html
<template>
    <section>
        <h2>正在进行 <span>{{todoList.length}}</span></h2>
        <ol class="demo-box">
            <li v-for="(item,index) in todoList" :key="index">
                <input type="checkbox" @click="e=>changeTodo(item.id,true)">
                <p>{{item.todo}}</p>
                <a @click="e=>deleteTodo(item.id)">-</a>
            </li>
        </ol>
        <h2>已经完成 <span>{{doneList.length}}</span> </h2>
        <ul>
            <li v-for="(item,index) in doneList" :key="index">
                <input type="checkbox" @click="e=>changeTodo(item.id,false)" checked='checked'>
                <p>{{item.todo}}</p>
```

```
            <a @click="e=>deleteTodo(item.id)">-</a>
          </li>
        </ul>
      </section>
</template>
```

script 脚本部分代码如下,通过 props 中定义的 list 变量接收父组件传入的任务列表,通过 computed 计算属性来计算任务列表中哪些任务是正在进行的任务,哪些任务是已完成的任务:

```
<script>
    export default {
        name:"List",
        props: {
            list: {
                type: Array,
                default: () => []
            }
        },
        computed:{
            todoList(){
                let todolist = []
                this.list.forEach(item=>{
                    !item.isDone?todolist.push(item):""
                })
                return todolist
            },
            doneList(){
                let donelist = []
                this.list.forEach(item=>{
                    item.isDone?donelist.push(item):""
                })
                return donelist
            }
        },
        data() {
            return {
            }
        },
        methods: {
            changeTodo(id,done){
                this.$emit("change",id,done)
            },
```

```
        deleteTodo(id){
            this.$emit("delete",id)
        }
      }
    };
</script>
```

style 样式部分代码如下，参考 www.todolist.cn 网站中的样式，设置 list 组件所需要的样式，在 List.vue 中添加以下内容：

```
<style>
    h2{ position:relative; }
    span{
        position:absolute;top:2px;right:5px;display:inline-block;
        padding:0 5px;height:20px;border-radius:20px;
        background:#E6E6FA;line-height:22px;text-align:center;
        color:#666;font-size:14px;
    }
    ol,ul{ padding:0;list-style:none; }
    li input{
        position:absolute;top:2px;left:10px;width:22px;
        height:22px;cursor:pointer;
    }
    p{ margin: 0; }
    li p input{
        top:3px;left:40px;width:70%;height:20px;line-height:14px;
        text-indent:5px;font-size:14px;
    }
    li{
        height:32px;line-height:32px;background: #fff;
        position:relative;margin-bottom: 10px;padding:0 45px;
        border-radius:3px;border-left: 5px solid #629A9C;
        box-shadow: 0 1px 2px rgba(0,0,0,0.07);
    }
    ol li{cursor:move;}
    ul li{border-left: 5px solid #999;opacity: 0.5;}
    li a{
        position:absolute;top:2px;right:5px;display:inline-block;width:14px;
        height:12px;border-radius:14px;border:6px double #FFF;
        background:#CCC;line-height:14px;text-align:center;color:#FFF;
        font-weight:bold;font-size:14px;cursor:pointer;
    }
</style>
```

4. 创建底部组件

在 src/components 目录下创建 Footer.vue 文件，Footer.vue 文件内容包含三部分：template（模板）、script（脚本）和 style（样式）。

template 部分代码如下，监听清空按钮的单击事件，单击触发声明的 clear 函数：

```
<template>
    <footer>
        Copyright © bifang <a @click="clear">clear</a>
    </footer>
</template>
```

script 脚本部分代码如下，声明 clear 函数，内部调用 $emit 方法通知父组件清除任务列表：

```
<script>
export default {
    name:"Footer",
    data(){
        return {
        }
    },
    methods:{
        clear(){
            this.$emit("clear")
        }
    }
}
</script>
```

style 样式部分代码如下，参考 www.todolist.cn 网站中的样式，设置 list 组件所需要的样式，在 Footer.vue 中添加以下内容：

```
<style>
    footer{
        color:#666;
        font-size:14px;
        text-align:center;
    }
    footer a{
        color:#666;
        text-decoration:none;
```

```
        color:#999;
    }
</style>
```

5. 创建主入口组件

通过上面的介绍，发现主入口组件是最重要的组件，它不仅负责存储任务，同时监听三个子组件发送的事件，同步更新任务列表；引用头部组件（Header.vue）、列表组件（List.vue）和底部组件（Footer.vue）并传递任务列表到列表组件。

template 部分代码如下，引用的三个子组件根据顺序挂载到 id 为 app 的容器中；监听 Header 组件的 add 事件，监听 List 组件的 change 和 delete 事件，监听 Footer 组件的 clear 事件；把任务列表 tasklist 传递给 List 组件：

```
<template>
    <div id="app">
        <Header @add="add" />
        <List :list="tasklist" @change="changeItem" @delete="deleteItem" />
        <Footer @clear="clear" />
    </div>
</template>
```

script 脚本部分代码如下，顶部引入三个子组件 Header、List 和 Footer 并通过 components 属性引入。声明监听子组件的处理函数：add（添加任务）、changeItem（处理任务的状态）、deleteItem（删除对应的任务）和 clear（清空任务列表）：

```
<script>
    import Header from "./components/Header";
    import List from "./components/List";
    import Footer from "./components/Footer";

    export default {
        name: "App",
        components: {
            Header,
            List,
            Footer
        },
        data() {
            return {
```

```
                    tasklist: []
                };
            },
            methods: {
                add(val) {
                    let data = {
                        name: val,
                        isDone: false,
                        id: new Date().getTime()
                    };
                    this.tasklist.push(data);
                },
                changeItem(id,done) {
                    let index = null;
                    this.tasklist.map((item, i) => {
                        if (item.id === id) index = i;
                    });
                    this.tasklist[index].isDone = done;
                },
                deleteItem(id){
                    let index = null;
                    this.tasklist.map((item, i) => {
                        if (item.id === id) index = i;
                    });
                    this.tasklist.splice(index,1)
                },
                clear(){
                    this.tasklist=[]
                }
            }
        };
</script>
```

主入口组件的 style 样式为空,但要设置全局样式。在 src 目录下添加 index.css,内容如下,并在 main.js 中通过 import 方式引入:

```
section{margin:0 auto;}
@media screen and (max-device-width: 620px)
{
    section{width:96%;padding:0 2%;}
}
@media screen and (min-width: 620px)
{
```

```
section{width:600px;padding:0 10px;}
}
```

6. 运行项目

通过启动命令运行项目：

```
npm run serve
```

运行后的页面截图如图 13-5 所示。

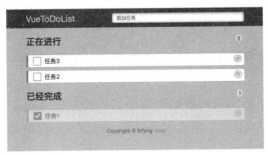

图 13-5　Vue ToDoList 实现

13.5　小　结

本章带领大家使用 Angular、React 和 Vue 框架完成了 ToDoList 的功能，介绍了使用各自的脚手架搭建基础项目，通过实战了解 Angular、React 以及 Vue 组件间传值方式的区别，相信没有了解过相关知识点的读者能大有收货，并在学习了解后能在后续对应的项目中得心应手地使用。

在实战 ToDoList 项目中，我没有使用状态管理工具，如 Angular 状态管理工具 service 或 rxjs、React 状态管理工具 redux、Vue 状态管理工具 Vuex，有兴趣的读者可以在 ToDoList 项目中添加状态管理，在大型项目中普遍使用状态管理工具作为中间数据层。

第 14 章将全局介绍前端开发模式，让读者有个全局的开发概念，接下来重点介绍 BiFang 项目使用的 Vue 相关的概念，话不多说，让我们继续前进！

每一个成功者都有一个开始。勇于开始，才能找到成功的路！

第 14 章 前端开发模式以及公共服务配置

业精于勤，荒于嬉，行成于思，毁于随。

——韩愈

本章 GitHub 代码地址：https://github.com/aguncn/bifang-book/tree/main/bifang-ch14。

第 13 章讲述了使用前端框架选型中的三大框架 Angular、React 和 Vue 实现 todolist 项目，详细讲了三者实现的区别，接下来讲前后端开发模式以及使用 Vue 框架搭建公共服务。

对于大型的项目会有各种角色参与项目，比如产品经理、项目经理、开发人员和测试人员等，项目的开发模式关系着项目是否能持续、顺畅地进行，目前主流的开发模式有两种：传统瀑布型开发模式和主流的敏捷开发。瀑布开发模式严格地把软件项目的开发分隔成各个开发阶段，阶段不重叠；敏捷开发模式强调快速迭代，拥抱变化，每个阶段都可重新根据优先级排序，任务可并行进行。前后端开发任务并行进行就需要把任务解耦，分离开发。

14.1 前后端分离开发架构简介

前后端分离已成为互联网项目开发的业界标准使用方式，通过 nginx+tomcat 的方式有效地进行解耦，并且前后端分离会为以后的大型分布式架构、微服务架构、多端化服务（PC、H5、安卓、IOS 等）打下坚实的基础，其核心思想是前端 HTML 页面通过 AJAX 调用后端的 RESTFUL API 接口并使用 JSON 数据进行交互。前后端分离这一步是系统架构"从猿进化成人"的必经之路。

14.1.1 开发流程

在未分离的时代，项目的开发流程如图 14-1 所示。

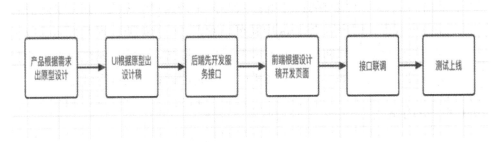

图 14-1　未分离开发流程

以 Java Web 项目为例，Java 后端开发人员"又当爹又当妈"，又做前端，又做后端；前端开发人员在开发过程中严重依赖后端，在后端没有完成的情况下，前端根本无法干活，开发效率低。这种一环套一环的传统开发流程，严重违背敏捷开发的流程，如何进一步提升开发效率，提前上线时间成为各大公司急需解决的痛点，于是前后端分离应运而生。前后端分离的开发流程如图 14-2 所示。

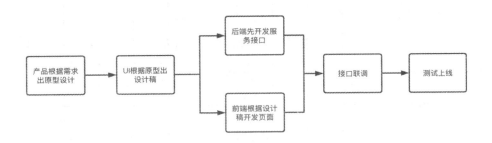

图 14-2　前后端分离开发流程

前后端分离对比未分离有以下优点：
- 缩短开发时间。前后端以 JSON 契约为准，可同时进入开发流程，从而缩短开发时间，提前上线时间。
- 适配性提升。在未分离时服务端经常会给 PC 端、H5、app 端各自研发一套 JSP，其实对于这三端来说，大部分端业务逻辑一样，唯一区别就是交互展现逻辑不同，而这恰恰是前端所擅长的。
- 页面响应速度提升。在未分离时，用户发起页面请求需要浏览器向服务器请求 JSP 文件，业务逻辑越复杂，服务器执行的时间越长，用户看到页面的时间也同步增加。谷歌公司研究表明，大多数用户期望的网站加载时间是 3s，如果时间长过 3s，大约 57% 的用户会流失。如果采用前后端分离流程，用户可以在 1s 内就可以看到页面骨架图，提升页面响应速度，给用户一个预期提示再去请求服务端数据。当然，现在有更好的渲染方式：server-side-render（SSR），有兴趣的读者可以自行研究。

14.1.2 契约格式

BiFang 项目采用前后端分离方式开发，前后端以 JSON 格式为契约以及对应状态为代表的事件以及处理方式，参见如下格式：

```
{
    "code":0,
    "message":"success",
    "data":{}
}
```

其中 code 值为 0 时表示数据返回成功，值为 1 时表示数据返回错误，需要提示对应的 message 信息。

请求服务采用 http 状态码，常用状态码如表 14-1 所列。

表 14-1　http 常用状态码

状态码	英文名称	描　述
200	OK	请求成功
401	Unauthorized	请求要求用户的身份认证
403	Forbidden	拒绝执行此请求
500	Internal Server Error	服务器内部错误

前后端服务根据以上契约格式达成协议，若出现问题，对应开发人员排查问题并解决，契约格式发生变更需要双方协商认可。契约约定好后，接下来就是前后端服务如何认证请求方式。

14.2　请求认证方式

前端与服务端的认证方式，是前后端分离带来的核心问题，下面介绍最常用的 5 种前端与服务端认证方式。

第 14 章 前端开发模式以及公共服务配置

1. Http Basic Auth

这是一种最古老的安全认证方式，这种方式就是在访问 API 的时候带上访问的 username 和 password，由于信息会暴露，所以现在也越来越少用了，现在都用更加安全保密的认证方式。

2. Cookie-Session 认证

Cookie-Session 认证机制是为一次请求认证在服务端创建一个 Session 对象，同时在客户端的浏览器端创建一个 Cookie 对象；通过客户端带上来 Cookie 对象来与服务器端的 Session 对象匹配来实现状态管理的。默认当关闭浏览器时，Cookie 会被删除，但可以通过修改 Cookie 的 expire time 使 Cookie 在一定时间内有效。

这种认证模式看似问题不大，但随着用户量的增加以及分布式部署的出现，Cookie-Session 认证应用的问题就会暴露出来：

- Session 增多会增加服务器开销。每个用户经过应用认证后，应用都要在服务端做一次记录，以方便用户下次请求的鉴别，通常 Session 都保存在内存中，而随着认证用户的增多，服务端的开销会明显增大。
- 在分布式或多服务器环境中适应性不好。用户认证后，服务端做认证记录，如果认证的记录被保存在内存中的话，这意味着用户下次还必须要在这台服务器上，请求才能拿到授权的资源，这样在分布式的应用上，相应地限制了负载均衡器的能力。这也意味着限制了应用的扩展能力。不过，现在某些服务器可以通过设置黏性 Session，来做到每台服务器之间的 Session 共享。
- 容易遭到 CSRF 攻击。因为该认证是基于 Cookie 来进行用户识别的，Cookie 如果被截获，用户就很容易受到跨站请求伪造的攻击。

3. OAuth 认证

OAuth 即 Open Authrization（开放授权），它是一个开放标准，允许用户让第三方应用访问该用户在某一网站上存储的私密资源，而无需将用户名和密码提供给第三方。OAuth2 在认证和授权的过程中有三个角色：

- 服务提供方，即提供受保护的服务和资源。
- 用户，在服务提供方存了资源（照片、资料等）的人。
- 客户端，服务调用方，它要访问服务提供方的资源，需要在服务提供方进行注册。

OAuth2 认证和授权的过程：
- 用户想操作存放在服务提供方的资源。
- 用户登录客户端，客户端向服务提供方请求一个临时 token。
- 服务提供方验证客户端的身份后，给它一个临时 token。
- 客户端获得临时 token 后，将用户引导至服务提供方的授权页面，并请求用户授权。在这个过程中会将临时 token 和客户端的回调链接 / 接口发送给服务提供方。
- 用户输入用户名密码登录，登录成功后，可授权客户端访问服务提供方的资源。
- 授权成功后，服务提供方将用户引导至客户端的网页。
- 客户端根据临时 token 从服务提供方那里获取正式的 access token。
- 服务提供方根据临时 token 以及用户的授权情况授予客户端 access token。
- 客户端使用 access token 访问用户存放在服务提供方受保护的资源。

有兴趣的读者可以仔细阅读一下 OAuth2.0 标准及其实现，我们熟知的，通过 QQ、微信、微博等登录第三方平台都是通过这种方式。

4. token 认证

基于 token 的鉴权机制类似于 http 协议也是无状态的，它不需要在服务端去保留用户的认证信息或会话信息。这就意味着基于 token 认证机制的应用不需要去考虑用户在哪一台服务器登录了，这就为应用的扩展提供了便利。

token 认证流程如下：
- 用户使用用户名密码来请求服务器。
- 服务器进行验证用户的信息。
- 服务器通过验证发送给用户一个 token。
- 客户端存储 token，并在每次请求时附送上这个 token 值。
- 服务端验证 token 值，并返回数据。

token 必须在每次请求时传递给服务端，它应该保存在请求里。另外，服务端要支持 CORS（跨域源资源共享）策略，需要在服务端添加如下配置：

```
Access-Control-Allow-Origin：*
```

5. JWT 认证

JSON Web Token（JWT）是一个开放标准（RFC 7519），它定义了一种紧凑的、自包含的方式，用于作为 JSON 对象在各方之间安全地传输信息。该信息可以被验证和信任，因为它是数字签名的。下列场景中使用 JWT 很有用：

- Authorization（授权）：这是使用 JWT 的最常见场景。一旦用户登录，后续每个请求都将包含 JWT，允许用户访问该令牌允许的路由、服务和资源。单点登录是现在广泛使用的 JWT 的一个特性，因为它的开销很小，并且可以轻松地跨域使用。
- Information Exchange（信息交换）：对于安全地在各方之间传输信息而言，JWT 无疑是一种很好的方式。因为 JWT 可以被签名，比如用公钥/私钥对，可以确定发送人就是它们所说的那个人。另外，由于签名是使用头和有效负载计算的，还可以验证内容是否被篡改。

在认证的时候，当用户用他们的凭证成功登录以后，一个 JWT 将会被返回。此后，token 就是用户凭证了，必须非常小心，以防出现安全问题。一般而言，保存令牌的时候不应该超过你所需要它的时间。无论何时用户想要访问受保护的路由或者资源时，用户代理（通常是浏览器）都应该带上 JWT，典型地，通常放在 Authorization header 中，用 Bearer schema。header 应该如下所示：

```
Authorization: Bearer <token>
```

服务器上受保护的路由将会检查 Authorization header 中的 JWT 是否有效，如果有效，则用户可以访问受保护的资源。如果 JWT 包含足够多的必需数据，那么就可以减少对某些操作的数据库查询的需要，尽管可能并不总是如此。如果 token 是在授权头（Authorization header）中发送，那么 CORS 将不会成为问题，因为它不使用 Cookie。

JWT 与其他认证的差异：

- JWT 与 Session 的差异：Session 在服务器端，而 JWT 在客户端。
- JWT 与 OAuth 的差异：OAuth2 是一种授权框架，JWT 是一种认证协议。
- JWT 与服务端 Token 的差异：JWT 包含足够多的必需数据，那么就可以减少对某些操作数据库查询的需要。

bifang 项目使用 JWT 验证模式，因为它的开销很小，并且可以轻松地跨域使用。

14.3 mock 数据模拟服务

基于前后端分离的开发模式，前后端可以同时工作，服务端没有部署接口，前端如何模拟 ajax 请求到接口数据？目前主流的有两种方案：

（1）搭建一套自己的 mock 服务，利用开源的 mock 服务，比如淘宝的 API 管理工具 RAP、搜车出品的 EasyMock、开源的 DocLever 等。对于开发团队来说，搭建一套公共的 API 管理工具还是很必要的，可以统一输出接口文档、易于维护接口等。

（2）通过开源插件 mockjs，本地模拟服务端数据，用于帮助前端独立于后端进行开发，其模拟 ajax 并返回相应的数据，从而使前端不必依赖于后端接口，方便开发。

mockjs 具备以下几个特点：

- 前后端分离，让前端工程师独立于后端进行开发。
- 开发无侵入，不需要修改既有代码就可以拦截 Ajax 请求，返回模拟的响应数据。
- 数据类型丰富，支持生成随机的文本、数字、布尔值、日期、邮箱、链接、图片和颜色等。
- 用法简单，符合直觉的接口。
- 增加单元测试的真实性，通过随机数据模拟各种场景。
- 方便扩展，支持扩展更多数据类型，支持自定义函数和正则。

针对以上两种方式的不同特点，bifang 项目采用第二种，本地引入 mockjs，模拟服务端接口并返回数据。

1. 引入 mockjs

通过命令安装 mockjs 插件：

```
npm install mockjs --save-dev
```

2. 构建 mock 服务层

在项目中构建 mock 服务层，src 目录下新增 mock 文件夹，在 mock 文件夹中读者可以根据各模块为维度来拆分 mock 接口的文件夹。比如在 BiFang 项目中，我根据发布单、用户、服务器等维度拆分不同的 mock 接口目录，方便维护，当然读者也可以根据后端接口归属的子系统维度拆分 mock 目录。

以发布单接口为例，在 mock 文件夹下新建 release 文件夹，在 release 文件夹下新建 index.js 文件。index 入口文件负责引入 mockjs 并模拟发布单请求接口，返回对应的数据，具体代码如下：

```
import Mock from 'mockjs'
const releaseList = Mock.mock({
    'list|100': [{
        'id|+1': 0,
        'name': '@GOODS',
        'orderId': `${current}-@integer(1,100)`,
        'status|1-4': 1,
        'send': '@BOOLEAN',
        'sendTime': '@DATETIME',
        'orderDate': '@DATE',
        'auditTime': '@TIME'
    }]
})
Mock.mock(RegExp(`${process.env.VUE_APP_API_BASE_URL}/api/releaseList` + '.*'),'get', ({url}) => {
    const params = parseUrlParams(decodeURI(url))
    let {page, pageSize} = params
    page = eval(page) - 1 || 0
    pageSize = eval(pageSize) || 10
    const total = releaseList.length
    let result = []
    if ((page) * pageSize > total) {
        result = []
    } else {
        result = releaseList.slice(page * pageSize, (page + 1) * pageSize)
    }
    return {
        code:0,
        message: 'success',
        data:{
            page: 1,
            pageSize: 10,
            total: 10,
            list: result
        }
    }
})
```

其他模块的 mock 数据也通过如上方式实现，待所有契约接口都实现后，需要聚合所有接口供项目使用。在 mock 目录下新增 index.js 文件，引入各模块定义的契约接口，文件内容如下：

```
import Mock from 'mockjs'
import '@/mock/release'
import '@/mock/project'
import '@/mock/user'
import '@/mock/workplace'
import '@/mock/server'
import '@/mock/goods'

// 设置全局延时
Mock.setup({
    timeout: '200-400'
})
```

3. 项目引入 mock 服务

在项目中引入 mock 层才能使本地模拟数据生效，在项目主文件 main.js 中引入 src/mock 目录下的 index.js，代码如下：

```
import '@/mock'
```

使用命令启动项目，测试 mock 接口是否生效，在页面中添加 ajax 请求，请求如下地址：

```
http://localhost:8080/mock/jwt_auth/
```

由于在浏览器中无法查看返回值，打断点显示返回值如图 14-3 所示，mock 层模拟接口添加成功。

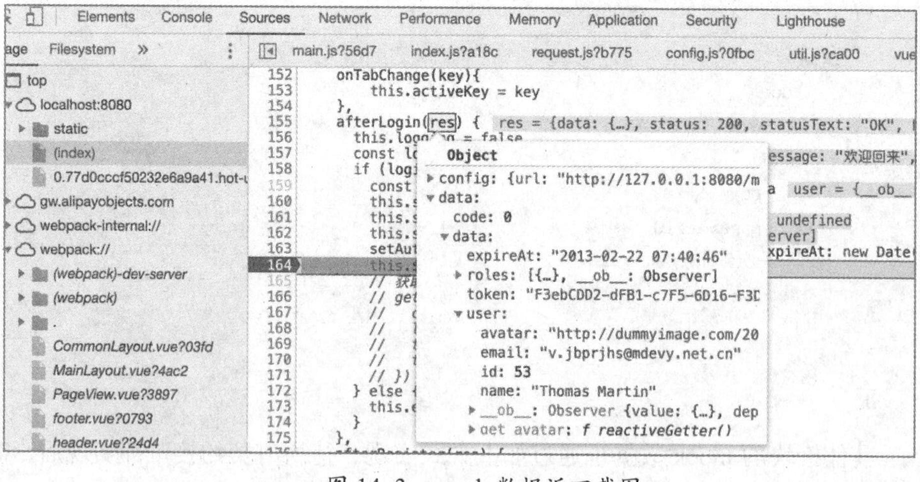

图 14-3 mock 数据返回截图

14.4 环境变量配置

细心的读者会发现在配置 Mock 接口路径时使用了 process.env.VUE_APP_API_BASE_URL 变量，这是从哪里来的？可以在你的项目根目录中放置下列文件来指定环境变量：

- .env # 在所有的环境中被载入。
- .env.local # 在所有的环境中被载入，但会被 git 忽略。
- .env.[mode] # 只在指定的模式中被载入。
- .env.[mode].local # 只在指定的模式中被载入，但会被 git 忽略。

默认情况下，一个 Vue CLI 项目有如下三个模式，当运行 vue-cli-service 命令时，所有的环境变量都从对应的环境文件中载入：

- development 模式用于 vue-cli-service serve。
- test 模式用于 vue-cli-service test:unit。
- production 模式用于 vue-cli-service build 和 vue-cli-service test:e2e。

为一个特定模式准备的环境文件（如 .env.production）将会比一般的环境文件（如 .env）拥有更高的优先级。此外，Vue CLI 启动时已经存在的环境变量拥有最高优先级，并不会被 .env 文件覆写。.env 环境文件是通过运行 vue-cli-service 命令载入的，因此环境文件发生变化，需要重启服务。

在本地开发环境可以通过配置 .env.development 文件配置接口访问的基础路径，如下。同时也可以在 .env 和 .env.development 文件中配置开发环境和生产环境不同的参数配置，如接口请求路径、存储的 Cookie 名称、RSA 加密公钥和 CDN 地址等：

```
//.env.development
VUE_APP_API_BASE_URL=/mock
//.env
VUE_APP_API_BASE_URL=/api
```

环境文件中配置的环境变量可以在项目中直接使用，但只有以 "VUE_APP_" 开头的变量会被 webpack.DefinePlugin 静态嵌入客户端侧的包中。你可以在应用的代码中以如下方式访问它们，在构建过程中，process.env.VUE_APP_BASE_URL 将被相应的值所取代。在 VUE_APP_BASE_URL=/api 的情况下，它会被替换为 "/api"。

除了 VUE_APP_* 变量外，在你的应用代码中始终可用的还有两个特殊的变量：

- NODE_ENV – "development"、"production" 或 "test" 中的一个，具体的值取决于应用运行的模式。
- BASE_URL – 会和 vue.config.js 中的 publicPath 选项相符，即你的应用会部署到的基础路径。

这两个变量在 webpack 配置中会经常用到，待讲到 webpack 配置，即 vue.config.js 配置时再详细讲述用法，读者可以暂时记下，在此作为一个伏笔。

14.5 Service 数据请求服务

本地 mock 服务搭建完毕，项目需要通过 ajax 请求接口地址访问 mock 服务，这就要搭建 service 数据请求服务。当然也有读者觉得可以在组件里直接写 ajax 请求，为什么要再单独定义 service 请求服务呢？搭建 service 数据请求服务有以下几点好处：

- 方便管理 API，各模块请求的接口分目录管理，项目请求的所有 API 接口都可以一目了然，方便查找和修改。
- 便于框架的替换，bifang 项目使用 axios 插件做 ajax 请求，以后可以方便地切换其他插件。

1. 封装公共 request 请求

BiFang 项目使用 axios 作为 ajax 请求的插件，axios 是一个开源免费的、基于 Promise 的 HTTP 客户端，可以工作于浏览器中，也可以在 node.js 中使用。它支持从浏览器中创建 XML 请求，支持 node.js 中发出 http 请求，支持拦截请求和响应，支持转换请求和响应数据，支持自动转换 JSON 数据。

在项目中安装 axios 模块：

```
npm install axios --save
```

在 src 新建 utils 目录，用户维护项目中使用的工具集，utils 目录下新增 request.js，引入 axios 模块，并对 axios 做配置：

```
import axios from 'axios'
```

```
import Cookie from 'js-cookie'
// 跨域认证信息 header 名
const authHeader = 'Authorization'
axios.defaults.timeout = 5000
axios.defaults.withCredentials= true
// http 请求方式
const METHOD = {
    GET: 'get',
    POST: 'post',
    DELETE: 'delete',
    PUT: 'put',
    PATCH: 'patch'
}
/**
 * axios 请求
 * @param url 请求地址
 * @param method {METHOD} http method
 * @param params 请求参数
 * @returns {Promise<AxiosResponse<T>>}
 */
async function request(url, method, params) {
    switch (method) {
        case METHOD.GET:
            return axios.get(url, {params})
        case METHOD.POST:
            return axios.post(url, params)
        case METHOD.DELETE:
            return axios.delete(url, params)
        case METHOD.PUT:
            return axios.put(url, params)
        case METHOD.PATCH:
            return axios.patch(url, params)
        default:
            return axios.get(url, {params})
    }
}
export {
    METHOD,
    request
}
```

添加 axios 拦截器，拦截器分为 request 请求拦截器和 response 响应拦截器。
- request 请求拦截器：发送请求前统一处理，如设置请求头 headers、应用的版本号、终端类型等。

- response 响应拦截器：有时候要根据响应的状态码来进行下一步操作，例如，由于当前的 token 过期，接口返回 401 未授权，那就要进行重新登录的操作。

拦截器代码如下，为了在拦截器中使用 $message 提示，需要开放 initInterceptor 方法：

```
//request 拦截器
const reqInterceptor = {
    onFulfilled(config,$message,router){
        const {url} = config;
        let token = Cookie.get(authHeader)
        config.headers.common[authHeader] = token
        config.headers['Content-Type'] = 'application/json';
        if (whiteRequstList.indexOf(url) === -1 && !Cookie.get(authHeader)) {
            $message.warning(' 认证 token 已过期，请重新登录 ');
            setTimeout(()=>{
                router.push('/login')
            },1000)
        }
        return config
    },
    onRejected(error, $message) {
        $message.error(error.message)
        return Promise.reject(error)
    }
}
//respone 拦截器
const resInterceptor = {
    onFulfilled(response, $message) {
        if (response.status === 401) {
            $message.error(' 无此接口权限 ')
        }
        return response
    },
    onRejected(error, $message) {
        $message.error(error.message)
        return Promise.reject(error)
    }
}
/**
 * 初始化 axios 拦截器
```

```
 * @param {*} $message
 */
function initInterceptor($message,router){
    axios.interceptors.request.use(
        config => reqInterceptor.onFulfilled(config, $message,router),
        error => reqInterceptor.onRejected(error, $message)
    )
    axios.interceptors.response.use(
        config => resInterceptor.onFulfilled(config, $message),
        error => resInterceptor.onRejected(error, $message)
    )
}
```

2. 封装接口路径集

在 src 目录下新建 service 目录，在目录中新增 api.js 文件，该文件用于管理项目接口的路径，代码如下：

```
const BASE_URL = process.env.VUE_APP_API_BASE_URL;
const REAL_URL = process.env.VUE_APP_API_REAL_URL;
const API = {
    LOGIN: `${REAL_URL}/jwt_auth/`,                          // 登录
    REGISTER: `${REAL_URL}/account/register/`,               // 注册
    RELEASELIST: `${REAL_URL}/release/list/`,                // 发布单列表
    CREATERELEASE: `${REAL_URL}/release/create/`,            // 新建发布单
    RELEASEDETAIL: `${REAL_URL}/release/detail/`,            // 发布单详情
    RELEASEBUILD: `${REAL_URL}/release/build/`,              // 发布单构建
    RELEASEBUILDSTATUS: `${REAL_URL}/release/build_status/`, // 发布单构建状态
}
export {
    API
}
```

其中：

- BASE_URL 为配置的真实请求地址。
- LOCAL_URL 为本地 mock 服务地址。

初始开发使用 mock 接口，待联调时可以把接口路径 LOCAL_URL 改为 BASE_URL，如此可对真实接口一个个联调，避免把所有接口访问路径全部修改为真实路径，导致联调复杂性骤升。

3. 新增各模块服务文件

接口访问路径维护在 api.js 后，接下来在 service 目录下新增各模块服务文

件，如 release（发布单服务）、project（项目服务）、application（应用服务）等，以发布单服务为例，新增 release.js，内容如下：

```javascript
import {request,METHOD} from '@/utils/request'
import {API} from './api'

/**
 * 获取发布单列表
 * @param {*} params
 */
async function ReleaseList(params){
    return request(
        API.RELEASELIST,
        METHOD.GET,
        params
    )
}
/**
 * 获取发布单详情
 * @param {*} params
 */
async function ReleaseDetail(params){
    URL = `${API.RELEASEDETAIL}${params}/`
    console.log(URL)
    return request(
        URL,
        METHOD.GET
    )
}
/**
 * 新建发布单
 * @param {*} params
 */
async function CreateRelease(params){
    return request(
        API.CREATERELEASE,
        METHOD.POST,
        params
    )
}
/**
 * 编译发布单
 * @param {*} params
```

```js
 */
async function BuildRelease(params){
    return request(
        API.RELEASEBUILD,
        METHOD.POST,
        params
    )
}
/**
 * 获取发布单编译进度和状态
 * @param {*} params
 */
async function BuildReleaseStatus(params){
    return request(
        API.RELEASEBUILDSTATUS,
        METHOD.POST,
        params
    )
}
export default {
    ReleaseList,
    ReleaseDetail,
    CreateRelease,
    BuildRelease,
    BuildReleaseStatus
}
```

添加各模块服务文件后，需要聚合各服务文件，提供统一出口。在 mock 目录下新增 index.js 文件，引入各模块服务，文件内容如下：

```js
import { requestWhitelistConfig } from '@/utils/request'
import account from './account'
import application from './application'
import env from './environment'
import git from './git'
import release from './release'
import deploy from './deploy'
import server from './server'
import history from './history'
import permission from './permission'
import stats from './stats'
import {API} from './api'
```

```
requestWhitelistConfig([
    API.LOGIN,
    API.REGISTER
])

export default {
    ...account,
    ...application,
    ...env,
    ...git,
    ...release,
    ...deploy,
    ...server,
    ...history,
    ...permission,
    ...stats
}
```

4. 服务层目录结构

添加各模块服务文件、接口集后 service 层的目录结构如下,每个服务文件都对应的是后续页面中的一个业务模块,有了一一对应的关系后,查找和修改都很方便:

```
>service
    >account.js
    >api.js
    >application.js
    >deploy.js
    >environment.js
    >git.js
    >history.js
    >index.js
    >permission.js
    >release.js
    >server.js
    >stats.js
```

在主项目文件 main.js 中引入 request 对象并挂载到 Vue 全局对象中并初始化拦截器,代码如下:

```
import {request,initInterceptor} from '@/utils/request'
Vue.prototype.$request = request;
initInterceptor(Vue.prototype.$message,router)
```

14.6 路由管理器 Vue router

在项目中需要有公共的数据仓库来存储全局数据，便于各功能使用这些公共数据，包括但不仅限于用户信息、项目信息等，而 Vuex 就是当仁不让的选择。

1. Vue router 简介

BiFang 部署平台前端使用 Vue 框架开发，是单页 Web 应用（single page web application，SPA），相对应的是多页 Web 应用（multiple page web applicaiton，MPA），两者区别如下：

- 应用构成不同：MPA 由多个完整页面构成；SPA 由一个外壳页面和多个页面片段构成。
- 跳转方式不同：MPA 页面之间的跳转是从一个页面到另一个页面；SPA 页面片段删除或隐藏，加载另一个页面片段并显示，片段间的模拟跳转没有开壳页面。
- URL 模式不同：MPA 采用 http://xxx/page1.html 和 http://xxx/page2.html 格式；SPA 使用 hash 方式，如 http://xxx/index.html#page1 和 http://xxx/index.html#page2。
- 用户体验对比：MPA 页面间切换加载慢、不流畅、用户体验差，尤其在移动端；SPA 页面片段间切换快，用户体验好，包括移动设备。

使用 Vue CLI 创建的项目是 SPA 项目，页面间的跳转采用 hash 跳转，而 Vue Router 是 Vue.js 官方的路由管理器。它和 Vue.js 的核心深度集成，让构建单页面应用变得易如反掌，其包含的功能包含：

- 嵌套的路由 / 视图表。
- 模块化的、基于组件的路由配置。
- 路由参数、查询、通配符。
- 基于 Vue.js 过渡系统的视图过渡效果。
- 细粒度的导航控制。
- 带有自动激活的 CSS class 的链接。
- HTML5 历史模式或 hash 模式，在 IE9 中自动降级。
- 自定义的滚动条行为。

当把 Vue Router 添加进来，我们需要做的是将组件 (components) 映射到路

由 (routes)，然后告诉 Vue Router 在哪里渲染它们。

2. 路由配置

使用 Vue CLI 创建的基础项目中包含 Vue router 模块，参见 router 目录下的 index.js 文件，内容如下（过滤注释）：

```javascript
import Vue from 'vue'
import VueRouter from 'vue-router'
import Home from '../views/Home.vue'
Vue.use(VueRouter)
const routes = [
    {
        path: '/',
        name: 'Home',
        component: Home
    },
    {
        path: '/about',
        name: 'About',
        component: () => import(/* webpackChunkName: "about" */ '../views/About.vue')
    }
]
const router = new VueRouter({
    routes
})
export default router
```

其中 path 为 '/' 和 '/about' 负责告诉 Vue 在遇到 hash 为 '/' 和 '/about' 时需要加载 Home 组件和 about 组件，加载 about 组件的方式为异步加载。

实际生活中的应用界面，通常由多层嵌套的组件组合而成，这就要使用嵌套路由。

3. 嵌套路由

在 BiFang 部署平台项目中，发布单需要包含发布单列表（/release/list）、创建发布单（/release/create）、发布单历史（/release/history），我们期望的 hash 值为发布单列表、创建发布单、发布单历史，如此一目了然，使用嵌套路由可以简单地表达这种关系，如图 14-4 所示。

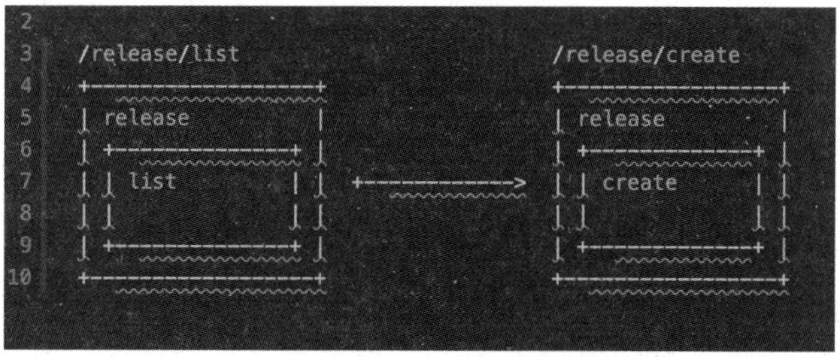

```
 2
 3    /release/list                    /release/create
 4   +--------------------+           +--------------------+
 5   | release            |           | release            |
 6   +--------------------+           +--------------------+
 7   | | list           | |  ----->   | | create         | |
 8   | |                | |           | |                | |
 9   | |                | |           | |                | |
10   +--------------------+           +--------------------+
```

图 14-4 嵌套路由说明

路由配置的相关代码如下：

```
const router = new VueRouter({
    routes:[
        {
            path: '/release',
            name: ' 发布单 ',
            Component:Release
            meta: {
                icon: 'profile'
            },
            children:[
                {
                    path: '/list',
                    name: ' 列表 ',
                    meta:{
                        title: " 发布单列表 "
                    },
                    component: () => import('@/views/release/list')
                },
                {
                    path: '/create',
                    name: ' 新建 ',
                    component: () => import('@/views/release/create')
                },
                {
                    path: '/history',
                    name: ' 发布单部署历史 ',
                    meta:{
                        isShowMenu:false
                    },
                    component: () => import('@/views/release/history')
                },
```

```
            ]
        }
    ]
})
```

通过以上配置，当访问路径的 hash 值为 /release/list 时，view 目录下的发布单列表组件（路径为 views/release/list.vue）会通过 <router-view></router-view> 标签渲染到 release.vue 组件中。

4. 配置文件

在 router 目录下新增 config.js 文件，该文件用于配置路由，同时提供主界面左侧菜单栏所需的配置，文件内容如下：

```
import MainLayout from '@/layouts/MainLayout.vue'
const routeConfig = [
    {
        path: '/login',
        name: '登录页',
        component: () => import('@/views/login/login')
    },
    {
        path: '/',
        name: 'Home',
        component: MainLayout,
        redirect: '/dashboard/workspace',
        children:[
            {
                path: '/dashboard',
                name: 'Dashboard',
                meta:{
                    icon: 'dashboard'
                },
                children:[
                ]
            },
            {
                path: '/release',
                name: '发布单',
                meta: {
                    icon: 'profile'
                },
                children:[
                ]
```

```
            },
            {
                path: '/environment',
                name: ' 环境 ',
                meta: {
                    icon: 'apartment'
                },
                children:[
                ]
            }
        ]
    }
]
export {
    routeConfig
}
```

在路由配置中，路由对象属性配置说明如下：
● path：字符串对应当前路由的路径，总是解析为绝对路径。
● name：当前路由的名称，用于展示主界面菜单栏显示菜单名称。
● meta：路由元信息，遍历 $route.matched 来检查路由记录中的 meta 字段。
● children：子路由配置。

5. 路由导航守卫

vue-router 提供的导航守卫主要用来通过跳转或取消的方式守卫导航。有多种机会植入路由导航过程：全局的、单个路由独享的或者组件级的。

在 BiFang 部署平台中，我们使用前置守卫监听用户未能验证身份时重定向到 /login 页面，在 router 目录下的 index.js 文件中添加以下代码：

```
router.beforeEach((to, from, next) => {
    if (to.path === '/login') {
        next();
    } else {
        if (!isLogin()) {
            next('/login');
        } else {
            next();
        }
    }
});
```

当用户访问非登录页面时，若用户登录状态未失效则允许用户继续访问，若已失效则重定向到登录页面。

6. router-link 及 router-view

<router-link> 组件支持用户在具有路由功能的应用中（单击）导航。通过 to 属性指定目标地址，默认渲染成带有正确链接的 <a> 标签，可以通过配置 tag 属性生成别的标签。另外，当目标路由成功激活时，链接元素自动设置一个表示激活的 CSS 类名。<router-link> 比起写死的 <a href="..."> 会好一些，理由如下：

- 无论是 HTML5 history 模式还是 hash 模式，它的表现行为一致，所以当切换路由模式或在 IE9 降级使用 hash 模式时，无需做任何变动。
- 在 HTML5 history 模式下，router-link 会守卫单击事件，让浏览器不再重新加载页面。
- 当你在 HTML5 history 模式下使用 base 选项之后，所有的 to 属性都不需要写（基路径）了。

<router-view> 组件是一个 functional 组件，渲染路径匹配到的视图组件。<router-view> 渲染的组件还可以内嵌自己的 <router-view>，根据嵌套路径渲染嵌套组件。其他属性（非 router-view 使用的属性）都直接传给渲染的组件，很多时候，每个路由的数据都包含在路由参数中。

因为它也是个组件，所以可以配合 <transition> 和 <keep-alive> 使用。如果两个结合在一起用，要确保在内层使用 <keep-alive>：

```
<transition>
    <keep-alive>
        <router-view></router-view>
    </keep-alive>
</transition>
```

14.7 小 结

本章带领读者了解了前端 Web 项目的开发模式，介绍了前后端分离开发架构下的流程、约定的基本契约格式，对其中的核心认证方式做了较详细的介绍，BiFang 项目采用了 JWT 认证方式，相信读者在做其他项目时也能根据需求采用合适的认证方式。

接下来深入项目搭建，介绍了公共服务 mock 数据模拟服务、service 数据请求服务以及路由管理配置的搭建步骤，随着我的思路能一步步构建前端项目的公共模块，相信读者能够结合自身经手的项目，合理地思考设计公共服务建设。

针对 BiFang 前端项目的搭建，在 15 章中我会带领大家深入 View 层，即页面结构的搭建，采用 ant-design-vue 封装的组件库，可以达到事半功倍的效果。

顽强的毅力能够征服世界上任何一座高峰！

第 15 章　登录页面设计与搭建

> 玉不琢、不成器，人不学、不知义。
>
> ——《三字经》

本章 GitHub 代码地址：https://github.com/aguncn/bifang-book/tree/main/bifang-ch15。

第 14 章主要介绍了前端开发模式，并在基于 Vue CLI 工具集创建的 Vue 基础项目中添加 mock 服务层和 service 数据请求层，接下来将带领读者进入页面设计和相关知识的介绍。

本章将在前端项目中新增登录页面，学习使用 Flex 布局构建页面框架，使用 vue-router 设置路由加载，使用第 14 章中的 service 层和 mock 数据完成页面的渲染和模拟数据访问。

15.1　前端登录界面设计

登录界面是用户进入管理平台必备的，验证用户身份信息的主入口，BiFang 部署平台的登录界面主要包含两个功能：登录和注册。

登录需要输入用户名和密码进行校验，前端通过正则表达式校验输入的信息是否符合规范，服务端校验是否为注册用户以及密码，通过后才能进入部署平台主界面。界面展示如图 15-1 所示。

注册需要用户输入用户名、设置登录密码、确认登录密码以及邮箱信息，前端通过正则表达式校验输入的信息是否符合规范，通过后提交服务端注册，注册成功返回登录界面进行登录。界面展示如图 15-2 所示。

第 15 章 登录页面设计与搭建

图 15-1 登录页面

图 15-2 注册页面

根据登录界面截图如图 15-2 所示，可以把页面分为三部分：

- 头部 header，展示标题以及说明、标识 BiFang 平台。
- 主内容 content，展示登录注册 tab 以及对应的表单，用于用户填写信息提交。
- 底部 footer，展示版权说明以及合作方信息，底部信息会出现在 BiFang 部署平台各页面中，所以这里把底部 footer 抽象为一个公共组件，供相关页面引用。

针对登录界面采用 flex 布局。可能有读者会产生疑问，什么是 flex 布局？传统布局和 flex 布局有什么区别？接下来讲 flex 布局。

15.2　flex 布局

flex 是 Flexible Box 的缩写，意为"弹性布局"，用来为盒状模型提供最大的灵活性。传统布局解决方案基于盒状模型，依赖"display 属性 + position 属性 + float 属性"。它对于那些特殊布局非常不方便，比如垂直居中就不容易实现。

2009 年，W3C 提出了一种新的方案——flex 布局，可以简便、完整、响应式地实现各种页面布局。目前它已得到了所有浏览器的支持，这意味着现在就能很安全地使用这项功能，flex 布局将成为未来布局的首选方案。

举个例子，要让多个 div 横向排列，传统做法是使用浮动，但浮空后因为脱离文档流的缘故，父元素会失去高度，这又涉及了清除浮动等一系列的问题。而 flex 布局相对简单很多，修改父元素 display:flex，会发现 div 自动排列成一行，而且没有浮动后的副作用。从回流角度考虑，flex 的性能更优于 float。

1. 基本概念

先来介绍 flex 布局的基本概念。采用 Flex 布局的元素称为 flex 容器（flex container），简称容器。它的所有子元素自动成为容器成员，称为 flex 项目（flex item），简称项目。

容器默认存在两根轴：水平的主轴（main axis）和垂直的交叉轴（cross axis）。主轴的开始位置（与边框的交叉点）叫 main start，结束位置叫 main end；交叉轴的开始位置叫 cross start，结束位置叫 cross end。项目默认沿主轴排列。单个项目占据的主轴空间叫 main size，占据的交叉轴空间叫 cross size。如图 15-3 所示。

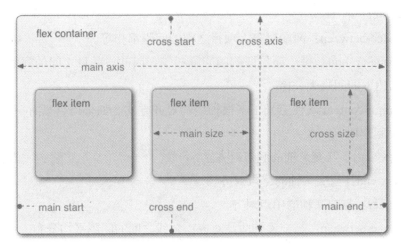

图 15-3　flex 容器示例

2. flex 容器属性

flex 容器有 6 个可设置的属性：

① flex-direction 属性决定主轴的方向，即项目的排列方向，有如下 4 个值：

- row（默认值）：主轴为水平方向，起点在左端。
- row-reverse：主轴为水平方向，起点在右端。
- column：主轴为垂直方向，起点在上沿。
- column-reverse：主轴为垂直方向，起点在下沿。

② flex-wrap 属性定义，如果一条轴线排不下，如何换行，有如下 3 个值：

- nowrap（默认）：不换行。
- wrap：换行，第一行在上方。
- wrap-reverse：换行，第一行在下方。

③ flex-flow 属性是 flex-direction 属性和 flex-wrap 属性的简写形式，默认值为 row nowrap：

```
.box {
    flex-flow: <flex-direction> <flex-wrap>;
}
```

④ justify-content 属性定义了项目在主轴上的对齐方式：

- flex-start（默认值）：左对齐。
- flex-end：右对齐。

- center：居中。
- space-between：两端对齐，项目之间的间隔都相等。
- space-around：每个项目两侧的间隔相等。因此，项目之间的间隔比项目与边框的间隔大一倍。

⑤ align-content 属性定义了多根轴线的对齐方式。如果项目只有一根轴线，该属性不起作用：

- flex-start：与交叉轴的起点对齐。
- flex-end：与交叉轴的终点对齐。
- center：与交叉轴的中点对齐。
- space-between：与交叉轴两端对齐，轴线之间的间隔平均分布。
- space-around：每根轴线两侧的间隔都相等。所以，轴线之间的间隔比轴线与边框的间隔大一倍。
- stretch（默认值）：轴线占满整个交叉轴。

3. 项目属性

容器属性是加在容器上的，那么项目属性呢？就是写在项目上的，这和容器属性给 ul、项目属性给 li 差不多一个意思。

- order 取值：默认 0，用于决定项目排列顺序，数值越小，项目排列越靠前。
- flex-grow 取值：默认 0，用于决定项目在有剩余空间的情况下是否放大，默认不放大；注意，即便设置了固定宽度，也会放大。假设默认三个项目中前两个项目都是 0，最后一个是 1，最后的项目会沾满剩余所有空间。
- flex-shrink 取值：默认 1，用于决定项目在空间不足时是否缩小，默认项目都是 1，即空间不足时大家一起等比缩小。注意，即便设置了固定宽度，也会缩小。但如果某个项目 flex-shrink 设置为 0，则即便空间不够，自身也不缩小。
- flex-basis 取值：默认 auto，用于设置项目宽度，默认 auto 时，项目会保持默认宽度，或者以 width 为自身的宽度，但如果设置了 flex-basis，权重 width 属性会高，因此会覆盖 widtn 属性。
- flex 取值：默认 0 1 auto，flex 属性是 flex-grow、flex-shrink 与 flex-basis 三个属性的简写，用于定义项目放大、缩小与宽度。该属性有两个快捷键值，分别是 auto（1 1 auto）等分放大缩小，与 none（0 0 auto）不放大不缩小。

- align-self 取值：auto（默认）| flex-start | flex-end | center | baseline | stretch，表示继承父容器的 align-items 属性。如果没父元素，则默认 stretch。用于让个别项目拥有与其他项目不同的对齐方式，各值的表现与父容器的 align-items 属性完全一致。

以上就是 flex 布局属性的介绍，BiFang 部署平台登录页面采用 flex 布局。

15.3　通用布局组件简介

在编写登录页面代码前，先创建登录页面所需的通用布局组件，即包含内容区和底部页脚的简单布局。布局组件页面的内容如图 15-4 所示。

图 15-4　通用布局示例

在 src 目录下创建 layouts 目录，该目录存储页面所需的通用布局，存放登录页面以及后续主页面框架所需要的组件，在 layouts 目录下创建 commonLayout.vue 文件。

1. 布局组件 template

在 commonLayout.vue 文件中添加 template 模块，内容如下：

```
<template>
    <div class="common-layout">
        <div class="content">
            <slot></slot>
        </div>
        <div class="footer">
            <div class="links">
                <a target="_blank" :key="index"
                    :href="item.link ? item.link : 'javascript: void(0)'"
                    v-for="(item, index) in linkList">
                    <a-icon v-if="item.icon" :type="item.icon"/>{{item.name}}
                </a>
            </div>
            <div class="copyright">
                Copyright<a-icon type="copyright" />{{copyright}}
            </div>
        </div>
    </div>
</template>
```

最外层 div.common-layout 使用 Flex 布局，内部包含 content（内容）和 footer（页脚）两个区域。content 中包含 slot 插槽标签，其实就相当于占位符。它在组件中给 HTML 模板占了一个位置，让你来传入一些东西。插槽又分为匿名插槽、具名插槽（name 属性）和作用域插槽（在插槽中使用数据），在 commonLayout 组件中使用匿名插槽。

footer 中包含 div.links 和 div.copyright 两块内容，分别用于展示网站链接以及版权说明。

2. 布局组件 style

在 commonLayout.vue 文件中添加 style 标签，样式代码如下：

```
<style scoped lang="less">
.common-layout{
    display: flex;
    flex-direction: column;
    height: 100vh;
    overflow: auto;
    background-color: @layout-body-background;
    background-image: url('./assets/TVYTbAXWheQpRcWDaDMu.svg');
    background-repeat: no-repeat;
    background-position-x: center;
    background-position-y: 110px;
```

```less
    background-size: 100%;
    .content{
        padding: 32px 0;
        flex: 1;
        @media (min-width: 768px){
            padding: 112px 0 24px;
        }
    }
    .footer{
        padding: 48px 16px 24px;
        text-align: center;
        .copyright{
            color: 'rgba(0, 0, 0, 0.45)';
            font-size: 14px;
        }
        .links{
            margin-bottom: 8px;
            a:not(:last-child) {
                margin-right: 40px;
            }
            a{
                color: 'rgba(0, 0, 0, 0.45)';
                -webkit-transition: all .3s;
                transition: all .3s;
            }
        }
    }
}
</style>
```

这块样式代码中包含几个知识点，简单介绍一下，对于初学者来说可以更方便地理解这块样式的含义，已了解的读者可以跳过。

- scoped 属性：在 Vue 组件中为了使样式私有化（模块化），不对全局造成污染，可以在 style 标签上添加 scoped 属性以表示它只属于当下的模块。编译后加载组件样式会带有标签区分作用域，如编译后的 common-layout class 类型会变为 common-layout[data-v-4312a6d]，对应 commonLayout.vue 编译后的组件 ID。

- @layout-body-background：该样式定义在 ant-design-vue 框架提供的公共样式文件中，具体的参考链接：https://github.com/vueComponent/ant-design-vue/blob/master/components/style/themes/default.less。

- 100vh：vh 是 CSS 单位，相对于视口的高度和宽度，而不是父元素的（CSS 百分比是相对于包含它的最近的父元素的高度和宽度），常用的像素 px 是相对于显示器屏幕分辨率而言的。
- @media 标签：使用 @media 查询，可以针对不同的媒体类型定义不同的样式。@media 可以针对不同的屏幕尺寸设置不同的样式，特别是当你需要设置、设计响应式的页面时，@media 是非常有用的。当重置浏览器大小的过程时，页面也会根据浏览器的宽度和高度重新渲染页面。

3. 布局组件 script

在 commonLayout.vue 文件中添加 script 标签，代码如下：

```
export default {
    name: 'CommonLayout',
    data(){
        return {
            copyright:'bifang@copyright',
            linkList:[{
                name:"Ant Design Vue",
                icon:"",
                link:"https://antdv.com/docs/vue/introduce-cn/"
            }]
        }
    }
}
```

通过 name 定义组件名称，data 中的 copyright 以及 links 用于维护版权以及网站链接。

4. 创建 PageFooter 组件

前文提到底部标签会在 BiFang 部署平台所有页面中展示，那就需要单独抽取一个单独的组件以供其他组件引用。

在 src/components 目录中新建 footer 目录，在目录下新增 footer.vue 文件，把 commonLayout.vue 中关于底部标签的代码抽取到 footer.vue 文件中，抽取后 footer.vue 中的代码如下：

```
<template>
    <div class="footer">
        <div class="links">
            <a target="_blank" :key="index" :href="item.link ? item.link : 'javascript: void(0)'"
                v-for="(item, index) in linkList">
```

```
                <a-icon v-if="item.icon" :type="item.icon"/>{{item.name}}
            </a>
        </div>
        <div class="copyright">
            Copyright<a-icon type="copyright" />{{copyright}}
        </div>
    </div>
</template>

<script>
export default {
    name: 'PageFooter',
    data(){
        return {
            copyright:'bifang@copyright',
            linkList:[{
                name:"Ant Design Vue",
                icon:"",
                link:"https://antdv.com/docs/vue/introduce-cn/"
            }]
        }
    }
}
</script>

<style lang="less" scoped>
    .footer{
        padding: 48px 16px 24px;
        text-align: center;
        .copyright{
            color: 'rgba(0, 0, 0, 0.45)';
            font-size: 14px;
        }
        .links{
            margin-bottom: 8px;
            a:not(:last-child) {
                margin-right: 40px;
            }
            a{
                color: 'rgba(0, 0, 0, 0.45)';
                -webkit-transition: all .3s;
                transition: all .3s;
            }
```

```
        }
    }
</style>
```

15.4 前端登录页面搭建

登录界面所需的通用布局组件创建完毕，接下来搭建登录界面组件，组件中引用 CommonLayout 通用布局组件，在其中嵌套登录界面设计中的 header（头部）以及 content（内容）部分。

在 src 目录创建 views 目录，该目录存放 MVVM 中的展现层，即 BiFang 部署平台所有业务页面都会放在该目录中。在 views 目录中创建 login 目录，目录下创建 login.vue 文件，该文件即为登录界面组件。

15.4.1 布局结构搭建

在登录组件中的 template 部分中，最外层使用 CommonLayout 组件，内部嵌套 class 为 top 和 login 的两个 div 标签，分别布局 header 和登录注册表单。具体的代码如下：

```
<template>
    <common-layout>
        <div class="top">
        </div>
        <div class="login">
        </div>
    </common-layout>
</template>
```

创建两个 div 标签对应的样式，代码如下：

```
<style lang="less" scoped>
    .common-layout{
        .top {
            text-align: center;
        }
        .login{
            width: 368px;
            margin: 0 auto;
```

```
            @media screen and (max-width: 576px) {
                width: 95%;
            }
        }
    }
</style>
```

在组件 template 下的 div.top 中创建 header（头部）部分，代码如下：

```
<div class="top">
    <div class="header">
        <img alt="logo" class="logo" src="@/assets/logo.png" />
        <span class="title"> 毕方部署平台 </span>
    </div>
    <div class="desc">BiFang 是最具影响力的服务部署平台 </div>
</div>
```

使用 less 语法在 style 标签中的 commonLayout>top 下添加头部对应的样式文件，代码如下：

```
.header {
    height: 44px;
    line-height: 44px;
    a {
        text-decoration: none;
    }
    .logo {
        height: 44px;
        vertical-align: top;
        margin-right: 16px;
    }
    .title {
        font-size: 33px;
        color: @bf-title-color;
        font-family: 'Myriad Pro', 'Helvetica Neue', Arial, Helvetica, sans-serif;
        font-weight: 600;
        position: relative;
        top: 2px;
    }
}
.desc {
    font-size: 14px;
    color: @bf-text-color-second;
    margin-top: 12px;
```

```
      margin-bottom: 40px;
}
```

header 部分布局搭建完毕，接下来搭建登录及注册的表单，这里用到 ant-design-vue 中的 a-tab、a-form 以及相关组件。因为在第 12 章项目搭建中使用的是全局引用方式，所以登录组件不需要额外单独引用这部分组件，ant-design 框架已经把相关组件注册到全局变量中，当然这种方式较为粗暴，毕竟项目有时只用到部分 ant-design 组件。为了页面的加载性能，还是建议读者能够按需引用。

登录以及注册表单代码如下：

```
<a-tabs size="large"
  :activeKey="activeKey"
  @change="onTabChange"
  :tabBarStyle="{textAlign: 'center'}"
  style="padding: 0 2px;">
    <a-tab-pane tab=" 登录 " key="1">
      <a-form @submit="onLogin" :form="loginForm">
        <a-alert type="error"
          :closable="true"
          v-show="error"
          :message="error"
          showIcon style="margin-bottom: 24px;" />
        <a-form-item>
          <a-input
            autocomplete="autocomplete"
            size="large"
            placeholder="admin"
            v-decorator="['name', {rules: [{ required: true,
              message: ' 请输入账户名 ', whitespace: true}]}]"
          >
            <a-icon slot="prefix" type="user" />
          </a-input>
        </a-form-item>
        <a-form-item>
          <a-input
            size="large"
            placeholder="password"
            autocomplete="autocomplete"
            type="password"
            v-decorator="['password', {rules: [{ required: true,
              message: ' 请输入密码 ', whitespace: true}]}]"
          >
```

```html
        <a-icon slot="prefix" type="lock" />
      </a-input>
    </a-form-item>
    <div>
      <a-checkbox :checked="true" > 自动登录 </a-checkbox>
        <a style="float: right"> 忘记密码 </a>
    </div>
    <a-button
      :loading="logging"
      style="width: 100%;margin-top: 24px"
      size="large"
      htmlType="submit" type="primary">
        登录 </a-button>
  </a-form>
</a-tab-pane>
<a-tab-pane tab=" 注册 " key="2">
  <a-form @submit="onRegister" :form="registerForm">
    <a-alert type="error"
      :closable="true"
      v-show="regError"
      :message="regError" showIcon style="margin-bottom: 24px;" />
    <a-form-item>
      <a-input
        autocomplete="autocomplete"
        size="large"
        placeholder=" 请输入用户名 "
        v-decorator="['username', {rules: [{ required: true,
          message: ' 请输入账户名 ', whitespace: true}]}]"
      >
          <a-icon slot="prefix" type="user" />
      </a-input>
    </a-form-item>
    <a-form-item>
      <a-input
        size="large"
        placeholder=" 请输入密码 "
        autocomplete="autocomplete"
        type="password"
        v-decorator="['password', {rules: [{ required: true,
          message: ' 请输入密码 ', whitespace: true}]}]"
      >
          <a-icon slot="prefix" type="lock" />
      </a-input>
```

```
            </a-form-item>
            <a-form-item>
              <a-input
                size="large"
                placeholder=" 请输入确认密码 "
                autocomplete="autocomplete"
                type="password"
                v-decorator="['confirmPassword', { rules: [{ required: true,
                  message: ' 请输入确认密码 !'}, { validator: compareToPassword}] }]"
              >
                <a-icon slot="prefix" type="lock" />
              </a-input>
            </a-form-item>
            <a-form-item>
              <a-input
                size="large"
                placeholder=" 请输入邮箱 "
                type="email"
                v-decorator="['email',{rules: [
                  { type: 'email',message: ' 输入的邮箱地址不正确！ '},
                  { required: true,message: ' 请输入邮箱 '}]}]">
                  <a-icon slot="prefix" type="mail" />
              </a-input>
            </a-form-item>
            <a-form-item>
              <a-button
               :loading="registing"
               style="width: 100%;margin-top: 24px"
               size="large" htmlType="submit" type="primary">
              注册 </a-button>
            </a-form-item>
          </a-form>
      </a-tab-pane>
    </a-tabs>
```

引用组件库的好处之一即不再需要创建额外的样式，除非有特殊需求。

到此处，登录界面所需的元素以及样式已搭建完毕，接下来要实现登录组件中的 script 部分。

15.4.2 功能实现

登录组件中实现的事件包含登录和注册按钮添加事件，注册表单中的两次输入框校验以及 tab 切换，相关代码如下：

第 15 章 登录页面设计与搭建

```
<script>
import CommonLayout from '@/layouts/CommonLayout'
import API from '@/service'
import {setAuthorization} from '@/utils/request'
export default {
    name: 'Login',
    components: {CommonLayout},
    data () {
        return {
            logging: false,              // 登录状态
            registing:false,             // 注册状态
            error: '',                   // 登录错误提示
            regError:'',                 // 注册错误提示
            activeKey:'1',               //tab 切换
            loginForm: this.$form.createForm(this),    // 登录表单
            registerForm: this.$form.createForm(this)  // 注册表单
        }
    },
    methods: {
        // 登录事件
        onLogin (e) {
            e.preventDefault()
            this.loginForm.validateFields((err) => {
                if (!err) {
                    this.logging = true
                    const username = this.loginForm.getFieldValue('name')
                    const password = this.loginForm.getFieldValue('password')
                    API.Login({username, password}).then(this.afterLogin)
                }
            })
        },
        // 注册事件
        onRegister (e) {
            e.preventDefault()
            this.registerForm.validateFields((err) => {
                if (!err) {
                    this.registing = true
                    const username = this.registerForm.getFieldValue('username')
                    const password = this.registerForm.getFieldValue('password')
                    const passwordConfirm = this.registerForm.getFieldValue('confirmPassword')
                    const email = this.registerForm.getFieldValue('email')
                    API.Register({username, password, passwordConfirm, email}).then(this.afterRegister)
                }
```

```
        })
    },
    // 密码校验
    compareToPassword(rule, value, callback) {
        const form = this.registerForm;
        if (value && value !== form.getFieldValue('password')) {
            callback(' 两次输入的密码不同 ');
        } else {
            callback();
        }
    },
    //tab 切换
    onTabChange(key){
        this.activeKey = key
    },
    // 登录回调
    afterLogin(res) {
        this.logging = false
        const loginRes = res.data
        if (loginRes.code == 0) {
            const {user, roles} = loginRes.data
            setAuthorization({
                token: loginRes.data.token,
                expireAt: new Date(loginRes.data.expireAt)
            })
            this.$router.push('/release/releaseList')
        } else {
            this.error = loginRes.data.message
        }
    },
    // 注册回调
    afterRegister(res) {
        const registerRes = res.data
        this.registing = false;
        if (registerRes.code == 0) {
            this.$message.success(' 恭喜，注册成功！ ', 3)
            const username = this.registerForm.getFieldValue('username')
            const password = this.registerForm.getFieldValue('password')
            this.registerForm.resetFields()
            // 回填
            this.loginForm.setFieldsValue({
                name:username,
                password
```

```
                })
                this.activeKey = "1"
            } else {
                this.regError =registerRes.data
            }
        }
    }
}
</script>
```

其中顶部通过 import 方法引入组件以及服务：

import commonLayout 通用布局组件并通过 components 注册为登录组件的子组件；import @service 层供登录和注册事件调用 mock 接口；Import setAuthorization，引用 request 中的方法，该方法用于把登录接口返回的 JWT Token 信息储存在 Cookie 中，便于设置登录态期限。

登录组件中的实现方法可参见代码以及备注，不再详细解释，接下来介绍 Vuex 中如何设置存储用户的相关信息。

15.5　全局数据仓库 Vuex

在项目中需要有公共的数据仓库来存储全局数据，便于各功能使用这些公共数据，包括但不仅限于用户信息、项目信息等，而 Vuex 就是当仁不让的选择。

1. Vuex 简介

Vuex 是一个专为 Vue.js 应用程序开发的状态管理模式。它采用集中式存储管理应用的所有组件的状态，并以相应的规则保证其状态以一种可预测的方式发生变化。Vuex 也集成到 Vue 的官方调试工具 devtools extension (opens new window) 上，提供了诸如零配置的 time-travel 调试、状态快照导入导出等高级调试功能。

状态自管理应用包含：

- state：驱动应用的数据源。
- view：以声明方式将 state 映射到视图。
- actions：响应在 view 上的用户输入导致的状态变化。

"单向数据流"理念的简单示意如图 15-5 所示。

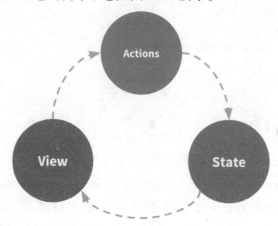

图 15-5 Vuex 单向数据流

Vuex 应用的核心就是 store（仓库），store 基本上就是一个容器，其包含应用中大部分的状态 (state)。Vuex 和单纯的全局对象有以下两点不同：

① Vuex 的状态存储是响应式的。当 Vue 组件从 store 中读取状态时，若 store 中的状态发生变化，那么相应的组件也会相应地得到高效更新。不能直接改变 store 中的状态。

② 改变 store 中状态的唯一途径就是显式地提交 (commit) mutation。这样使得我们可以方便地跟踪每一个状态的变化，从而能够实现让一些工具帮助我们更好地了解应用。

2. 用户信息存储

BiFang 部署平台项目使用 Vuex 存储用户信息以及角色信息，Vuex 是在我搭建项目时默认创建的，路径为 src 目录下的 store 目录，修改 index.js 文件实现用户相关信息的存储，具体的代码如下：

```
import Vue from 'vue'
import Vuex from 'vuex'
Vue.use(Vuex)
const USER_KEY = "bifang.user"           // 存储用户的 key
const ROLES_KEY = "bifang.roles"         // 存储角色的 key
export default new Vuex.Store({
    state: {
        user: null,
        roles: null,
    },
```

```
    getters:{
        getUser: state => {
            if (!state.user) {
                try {
                    const user = localStorage.getItem(USER_KEY)
                    state.user = JSON.parse(user)
                } catch (e) {
                    console.error(e)
                }
            }
            return state.user
        },
        getRoles: state => {
            if (!state.roles) {
                try {
                    const roles = localStorage.getItem(ROLES_KEY)
                    state.roles = JSON.parse(roles)
                    state.roles = state.roles ? state.roles : []
                } catch (e) {
                    console.error(e.message)
                }
            }
            return state.roles
        }
    },
    mutations: {
        setUser (state, user) {
            state.user = user
            localStorage.setItem(USER_KEY, JSON.stringify(user))
        },
        setRoles(state, roles) {
            state.roles = roles
            localStorage.setItem(ROLES_KEY, JSON.stringify(roles))
        },
    }
})
```

Vuex 中使用 localStorage 做本地化存储，便于用户在登录态未过期时打开 BiFang 部署平台页面，而不是每次都需要用户登录。

store 中定义 getter，就像计算机属性一样，getter 的返回值会根据它的依赖被缓存起来，且只有当它的依赖值发生改变才会被重新计算。

更改 Vuex 的 store 中状态的唯一方法是提交 mutation。Vuex 中的 mutation

非常类似于事件：每个 mutation 都有一个字符串的"事件类型（type）"和一个"回调函数（handler）"，这个回调函数就是实际进行状态更改的地方，并且它会接受 state 作为第一个参数。

3. 登录组件引入仓库

在登录组件中通过 import 引入 Vuex 中的 mapMutation 函数，该辅助函数用于将组件中的 methods 映射为 store.commit 调用，在登录成功的回调函数 afterLogin 中调用 Vuex mutation 对应的方法，代码如下：

```js
import {mapMutations} from 'vuex'
methods: {
    ...mapMutations(['setUser', 'setPermissions', 'setRoles']),
    // 其他函数略过
    // 登录回调
    afterLogin(res) {
        this.logging = false
        const loginRes = res.data
        if (loginRes.code == 0) {
            const {user, roles} = loginRes.data
            this.setUser(user)
            this.setRoles(roles)
            setAuthorization({token: loginRes.data.token, expireAt: new Date(loginRes.data.expireAt)})
            this.$router.push('/release/releaseList')
        } else {
            this.error = loginRes.data.message
        }
    }
}
```

以上代码便完成了 Vuex 的引入，到此登录组件的代码开发告一段落，接下来要让登录页面跑起来能展示。

15.6　路由以及 mock 服务配置

登录页面开发完成后，就需要在浏览器中测试开发的页面展示以及功能是否有问题，即进入开发本地测试阶段，这时需要用到第 14 章介绍的路由以及 mock 服务。

第15章 登录页面设计与搭建

1. 路由配置

根据第 14 章的介绍，读者已了解 Vue router 的作用以及 Vue CLI 工具集创建的默认路由配置，对此还不清楚的读者可再回顾一下。若要在浏览器能够展示登录页面，只需要把登录页面的组件配置到 router 目录下的 config.js 文件中，具体代码如下：

```
const routeConfig = [
    {
        path: '/login',
        name: ' 登录页 ',
        component: () => import('@/views/login/login')
    },
]
```

通过在浏览器地址中添加 hash 值为 login，即访问 http://localhost:8080/#/login，我们便可以在浏览器中看到开发完的登录页面。

2. mock 服务配置

项目中的本地 mock 服务用于模拟服务端请求，并返回自定义的模拟数据。登录页面请求两个服务端接口：登录和注册，所以需要在 mock 目录下新增 login 目录，在目录中新增 index.js 文件，文件内容如下：

```
import Mock from 'mockjs'
const loginInfo = Mock.mock({
    'token': '@guid',
    'expireAt': '@DATETIME',
    'user_id': '@integer(1,100)',
    'user': {
        'id':'@integer(1,100)',
        'name':'@name',
        'email':'@email',
        'avatar':"@image('200x100', '#4A7BF7', 'bifang')"
    },
    "roles": [
        {
            "id": '@name',
            "operation|2": [
                "add",
                "edit",
                "delete"
            ]
```

```js
            }
        ]
    })
    Mock.mock(`${process.env.VUE_APP_API_BASE_URL}/jwt_auth/`, 'post', ({body}) => {
        return {
            "code": 0,
            "message": " 欢迎回来 ",
            "data": loginInfo
        }
    })
    const registerInfo = Mock.mock({
        'username': '@name',
        'email': '@email'
    })
    Mock.mock(`${process.env.VUE_APP_API_BASE_URL}/account/register`, 'post', ({body}) => {
        return {
            "code": 0,
            "message": " 注册成功 ",
            "data": registerInfo
        }
    })
```

在 mock 服务的入口文件 index.js 中引入 login 目录，既完成了登录 mock 服务的注册；当登录组件访问 /jwt_auth/ 和 /account/register 两个路径时，mock 服务层会返回配置的数据。登录组件访问 mock 服务层的登录数据需要通过 service 层，所以我们还需要在 service 层配置登录模块。

3. service 服务配置

service 服务统一封装了毕方部署平台项目所需要的所有接口路径以及请求方式，并通过 axios 调用下游的 mock 数据，所以 service 服务需要新增登录模块。

在 services 目录下的 api.js 路径集中新增登录和注册的访问路径，代码如下：

```js
const API = {
    LOGIN: `${BASE_URL}/jwt_auth/`,
    REGISTER: `${BASE_URL}/account/register/`
}
```

在 services 目录下的新增 login.js 文件，用于配置登录组件所需服务，代码如下：

```
import {request,METHOD} from '@/utils/request'
import {API} from './api'
/**
 * 登录
 * @param {*} params
 */
async function Login(params){
    let {username,password} = params
    return request(
        API.LOGIN,
        METHOD.POST,
        {
            username,
            password
        }
    )
}

/**
 * 注册
 * @param {*} params
 */
async function Register(params){
    let {
        username = '',
        password = '',
        passwordConfirm = '',
        email = ''
    } = params;

    return request(
        API.REGISTER,
        METHOD.POST,
        {
            username,
            password,
            passwordConfirm,
            email
        }
    )
}
export default {
    Login,
```

```
    Register
}
```

登录和注册的接口都通过 POST 方式请求服务端接口，在登录组件 login.vue 中调用，并通过 promise.then(response=>{}) 的回调方式把获取的数据返回给 afterLogin 处理函数。在 afterLogin 函数中通过 Vuex 中的 setUser、setRole 方法把用户角色数据存储在全局仓库 store 中，同时调用 request.js 中的 setAuthorition 方法把登录校验返回的 token 信息存储在 Cookie 中，用于后续接口调用。

到此为止，登录组件以及相对应的 mock 服务、service 服务代码都已全部编写完毕，启动项目，在浏览器输入 http://localhost:8080/#/login，登录页面即可呈现在读者面前，在登录表单和注册表单输入相关信息，即可调用接口返回成功标识。

在 service 服务层中的 api.js 中改变 BASE_UR 和 REAL_URL 即可完成 mock 数据和真实环境的请求切换，方便项目的调试。

15.7 小　　结

本章以登录界面为例，带领读者熟悉了前端 Web 项目中的页面设计、组件布局、页面搭建到功能实现流程，掌握了如何配置公共服务路由、mock 服务以及 service 服务，中间也穿插了 Vuex、组件 scope 以及 CSS 相关的知识点，相信没有了解过相关知识点的读者能大有收获。

开发登录界面的流程贯穿在整个毕方部署平台项目中，所有页面的搭建都万变不离其宗，也是需要读者深入思考以及掌握的。掌握了这套流程方法论，在接下来各项目模块的构造中，读者学习也能达到事半功倍的效果。

第 16 章将带领大家进入部署项目所需的基础配置——服务器以及环境界面的搭建。服务器可配置哪些信息？为何需要区分多环境，如何配置？让我们带着这些问题，前进！

行动不一定带来快乐，而无行动则绝无快乐！

第 16 章　主界面及管理员模块设计与搭建

学而不思则罔，思而不学则殆。

——孔子

本章 GitHub 代码地址：https://github.com/aguncn/bifang-book/tree/main/bifang-ch16。

第 15 章主要介绍了前端登录页面的设计与搭建，开发通用布局组件和底部标签组件并了解组件的搭配使用，在 mock 服务层、service 服务层以及路由层添加登录页面专属的登录模块，熟悉整个流程配置方法，接下来将带领大家进入 BiFang 管理平台主页面的搭建，并在其中以用户模块为例，介绍如何使用主页面组件嵌套子模块，再次实践流程配置方法。

本章将在前端项目中设计搭建主页面框架，学习使用 ant-design-vue 提供的组件构建主页面通用框架组件，拆分布局组件；新增用户配置页面，在 mock 服务、service 服务和路由层添加页面对应的用户模块。

16.1　主界面框架设计

管理平台是各公司管理内部信息和资源的重要系统，优秀的管理系统在灵活性、维护性和可操作性上能让用户满意。对于可操作性，经过近十几年的发展，页面的设计风格一变再变，而页面的布局反而没有大的变化。管理平台的界面设计如图 16-1 所示。

根据界面截图 16-1，可以把页面布局分为四部分：
- 侧边导航栏，展示平台名称图标信息以及所有功能的菜单列表，单击菜单可以展开子菜单，方便用户一览无余地了解平台具有哪些功能。
- 顶部导航栏，左侧展示菜单折叠按钮，单击可折叠侧边导航栏；右侧展示用户相关信息以及可操作导航，如全局搜索、帮助信息等。

- 主内容区，单击侧边导航栏菜单展示的页面具体信息，是信息的主要展示操作区域，搭配面包屑导航，方便用户快速识别入口路径。
- 底栏页脚，展示版权说明以及合作方信息，底部信息会出现在 BiFang 部署平台各页面中，这部分可以使用第 15 章中封装的底部组件。

图 16-1 通用布局示意图

针对 BiFang 管理平台主界面，这里采用 Ant-design-vue 中提供的通用 Layout 布局组件，接下来具体讲解如何通过 Layout 布局组件搭建主界面页面。

16.2 主界面布局组件

第 15 章讲到在 src 目录下的 layouts 目录存储页面所需要的布局组件，已创建了登录页面所需要的通用组件 CommonLayout，在 layouts 目录下再创建 MainLayout.vue 文件，该文件即为主界面布局组件。

在进入组件代码开发前，有必要介绍 ant-design-vue 中的 Layout 布局组件是如何工作的。

1. Layout 布局简介

在 ant-design-vue 框架中封装了一系列高质量开箱即用的 Vue 组件，组件风格与 Ant Design 保持同步，服务于企业级后台平台产品，而其中的 Layout 布局组件协助进行页面级整体布局。

Layout 布局组件包含：
- Layout：布局容器，其下可嵌套 Header Sider Content Footer 或 Layout 本身，可以放在任何父容器中。
- Header：顶部布局，自带默认样式，其下可嵌套任何元素，只能放在 Layout 中。
- Sider：侧边栏，自带默认样式及基本功能，其下可嵌套任何元素，只能放在 Layout 中。
- Content：内容部分，自带默认样式，其下可嵌套任何元素，只能放在 Layout 中。
- Footer：底部布局，自带默认样式，其下可嵌套任何元素，只能放在 Layout 中。

此处用几个示例来详细演示如何用短短几行代码即可完成页面的布局。

（1）页面包含 Header、Content、Footer 部分，界面展示如图 16-2 所示。

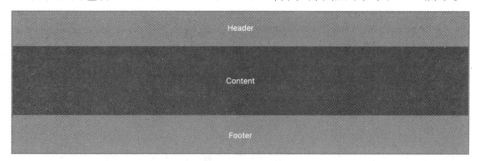

图 16-2　Layout 布局 1

代码如下：

```
<a-layout>
    <a-layout-header>Header</a-layout-header>
    <a-layout-content>Content</a-layout-content>
    <a-layout-footer>Footer</a-layout-footer>
</a-layout>
```

（2）页面包含 Header、Sider、Content 和 Footer，界面展示如图 16-3 所示。

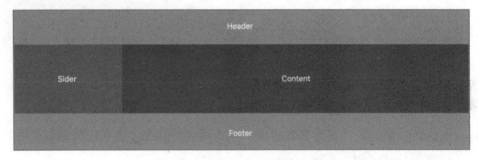

图 16-3　Layout 布局示例 2

代码如下：

```
<a-layout>
    <a-layout-header>Header</a-layout-header>
    <a-layout>
        <a-layout-sider>Sider</a-layout-sider>
        <a-layout-content>Content</a-layout-content>
    </a-layout>
    <a-layout-footer>Footer</a-layout-footer>
</a-layout>
```

（3）页面左侧为 Sider，右侧自上至下分别为 Header、Content 和 Footer，界面展示如图 16-4 所示。

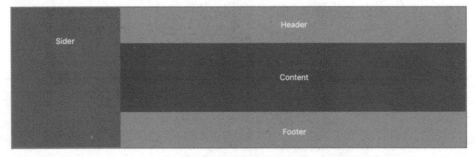

图 16-4　Layout 布局示例 3

代码如下：

```
<a-layout>
    <a-layout-sider>Sider</a-layout-sider>
    <a-layout>
        <a-layout-header>Header</a-layout-header>
        <a-layout-content>Content</a-layout-content>
        <a-layout-footer>Footer</a-layout-footer>
```

```
        </a-layout>
    </a-layout>
```

2. 主界面组件 template

第 1 小节中的第三个示例布局符合 BiFang 部署平台的设计，这里采用第三种布局方式，但稍有不同的是，为了保证组件的可复用和可维护性，侧边导航栏、右侧顶部导航栏、右侧内容区域以及右侧底部页脚封装为 4 个组件，分别对应 sideMenu、adminHeader、pageView 以及第 15 章封装的 pageFooter 组件。

在 MainLayout.vue 文件中添加 template 标签，内容如下：

```
<template>
    <a-layout class="admin-layout beauty-scroll">
        <!-- 左侧导航栏 -->
        <side-menu class="fixed-side"
            :collapsed="collapsed"
            :collapsible="true" />
        <!-- 左侧导航栏收起展示效果 -->
        <div
            :style="`width: ${sideMenuWidth};
                min-width: ${sideMenuWidth};max-width: ${sideMenuWidth};`"
            class="virtual-side"></div>
        <a-layout class="admin-layout-main beauty-scroll">
            <!-- 右侧顶部导航栏 -->
            <admin-header :style="headerStyle"
                :collapsed="collapsed" @toggleCollapse="toggleCollapse"/>
            <a-layout-content class="admin-layout-content" :style="`min-height: ${minHeight}px;`">
                <div style="position: relative">
                    <!-- 右侧内容区域 -->
                    <page-view></page-view>
                </div>
            </a-layout-content>
            <!-- 右侧底部页脚 -->
            <a-layout-footer style="padding: 0px">
                <page-footer />
            </a-layout-footer>
        </a-layout>
    </a-layout>
</template>
```

布局结构一目了然，参见代码注释，涉及自定义的 4 个组件。

3. 主界面组件 style

在 MainLayout.vue 文件中添加 style 标签，设置左侧导航栏被收起时的样式和顶部导航样式，代码如下：

```less
<style lang="less" scoped>
    .admin-layout{
        .side-menu{
            &.fixed-side{
                position: fixed;
                height: 100vh;
                left: 0;
                top: 0;
            }
        }
        .virtual-side{
            transition: all 0.2s;
        }
        .admin-layout-main{
            .admin-header{
                top: 0;
                right: 0;
                overflow: hidden;
                transition: all 0.2s;
            }
        }
        .admin-layout-content{
            padding: 24px 24px 0;
        }
    }
</style>
```

其中 virtual-side 的宽度不固定，会随着用户单击展开、收起发生变动，遂放在组件数据中，动态变更。

4. 主界面组件 script

在 MainLayout.vue 文件中添加 script 标签，通过 import 方式引入 4 个组件：sideMenu、adminHeader、pageView 以及 pageFooter，并使用 components 属性注册为子组件，代码如下：

```
<script>
import AdminHeader from '@/components/header/header'
import PageFooter from '@/components/footer/footer'
```

```
import SideMenu from '@/components/menu/sideMenu'
import PageView from './pageView/PageView'
export default {
    name: 'MainLayout',
    components: { AdminHeader, SideMenu, PageFooter, PageView},
    data () {
        return {
            minHeight: window.innerHeight - 64 - 122,
            collapsed: false
        }
    },
    created(){
        console.log("MainLayout",this.$route.matched)
    },
    computed: {
        sideMenuWidth() {
            return this.collapsed ? '80px' : '256px'
        },
        headerStyle() {
            return `width: 100%; position:static`
        }
    },
    methods: {
        toggleCollapse () {
            this.collapsed = !this.collapsed
        },
        onMenuSelect () {
            this.toggleCollapse()
        }
    },
}
</script>
```

在主布局组件中使用两个属性：collapsed 和 minHeight，collapsed 用于存储顶部导航和左侧导航栏的值通信，当用户单击收起按钮时，触发收起效果并传递值给左侧导航组件；minHeight 用于动态计算右侧内容展示区域的大小，页面总高度减去顶部导航高度（64px）和底部页脚的高度（122px）即为展示区域高度。

5. 主界面组件引入

在后续小节中将介绍主界面布局所需的侧边导航组件 SideMenu、顶部导航组件 AdminHeader、内容组件 PageView 的开发，加上第 15 章提到的

PageFooter 组件，若已全部开发完毕，在路由配置文件 route/config.js 文件中引用 MainLayout 组件，在根路径下配置 component 为 MainLayout，后续所有路由页面都会使用该主布局组件渲染页面，配置文件中的引用代码如下：

```js
import MainLayout from '@/layouts/MainLayout.vue'
const routeConfig = [
    {
        path: '/',
        name: 'Home',
        component: MainLayout,
        redirect: '/dashboard/workspace',
        children:[
            {
                path: '/dashboard',
                name: 'Dashboard',
                meta:{
                    icon: 'dashboard'
                },
                children:[
                    {
                        path: '/workspace',
                        name: ' 工作台 ',
                        meta:{
                            title: " 工作台 "
                        },
                        component: () => import('@/views/dashboard/workspace')
                    }
                ]
            }
        ]
    }
]
```

介绍完主界面组件的代码、结构以及使用方式，接下来介绍 3 个子组件：SideMenu、AdminHeader 以及 PageView。

16.3 侧边导航组件 SideMenu

主界面组件 MainLayout 中引用的子组件 SideMenu 用于封装左侧菜单导航

栏，其中包含：
- 导航栏顶部展示项目名称以及 logo，单击该区域回到指定的页面，如 dashboard 面板页面。
- 渲染路由文件中配置的菜单，根据配置的层级展示对应的菜单和子菜单，单击子菜单使用 router-view 标签展示对应的页面。
- 根据登录用户的角色展示不同的菜单，在 BiFang 部署平台项目中分为两种角色：管理员和普通用户。管理员比普通用户多了用户模块的权限，可以添加用户，设置用户角色，这就是本章后续提到的用户模块。

在 src 下的 components 目录中创建 menu 文件夹，在 menu 文件夹中创建 sideMenu.vue 文件，该文件即为侧边导航组件。

1. 侧边导航组件 template

在侧边导航组件的 template 部分中，最外层使用 ant-design-vue Layout 布局中的 Sider 组件，内部嵌套两层，第一层是 class 为 logo 的 div 标签，第二层使用 Antd 中的 menu 组件，这两层分别表示布局顶部项目信息和菜单列表，代码如下：

```
<template>
    <a-layout-sider :class="['side-menu', 'beauty-scroll', 'shadow']"
        width="256px" :collapsible="collapsible"
        v-model="collapsed" :trigger="null">
        <div :class="['logo']">
            <router-link to="/dashboard">
                <img src="@/assets/logo.png">
                <h1>{{systemName}}</h1>
            </router-link>
        </div>
        <a-menu>
        </a-menu>
    </a-layout-sider>
</template>
```

在 div.logo 项目信息模块中使用第 14 章提到的 router-link 嵌套 img 和 h1 标签，当用户单击时触发路由变化，跳转至 dashboard 面板页面。

SideMenu 组件 template 中的 a-menu 标签中使用 Antd 框架中的示例编写菜单栏代码：

```
<a-menu theme="dark" mode="inline"
```

```
            @select="onSelect"
            :default-selected-keys="defaultKey"
            :open-keys.sync="openKey">
                <a-sub-menu v-for="menu in menuList"
                    v-if="isVisible(menu.meta.role||[])" :key="menu.path">
                    <span slot="title">
                        <a-icon :type="menu.meta.icon||'form'" />
                        <span>{{menu.name}}</span>
                    </span>
                    <a-menu-item
                        v-show="!(sub.meta&&(sub.meta.isShowMenu==false))"
                        v-for="sub in menu.children" :key="sub.path">
                        <router-link :to="menu.path+sub.path">
                            {{sub.name}}</router-link>
                    </a-menu-item>
                </a-sub-menu>
</a-menu>
```

遍历 menuList 数组，根据层级结构使用 a-sub-menu 以及 a-menu-item 生成菜单标签，menuList 是经过处理的路由配置，即处理项目中 router 目录下的 config 配置文件。

在项目中存在这样一种场景，某些特定页面由业务页面中的按钮触发，不需要在菜单导航中展示，这就需要在配置文件中设计该路由是否展示，所以我在路由的 meta 对象下添加 isShowMenu 属性，用于处理该场景。当 isShowMenu 不存在或值为 true 时，展示该路由菜单导航中；当值为 false 时，则不展示该路由信息。

针对该场景，路由配置文件中的配置参考如下：

```
{
    path: '/deployment',
    name: ' 部署 ',
    meta: {
        icon: 'cloud-upload'
    },
    children:[
        {
            path: '/deployList',
            name: ' 服务部署 ',
            component: () =>
                import('@/views/deployment/deployList')
        },
```

```
        {
            path: '/deploy',
            name: '发布单部署',
            meta:{
                isShowMenu:false
            },
            component: () => import('@/views/deployment/deploy')
        }
    ]
}
```

2. 侧边导航组件 style

在 sideMenu.vue 文件中添加 style 标签,设置左侧导航栏的样式,代码如下:

```
<style lang="less" scoped>
    .shadow{
        box-shadow: 2px 0 6px rgba(0, 21, 41, .35);
    }
    .side-menu{
        min-height: 100vh;
        overflow-y: auto;
        z-index: 10;
        .logo{
            height: 64px;
            position: relative;
            line-height: 64px;
            padding-left: 24px;
            -webkit-transition: all .3s;
            transition: all .3s;
            overflow: hidden;
            background-color: @bf-layout-trigger-background;
            &.light{
                background-color: #fff;
                h1{
                    color: @bf-primary-color;
                }
            }
            h1{
                color: #fff;
                font-size: 20px;
                margin: 0 0 0 12px;
                display: inline-block;
```

```
                vertical-align: middle;
            }
            img{
                width: 32px;
                vertical-align: middle;
            }
        }
    }
    .menu{
        padding: 16px 0;
    }
</style>
```

侧边导航组件所需的 DOM 元素以及样式已搭建完毕，接下来要实现侧边导航组件中的 script 部分。

3. 侧边导航组件 script

在 sideMenu.vue 文件中添加 script 标签，侧边导航组件中实现的事件包含：
- 根据用户角色判断是否展示菜单，引入 vuex 中的数据，针对用户角色判断菜单的展示。
- 菜单被单击时通知父组件，父组件监听菜单变更做相应的处理。

相关代码如下：

```
<script>
import {mapGetters} from 'vuex'
import {routeConfig} from '@/router/config'         // 引入路由配置文件
export default {
    name: 'SideMenu',
    data(){
        return {
            systemName:" 毕方部署平台 ",
            menuList:[],                              // 菜单列表
            defaultKey:['/createRelease'],            // 默认选中
            openKey:['/release']
        }
    },
    computed:{
        ...mapGetters({
            'userRole':'getRoles'
        }),
    },
    created(){
```

```
            let match = this.$route.path.slice(1).split('/');
            // 匹配链接中携带的路由数据
            if(match && match.length > 1){
                this.openKey = ["/"+match[0]]
                this.defaultKey = ["/"+match[1]]
            }
            let item = routeConfig.find(item=>item.path == '/')
            this.menuList = item?item.children:[]
        },
        props: {
            collapsible: {
                type: Boolean,
                required: false,
                default: false
            },
            collapsed: {
                type: Boolean,
                required: false,
                default: false
            },
            // todo-- 动态生成菜单
            // menuData: {
            //   type: Array,
            //   required: true
            // }
        },
        methods: {
            isVisible(roleList){
                if(roleList.length == 0) return true;
                let isVisible = false
                // 根据角色过滤
                this.userRole.forEach((item)=>{
                    roleList.indexOf(item.id) > -1?isVisible=true:""
                })
                return isVisible
            },
            onSelect (obj) {
                this.$emit('menuSelect', obj)           // 通知父组件
            }
        }
    }
</script>
```

针对用户菜单展示的方案，我认为有更好的方案：

- 用户登录时,根据用户角色在 Vuex 中匹配过滤路由配置,在 sideMenu 中通过属性传入。
- 服务端根据用户角色生成菜单路由数据发送给前端,根据服务端数据初始化菜单导航,相对来说,服务端方案更安全。

侧边导航组件中的实现方法可参见代码及备注,不再详细解释,接下来介绍顶部导航组件 AdminHeader。

16.4 顶部导航组件 AdminHeader

主界面组件 MainLayout 中引用的子组件 AdminHeader 封装顶部导航栏,其中包含:
- 左侧菜单栏展开收起图标,单击该图标触发菜单栏展开收起。
- 右侧展示用户名称和图标,单击用户图标出现下拉菜单,展示退出按钮。

在 src 下的 components 目录中创建 header 文件夹,在 header 文件夹中创建 header.vue 文件,该文件即为顶部导航组件。

1. 顶部导航组件 template

在顶部导航组件的 template 部分中,最外层使用 ant-design-vue Layout 布局中的 Header 组件,嵌套 a-icon 和以 class 为 admin-header-right 的 div 标签。在右侧 div 标签中使用 a-dropdown 布局用户信息以及下拉菜单。代码如下:

```html
<template>
    <a-layout-header :class="['light', 'admin-header']">
        <div :class="['admin-header-wide']">
            <a-icon class="trigger"
                :type="collapsed ? 'menu-unfold' : 'menu-fold'"
                @click="toggleCollapse"/>
            <div :class="['admin-header-right', 'dark']">
                <a-tooltip class="header-item"
                    title=" 帮助文档 " placement="bottom" >
                    <a href="https://iczer.gitee.io/vue-antd-admin-docs/"
                        target="_blank">
                        <a-icon type="question-circle-o" />
                    </a>
                </a-tooltip>
                <a-dropdown>
```

```html
            <div class="header-avatar" style="cursor: pointer">
                <a-avatar class="avatar" size="small"
                    shape="circle" :src="avatar"/>
                <span class="name">{{user.name}}</span>
            </div>
            <a-menu :class="['avatar-menu']" slot="overlay">
                <a-menu-item @click="logout">
                    <a-icon style="margin-right: 8px;" type="poweroff" />
                    <span> 退出登录 </span>
                </a-menu-item>
            </a-menu>
          </a-dropdown>
        </div>
      </div>
  </a-layout-header>
</template>
```

顶部导航组件布局较简单，主要使用 Antd 框架中的 a-layout-header、a-icon、a-tooltip、a-dropdown 以及 a-menu 组件，详细的使用方法可参见官网，在此不再赘述。

2. 顶部导航组件 style

在 header.vue 文件中添加 style 标签，设置顶部导航栏的样式，代码如下：

```less
<style lang="less" scoped>
    .admin-header{
        padding: 0;
        z-index: 2;
        box-shadow: @bf-shadow-down;
        position: relative;
        background: @bf-base-bg-color;
        .admin-header-wide{
            padding: 0 24px;
            &.head.fixed{
                max-width: 1400px;
                margin: auto;
                padding-left: 0;
            }
            &.side{
                padding-right: 12px;
            }
            .logo {
                height: 64px;
```

```
            line-height: 58px;
            vertical-align: top;
            display: inline-block;
            padding: 0 12px 0 24px;
            cursor: pointer;
            font-size: 20px;
            color: inherit;
            &.pc{
                padding: 0 12px 0 0;
            }
            img {
                vertical-align: middle;
            }
            h1{
                color: inherit;
                display: inline-block;
                font-size: 16px;
            }
        }
        .trigger {
            font-size: 20px;
            line-height: 64px;
            padding: 0 24px;
            cursor: pointer;
            transition: color .3s;
            &:hover{
                color: @primary-color;
            }
        }
        .admin-header-menu{
            display: inline-block;
        }
        .admin-header-right{
            float: right;
            display: flex;
            color: inherit;
            .header-item{
                color: inherit;
                padding: 0 12px;
                cursor: pointer;
                align-self: center;
                a{
                    color: inherit;
```

```
            i{
                font-size: 16px;
            }
          }
        }
      }
    }
  }
}
.header-avatar{
    display: inline-flex;
    .avatar, .name{
        align-self: center;
    }
    .avatar{
        margin-right: 8px;
    }
    .name{
        font-weight: 500;
    }
}
.avatar-menu{
    width: 150px;
}
</style>
```

顶部导航组件所需的 DOM 元素以及样式已搭建完毕，接下来要实现顶部导航组件中的 script 部分。

3. 顶部导航组件 script

在 header.vue 文件中添加 script 标签，顶部导航组件中实现的事件包含：

- 单击左侧 a-icon 图标，触发左侧导航栏的切换事件，通知父组件用 sideMenu 组件处理对应的事件。
- 单击用户名称，在出现的下拉菜单"退出登录"中绑定 logout 事件，清除登录返回的的 token 信息，并跳转至登录页面。

相关代码如下：

```
<script>
import {mapGetters} from 'vuex'
import {logout} from '@/utils/request'
export default {
```

```
        name: 'AdminHeader',
        props: ['collapsed'],
        data() {
            return {
                avatar: './asset/user.png',
                searchActive: false
            }
        },
        computed: {
            ...mapGetters({
                'user':'getUser'
            }),
        },
        methods: {
            toggleCollapse () {
                this.$emit('toggleCollapse')
            },
            logout() {
                logout()
                this.$router.push('/login')
            }
        }
    }
</script>
```

引入 Vuex 中的 getUser 方法获取用户信息，并绑定到导航右侧。引入 request.js 中的 logout 方法，在该方法中清理 cookie，代码如下：

```
/*** 登出 */
function logout(){
    Cookie.remove(authHeader);
}
```

顶部导航组件中的其他实现方法可参见代码以及备注，不再详细解释，接下来介绍内容区域组件 PageView。

16.5　内容组件 PageView

主界面组件 MainLayout 中引用的子组件 PageView 封装内容区域，它是加载业务页面的载体，在其中使用第 14 章中提到的 router-view 加载路由配置的子

业务组件。其中包含:
- 顶部使用面包屑组件 a-breadcrumb 展示页面路径,下方展示加载的业务页面名称。
- 下方使用 router-view 加载路由配置的子业务组件。

在 src 下的 components 目录中创建 page 文件夹,在 page 文件夹中创建 page.vue 文件,该文件即为顶部导航组件。

1. 内容组件 template

在内容组件的 template 标签中,最外层创建以 page-layout 为 class 类名的 div 标签,嵌套两个 div 标签,class 类名分别为 page-header 以及 page-content,在 div.page-content 中使用 router-view 标签加载子业务组件。代码如下:

```html
<template>
    <div class="page-layout">
        <div :class="['page-header', layout]">
            <div class="page-header-wide">
                <div class="breadcrumb">
                    <a-breadcrumb>
                        <a-breadcrumb-item
                            :key="index" v-for="(item, index) in breadcrumb">
                            <span>{{item}}</span>
                        </a-breadcrumb-item>
                    </a-breadcrumb>
                </div>
            </div>
        </div>
        <div ref="page" :class="['page-content', layout]" >
            <router-view ref="page" />
        </div>
    </div>
</template>
```

内容组件布局使用 Antd 框架中的 a-breadcrumb 组件,其他使用原生 DOM 标签构建,在此不再赘述。

2. 内容组件 style

在 page.vue 文件中添加 style 标签,设置内容组件的样式,代码如下:

```less
<style lang="less" scoped>
    .page-header{
        background: @bf-base-bg-color;
```

```
                padding: 16px 24px;
                &.head.fixed{
                    margin: auto;
                    max-width: 1400px;
                }
            }
            .page-header-wide{
                .breadcrumb{
                    margin-bottom: 20px;
                }
            }
        }
        .page-content{
        position: relative;
        padding: 24px 0 0;
        &.head.fixed{
            margin: 0 auto;
            max-width: 1400px;
        }
    }
}
</style>
```

内容组件 page.vue 所需的 DOM 元素以及样式已搭建完毕，接下来实现区域组件中的 script 部分。

3. 内容组件 script

在 page.vue 文件中添加 script 标签，内容组件中解析 $route 对象中传入的路由对象，从中获取当前路由组件名称以及父路由组件名称。代码如下：

```
<script>
export default {
    name: 'PageView',
    data(){
        return {
            layout:'side'
        }
    },
    computed: {
        breadcrumb(){
            const list = [" 首页 "]
            if(this.$route.matched.length >1){
                list.push(this.$route.matched[1].meta.parent.name)
                list.push(this.$route.matched[1].name)
            }
```

```
                return list
        },
        title(){
                return this.$route.matched.length>1?
(this.$route.matched[1].meta.title ||
this.$route.matched[1].name):" 标题 "
        }
    }
}
</script>
```

内容组件 pageView 中的实现方法可参见代码以及备注，不再详细解释。

16.6　管理员模块

介绍完主界面布局，接下来正式进入业务组件的开发。本书开篇已介绍管理员模块，主要提供用户组和用户增删查改的管理功能。

管理员模块是只有具有管理权限用户才能看到的功能，用于为 BiFang 部署平台增加用户并分配对应权限，用户本身是没有权限属性的，只有分配给特定的角色（即用户组），才能拥有权限，这是在大型系统设计时遵循的一项规则。

1. 功能说明与清单

（1）管理员模块功能清单如表 16-1 所列。

表 16-1　管理员模块功能表

模块名称	功能名称	所属页面	功能描述
管理员	用户组管理	用户组管理页面	新增用户组 更改组名 删除用户组 搜索用户组
	用户管理	用户管理页面	新增用户 修改用户所属用户组 删除用户

（2）管理员模块功能说明。

① 只有在 admin 用户组中的用户，登录之后才能看到这个侧边管理菜单。

② 要所有用户均退出一个用户组后，才能删除这个用户组。

③ 用户组权限分菜单权限和操作权限，菜单权限用于管理用户可看到的左侧菜单栏；操作权限用于是否具有增删改查操作权限。

（3）管理员模块截图如图 16-5 ～图 16-7 所示。

图 16-5　BiFang 用户组管理

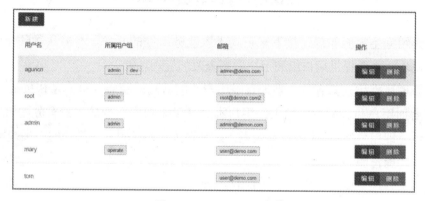

图 16-6　BiFang 用户管理

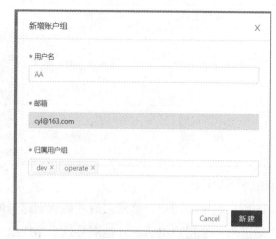

图 16-7　BiFang 新增用户

2. 功能技术要点

在管理员模块用户组和用户页面搭建中，本书主要使用 Antd 框架中的组件进行搭建，详细功能点使用的组件如下：

- 查询操作：使用 a-form 表单、a-table 表格、a-pagination 分页组件。
- 新增操作：使用 a-modal 对话框、a-form 表单。
- 编辑操作：使用 a-button 按钮。
- 删除操作：使用 a-button 按钮、a-popconfirm 确认框。

Antd 框架设计的组件开放很多属性供开发者灵活配置，接下来介绍上述使用的组件常用属性，以使读者能清晰地知道各属性的作用。

Form 表单是具有数据收集、校验和提交功能的表单，包含复选框、单选框、输入框、下拉选择框等元素，常用的 API 如下：

- layout：表单元素排列方式，水平排列（vertical）、垂直排列（horizontal，默认）、行内排列（inline）。
- labelAlign：label 标签的文本对齐方式，左对齐（left）、右对齐（right）。
- labelCol：label 标签布局，设置 span offset 值，如 {span: 3, offset:12}。
- wrapperCol：需要为输入控件设置布局样式时，使用该属性，用法同 labelCol。
- getFieldsValue：获取一组输入控件的值，如不传入参数，则获取全部组件的值。
- setFieldsValue：设置一组输入控件的值。
- validateFields：校验并获取一组输入域的值与 error，若 fieldNames 参数为空，则校验全部组件。

Table 表格用于当需要对数据进行排序、搜索、分页、自定义操作等复杂行为时，展示行列数据，常用的 API 如下：

- dataSource：表格数据源，对应服务端返回的数据。
- columns：表格列的配置，Object 类型数组。
- title：列头显示文字。
- align：内容的对齐方式，'left' | 'right' | 'center'。
- dataIndex：列数据在数据项中对应的 key。
- customRender：生成复杂数据的渲染函数，参数分别为当前行的值，当前行数据，行索引，@return 里面可以设置表格行/列合并。

- scopedSlots：使用 columns 时，可以通过该属性配置支持 slot-scope 的属性，如 scopedSlots: { customRender: 'XXX'}。
- loading：页面是否加载中。
- pagination：分页器，参考配置项或 pagination 文档，设为 false 时不展示和进行分页。
- rowKey：表格行 key 的取值，可以是字符串或一个函数。
- change：分页、排序、筛选变化时触发。

Pagination 采用分页的形式分隔长列表，每次只加载一个页面，常用的 API 如下：

- current：当前页数。
- pageSize：每页条数。
- total：数据总数。
- showSizeChanger：是否可以改变 pageSize。
- showQuickJumper：是否可以快速跳转至某页。
- showSizeChange：pageSize 变化的回调。

Button 按钮标记了一个（或封装一组）操作命令，响应用户单击行为，触发相应的业务逻辑，常用的 API 如下：

- disabled：按钮失效状态。
- icon：设置按钮的图标类型。
- loading：设置按钮载入状态。
- type：设置按钮类型，可选值为 primary dashed danger link。
- click：单击按钮时的回调。

Modal 模态对话框在当前页面正中打开一个浮层，承载用户处理事务，又不希望跳转页面以致打断工作流程，常用 API 如下：

- cancelText：取消按钮文字。
- closable：是否显示右上角的关闭按钮。
- destroyOnClose：关闭时销毁 Modal 里的子元素 footer 底部内容，当不需要默认底部按钮时，可以设为 ":footer=\"null\""。
- mask：是否展示遮罩 maskClosable，单击蒙层是否允许关闭。
- okText：确认按钮文字。
- okType：确认按钮类型。

- title：标题。
- width：宽度。

Popconfirm 气泡框在目标元素附近弹出浮层提示，询问用户进一步确认操作，常用 API 如下：

- cancelText：取消按钮文字。
- okText：确认按钮文字。
- title：确认框的描述。
- confirm：单击确认的回调。
- cancel：单击取消的回调。

在实践代码之前建议读者阅读 Antd 框架官网的样例，这样在正式开发时就能知道哪些功能可以使用组件实现。话不多说，直接进入用户组和用户页面的搭建。

16.7 用户组页面搭建

用户组页面包含查询用户组列表、新增用户组、删除用户组以及修改用户组名称，这里将以实现每个功能的步骤来呈现。

1. 查询列表

用户组列表包含组名称列、链接列以及操作列，其中操作列包含编辑和删除两个按钮，单击按钮执行对应的操作。同时列表顶部包含过滤条件，根据用户输入的组名称和创建事件进行过滤。

在 src 中的 views 目录下创建 account 目录，并在该目录下创建 group.vue 文件，此文件为用户组组件。具体的 template 代码如下：

```
<template>
    <a-card>
        <div class="search">
            <a-form layout="inline" :form="form"
                @submit="submitHandler">
                <a-form-item label=" 组名称 " >
                    <a-input
                        placeholder=" 请输入组名称 "
                        v-decorator="['name', { rules: [
                            { required: false, message: ' 请输入组名称 !' }] }]" />
```

```html
                    </a-form-item>
                    <a-form-item label=" 时间 ">
                        <a-range-picker
                        v-decorator="['timePicker', {rules: [
                            { type: 'array', required: false }]}]" />
                    </a-form-item>
                    <a-form-item>
                        <a-button type="primary" html-type="submit">
                            查询
                        </a-button>
                    </a-form-item>
                    <a-form-item>
                        <a-button type="primary" @click.prevent="createDialog">
                            新建
                        </a-button>
                    </a-form-item>
                </a-form>
            </div>
            <div>
                <a-table
                    :columns="columns"
                    :dataSource="dataSource"
                    rowKey="name"
                    @clear="onClear"
                    @change="onChange"
                    :pagination="{
                        current: params.currentPage,
                        pageSize: params.pageSize,
                        total: total,
                        showSizeChanger: true,
                        showLessItems: true,
                        showQuickJumper: true,
                        showTotal: (total, range) => `第 ${range[0]}-${range[1]} 条，总计 ${total} 条`
                    }">
                    <div slot="action" slot-scope="text, record">
                        <a-button-group>
                            <a-button type="primary" @click.prevent="onShowEdit(record)">
                                编辑
                            </a-button>
                            <a-popconfirm
                                title=" 确定执行删除操作么？"
                                ok-text=" 是 "
                                cancel-text=" 否 "
```

```
                            @confirm="onDelete(record)"
                        >
                            <a-button type="danger">
                                删除
                            </a-button>
                        </a-popconfirm>
                    </a-button-group>
                </div>
            </a-table>
        </div>
    </a-card>
</template>
```

用户组组件以 a-card 作为根节点，它是通用的卡片容器，其中内嵌 a-form 和 a-table 组件，在 a-table 组件中嵌套了 slot 为 action 的 template 模板，在 columns 中使用 scopedSlots: { customRender: 'action' } 可指定该列采用模板方式渲染，其他详细的属性已在技术要点中介绍，在此就不再赘述。

在 group.vue 文件中添加 script 标签，用户组组件中实现的事件包含：

- fetchData：请求接口获取用户组列表信息。
- onReset：重置顶部 form 表单数据。
- onChange：用户单击分页或表格个数 pageSize 变更时重新请求数据。
- submitHandler：用户单击查询按钮根据 form 表单输入的条件查询用户组列表信息。

相关代码如下：

```
<script>
import API from '@/service'
const columns = [
    {
        title: ' 组名称 ',
        dataIndex: 'name'
    },
    {
        title: ' 链接 ',
        dataIndex: 'url'
    },
    {
        title: ' 操作 ',
        scopedSlots: { customRender: 'action' }
    }
```

```js
    ]
    export default {
        name: 'group',
            data () {
                return {
                    total:0, // 数据总数
                    columns: columns,           // 表格列数据
                    dataSource: [],             // 数据源
                    visible:false,              // 弹出框是否展示
                    isEdit:false,               // 是否编辑模式
                    params:{                    // 分页数据
                        name:"",
                        currentPage:1,
                        pageSize:20,
                        begin_time:"",
                        end_time:"",
                        sorter:""
                    }
                }
            },
            beforeCreate() {
                this.form = this.$form.createForm(this, { name: 'groupList' });
                this.formDialog = this.$form.createForm(this, { name: 'formDialog' });
            },
            created(){
                this.fetchData()
            },
            methods: {
                submitHandler(e){
                    e.preventDefault()
                    this.form.validateFields((err, fieldsValue) => {
                        if (err) {
                            return;
                        }
                        const rangeValue = fieldsValue['timePicker']
                        this.params.name = fieldsValue["name"]
                        this.params.begin_time =
                            rangeValue?rangeValue[0].format("YYYY-MM-DD"):""
                        this.params.end_time =
                            rangeValue?rangeValue[1].format("YYYY-MM-DD"):""
                        this.fetchData()
                    });
```

```
        },
        fetchData(){
            API.GroupList(this.params).then((res)=>{
                let result = res.data
                if(res.status == 200 ){
                    this.total = result.count
                    this.dataSource = result.results;
                }
                else{
                    this.dataSource = []
                }
            })
        },
        onReset(){
            this.visible = false
            this.isEdit = false
            this.formDialog.resetFields()
        },
        onChange(pagination, filters, sorter) {
            let { current, pageSize } = pagination
            let { order, field } = sorter
            this.params.currentPage = current
            this.params.pageSize = pageSize
            this.params.sorter = (field?field:"")
            this.fetchData()
        }
    }
}
</script>
```

值得一提的是，在 group 用户组组件创建前使用 this.$form.createForm 收集 form 表单对象供后续做表单校验、获取表单内组件数据以及表单内部组件的赋值操作。API.GroupList 方法为 service 服务配置的请求用户组列表的 request 请求，在 service 服务配置中介绍具体的实现。

2. 编辑新增数据

用户组数据的新增和编辑采用模态框的形式实现，其中嵌套 form 表单、组名称输入框以及隐藏的 id 输入框，用于在编辑时自动提交对应的用户组 id。在 template 中新增代码如下：

```
<a-modal
    :visible="visible"
```

```
        :title="isEdit?' 编辑账户组 ':' 新增账户组 '"
        :okText="isEdit?' 更新 ':' 新建 '"
        @cancel="onReset"
        @ok="isEdit?onUpdateGroup():onCreateGroup()"
    >
        <a-form layout='vertical' :form="formDialog">
            <a-form-item
                label="ID"
                v-show="false"
            >
                <a-input placeholder=" 自动生成 "
                    v-decorator="['id']"
                />
            </a-form-item>
            <a-form-item label=' 组名称 '>
                <a-input
                    v-decorator="[
                        'groupName',
                        {
                            rules: [{ required: true, message: ' 请输入组名称 ' }],
                        }
                    ]"
                />
            </a-form-item>
        </a-form>
    </a-modal>
```

模态框使用 a-modal 组件，其中内嵌 a-form 和 a-input 组件，详细的属性已在技术要点中介绍，在此不再赘述。

在 script 标签中新增事件包含：

- onUpdateGroup：请求接口更新单个用户组信息。
- onCreateGroup：请求接口新增用户组信息。
- onShowEdit：控制模态框变更为编辑模式并使用行数据初始化表单。
- createDialog：控制模态框变更为新增模式。

新增的事件代码如下：

```
<script>
    ... // 其他忽略
    methods: {
        ...// 其他忽略
        onShowEdit(record){
```

```js
            this.visible = true
            this.isEdit = true
            this.$nextTick(()=>{
                this.formDialog.setFieldsValue({
                    id:record.id,
                    groupName:record.name
                })
            })
        },
        onUpdateGroup(){
            this.formDialog.validateFields((err, fieldsValue) => {
                if (err) {
                    return;
                }
                let id = fieldsValue["id"],
                    name = fieldsValue["groupName"];
                let data = {
                    id, name
                }
                API.UpdateGroup(data).then((res)=>{
                    let result = res.data
                    if(res.status == 200){
                        this.$message.success(" 用户组更新成功 ~")
                        this.onReset()
                        this.fetchData()
                    }
                    else{
                        this.$message.error(" 用户组更新失败 :"+result.message)
                    }
                })
            })
        },
        createDialog(){
            this.visible = true
        },
        onCreateGroup () {
            this.formDialog.validateFields((err, fieldsValue) => {
                if (err) {
                    return;
                }
                let name = fieldsValue["groupName"];
                let data = {
                    name
```

```
            }
            API.CreateGroup(data).then((res)=>{
                let result = res.data
                if(res.status == 201){
                    this.$message.success(" 用户组新增成功 ~")
                    this.visible = false
                    this.fetchData()
                }
                else{
                    this.$message.error(" 用户组新增失败 :"+result.message)
                }
            })
        })
    }
}
</script>
```

3. 删除数据

在 a-table 操作列中使用模板渲染了删除按钮，单击按钮可通过 slot-scope 获取当前行数据 record，在单击删除按钮时，可通过该变量获取对应的行数据的 id 进行删除操作。在 script 中添加删除事件 onDelete，代码如下：

```
<script>
    ... // 其他忽略
    methods: {
        ... // 其他忽略
        onDelete(record){
            let {id} = record;
            if(id == undefined){
                this.$message.error(" 操作参数非法！ ")
                return false
            }
            API.DeleteGroup({id}).then((res)=>{
                let result = res.data
                if(res.status == 204){
                    this.$message.success(" 删除成功 ~")
                    this.fetchData()
                }
                else{
                    this.$message.error(" 操作错误 ~:"+result.message)
                }
            })
        },
```

```
        }
    }
</script>
```

至此用户组组件已全部实现完毕，method 方法中调用的 API 函数将在 service 服务配置中介绍。

16.8 用户页面搭建

用户页面包含查询用户列表、新增用户、删除用户以及修改用户信息，这里以实现每个功能的步骤来呈现。

1. 查询列表

用户列表包含用户名列、所属用户组列、用户邮箱列以及操作列，其中操作列包含编辑和删除两个按钮，单击按钮执行对应的操作。在 account 目录下创建 user.vue 文件，此文件为用户组件。具体的 template 代码如下：

```
<template>
    <a-card>
        <div class="search">
            <a-form layout="inline" :form="form" >
                <a-form-item>
                    <a-button type="primary"
@click.prevent="createDialog">
                        新建
                    </a-button>
                </a-form-item>
            </a-form>
        </div>
        <div>
            <a-table
                :columns="columns"
                :dataSource="dataSource"
                rowKey="name"
                @change="onChange"
                :pagination="{
                    current: params.currentPage,
                    pageSize: params.pageSize,
                    total: total,
                    showSizeChanger: true,
```

```html
                    showLessItems: true,
                    showQuickJumper: true,
                    showTotal: (total, range) => `第 ${range[0]}-${range[1]} 条，总计 ${total} 条`
                }"
            >
                <div slot="username" slot-scope="text, record">
                    <a-tooltip>
                        <template slot="title">
                            {{record.url}}
                        </template>
                            {{text}}
                    </a-tooltip>
                </div>
                <template slot="groups_names" slot-scope="text,record">
                    <a-tag color="blue" v-for="item in text">
                        {{item.name}}
                    </a-tag>
                </template>
                <template slot="email" slot-scope="text,record">
                    <a-tag color='blue'>{{text}}</a-tag>
                </template>
                <div slot="action" slot-scope="{text, record}">
                    <a-button-group>
                        <a-button type="primary" @click.prevent="onShowEdit(record)">
                            编辑
                        </a-button>
                        <a-popconfirm
                            title=" 确定执行删除操作么？"
                            ok-text=" 是 "
                            cancel-text=" 否 "
                            @confirm="onDelete(record)"
                        >
                            <a-button type="danger">
                                删除
                            </a-button>
                        </a-popconfirm>
                    </a-button-group>
                </div>
            </a-table>
        </div>
    </a-card>
</template>
```

用户组件使用通用的卡片容器 a-card 作为根节点，其中内嵌 a-form 和

a-table 组件，在 a-table 组件中嵌套了 slot 名为 username、groups_names、email 和 action 的 template 模板，分别用于渲染用户名列、所属用户组列、邮箱列以及操作列，使用的 Antd 组件的详细属性已在技术要点中介绍，在此不再赘述。

在 user.vue 文件中添加 script 标签，用户组件中实现的事件包含：
- fetchData：请求接口获取用户列表信息。
- fetchGroupList：请求用户组列表，用于分配用户的归属组。
- onReset：重置 form 表单数据以及控制开关。
- onChange：用户单击分页或表格个数 pageSize 变更时触发重新请求数据。

相关代码如下：

```
<script>
import API from '@/service'
const columns = [
    {
        title: ' 用户名 ',
        dataIndex: 'username',
        scopedSlots: {customRender: 'username'}
    },
    {
        title: ' 所属用户组 ',
        dataIndex: 'groups_names',
        scopedSlots: {customRender: 'groups_names'}
    },
    {
        title: ' 邮箱 ',
        dataIndex: 'email',
        scopedSlots: { customRender: 'email' }
    },
    {
        title: ' 操作 ',
        scopedSlots: { customRender: 'action' }
    }
]

export default {
    name: 'user',
    data () {
        return {
            total:0,
            columns: columns,
```

```
                    dataSource: [],
                    visible:false,
                    options:[],
                    isEdit:false,
                    params:{
                        name:"",
                        currentPage:1,
                        pageSize:20,
                        begin_time:"",
                        end_time:"",
                        sorter:""
                    }

            }
        },
        beforeCreate() {
            this.formDialog = this.$form.createForm(this, { name: 'formDialog' });
        },
        created(){
            this.fetchData()
            this.fetchGroupList()
        },
        methods: {
            fetchData(){
                API.UserList(this.params).then((res)=>{
                    let result = res.data
                    if(res.status == 200 ){
                        this.total = result.count
                        this.dataSource = result.results;
                        console.log(this.dataSource)
                    }
                    else{
                        this.dataSource = []
                    }
                })
            },
            onReset(){
                this.visible = false
                this.isEdit = false
                this.formDialog.resetFields()
            },
            onChange(pagination, filters, sorter) {
```

```
        let {
            current, pageSize
        } = pagination
        let {
            order,
            field
        } = sorter
        this.params.currentPage = current
        this.params.pageSize = pageSize
        this.params.sorter = (field?field:"")
        this.fetchData()
    },
    fetchGroupList(){
        API.GroupList({}).then((res)=>{
            let result = res.data
            if(res.status == 200){
                result.results.forEach(item=>{
                    this.options.push({
                        label:item.name,
                        value:item.id
                    })
                })
            }
            else{
                this.$message.error(" 无法获取用户组列表 ~")
            }
        })
    }
}
</script>
```

在 user 用户组件创建前使用 this.$form.createForm 收集 form 表单对象，供后续做表单校验、获取表单内组件数据以及表单内部组件的赋值操作。

2. 编辑新增数据

用户数据的新增和编辑采用模态框的形式实现，其中嵌套 form 表单、用户名称输入框、邮箱输入框、用户组多选列表以及隐藏的 id 输入框，用于在编辑时自动提交对应的用户 id。在 template 中新增代码如下：

```
<a-modal
    :visible="visible"
    :title="isEdit?' 编辑账户组 ':' 新增账户组 '"
```

```html
        :okText="isEdit?' 更新 ':' 新建 '"
        @cancel="onReset"
        @ok="isEdit?onUpdateUser():onCreateUser()"
>

<a-form layout='vertical' :form="formDialog">
    <a-form-item
        label="ID"
        v-show="false"
    >
        <a-input placeholder=" 自动生成 "
            v-decorator="['id']"
        />
    </a-form-item>
    <a-form-item label=' 用户名 '>
        <a-input
            v-decorator="[
                'username',
                {
                    rules: [{ required: true, message: ' 请输入用户名 ' }],
                }
            ]"
        />
    </a-form-item>
    <a-form-item label=' 邮箱 '>
        <a-input
            v-decorator="[
                'email',
                {
                    rules: [{ required: true, message: ' 请输入邮箱 ' }],
                }
            ]"
        />
    </a-form-item>
    <a-form-item
        label=" 归属用户组 "
    >
    <a-select
        show-search
        mode="multiple"
        placeholder=" 请选择归属用户组 "
        option-filter-prop="children"
        style="min-width: 200px;width:100%"
        :filter-option="filterOption"
```

```
                v-decorator="['groups', { rules: [
{ required: true, message: ' 请选择归属用户组 !' }] }]"
                >
                    <a-select-option v-for="d in options" :key="d.value">
                        {{ d.label }}
                    </a-select-option>
                </a-select>
            </a-form-item>
        </a-form>
    </a-modal>
```

模态框使用 a-modal 组件，其中内嵌 a-form、a-input 以及 a-select 组件，详细的属性已在技术要点中介绍，在此不再赘述。

在 script 标签中新增事件如下：

- onUpdateUser：请求接口更新单个用户组信息。
- onCreateUser：请求接口新增用户组信息。
- onShowEdit：控制模态框变更为编辑模式并使用行数据初始化表单。
- createDialog：控制模态框变更为新增模式。

新增的事件代码如下：

```
<script>
... // 其他忽略
    methods: {
        ...// 其他忽略
        onShowEdit(record){
            this.visible = true
            this.isEdit = true
            this.$nextTick(()=>{
                this.formDialog.setFieldsValue(record)
            })
        },
        onUpdateUser(){
            this.formDialog.validateFields((err, fieldsValue) => {
                if (err) {
                    return;
                }
                let id = fieldsValue["id"],
                    username = fieldsValue["username"],
                    email = fieldsValue["email"],
                    groups = fieldsValue["groups"];
                let data = {
```

```
                    id,
                    username,
                    email,
                    groups
                }
                API.UpdateUser(data).then((res)=>{
                    let result = res.data
                    if(res.status == 200){
                        this.$message.success(" 用户组更新成功 ~")
                        this.onReset()
                        this.fetchData()
                    }
                    else{
                        this.$message.error(" 用户组更新失败 :"+result.message)
                    }
                })
            })
        },
        createDialog(){
            this.visible = true
        },
        onCreateUser () {
            this.formDialog.validateFields((err, fieldsValue) => {
                if (err) {
                    return;
                }
                let data = fieldsValue;
                API.CreateUser(data).then((res)=>{
                    let result = res.data
                    if(res.status == 201){
                        this.$message.success(" 用户组新增成功 ~")
                        this.visible = false
                        this.fetchData()
                    }
                    else{
                        this.$message.error(" 用户组新增失败 :"+result.message)
                    }
                })
            })
        }
    }
}
</script>
```

3. 删除数据

在 a-table 操作列中使用 action 模板渲染了删除按钮，单击按钮可通过 slot-scope 获取当前行数据 record，在单击删除按钮时，可通过该变量获取对应的行数据的 id 进行删除操作。在 script 中添加删除事件 onDelete，代码如下：

```
<script>
    ... // 其他忽略
    methods: {
        ... // 其他忽略
        onDelete(record){
            let {id} = record;
            if(id == undefined){
                this.$message.error(" 操作参数非法！")
                return false
            }
            API.DeleteUser({id}).then((res)=>{
                let result = res.data
                if(res.status == 204){
                    this.$message.success(" 删除成功 ~")
                    this.fetchData()
                }
                else{
                    this.$message.error(" 操作错误 ~:"+result.message)
                }
            })
        },
    }
}
</script>
```

至此，用户组件已全部实现完毕，接下来增加配置路由以访问管理员模块、在 service 服务配置 method 方法中调用的 API 函数以及配置 mock 服务模拟接口数据。

16.9 管理员基础服务配置

管理员模块用户组以及用户页面开发完成后，就需要在浏览器中测试开发的页面展示以及功能是否有问题，即进入开发本地测试阶段。这时需要用到第 14 章介绍的路由配置、service 服务以及 mock 服务。

1. 路由配置

根据第 14 章的介绍，相信读者已了解 Vue router 的作用以及 Vue CLI 工具集创建的默认路由配置，对此还不清楚的读者可再回顾一下。若要在浏览器能够展示登录页面，只需要把用户组和用户页面的组件配置到 router 目录下的 config.js 文件中，代码如下：

```
const routeConfig = [
    {
        path: '/account',
        name: '管理',
        meta: {
            icon: 'appstore',
            role: ['admin']
        },
        children:[
            {
                path: '/group',
                name: '用户组',
                component: () => import('@/views/account/group')
            },
            {
                path: '/user',
                name: '用户',
                component: () => import('@/views/account/user')
            },
        ]
    }
]
```

通过在浏览器地址中添加 hash 值为 /account/group 和 /account/user，即访问 http://localhost:8080/#/account/group 和 http://localhost:8080/#/account/user，便可以在浏览器中看到开发完的用户组页面和用户页面。

2. mock 服务配置

项目中的本地 mock 服务用于模拟服务端请求，并返回自定义的模拟数据。用户和用户组页面分别请求四个服务端接口：

- 查询列表：GroupList（用户组列表）、UserList（用户列表）。
- 新增数据：CreateGroup（用户组）、CreateUser（用户）。
- 编辑数据：UpdateGroup（用户组）、UpdateUser（用户）。
- 删除数据：DeleteGroup（用户组）、DeleteUser（用户）。

所以需要在 mock 目录下新增 account 目录，目录中新增 index.js 文件，文件内容如下：

```js
//todo
import Mock from 'mockjs'
const groupList = Mock.mock({
    "count":"@integer(1, 100)",
    "next":null,
    "previous":null,
    "results|100":[
        {
            "url":"@url",
            "name":"@name",
            "id":"@id"
        }
    ]
})

Mock.mock(`${process.env.VUE_APP_API_BASE_URL}/account/groups/`, 'get', ({body}) => {
    return {
        "code": 0,
        "message": "success",
        "data": groupList
    }
})
const editGroup = Mock.mock({
    "url":"@url",
    "name":"@name",
    "id":"@id"
})

Mock.mock(`${process.env.VUE_APP_API_BASE_URL}/account/groups/`, 'put', ({body}) => {
    return {
        "code": 0,
        "message": "success",
        "data": editGroup
    }
})

const addGroup = Mock.mock({
    "url":"@url",
    "name":"@name",
    "id":"@id"
```

```js
    })

    Mock.mock(`${process.env.VUE_APP_API_BASE_URL}/account/groups/`, 'post', ({body}) => {
        return {
            "code": 0,
            "message": "success",
            "data": addGroup
        }
    })

    Mock.mock(`${process.env.VUE_APP_API_BASE_URL}/account/groups/`, 'delete', ({body}) => {
        return {
            "code": 0,
            "message": "success",
            "data": {}
        }
    })

    const accountList = Mock.mock({
        "count":"@integer(1, 100)",
        "results":[
            {
                "id":"@id",
                "url":"@url",
                "username":"@name",
                "email":"@email",
                "groups":"@range(1,2)",
                "groups_names"|1-5:[
                    {
                        "id":"@id",
                        "name":"@name"
                    }
                ]
            }
        ]
    })

    Mock.mock(`${process.env.VUE_APP_API_BASE_URL}/account/users/`, 'get', ({body}) => {
        return {
            "code": 0,
            "message": "success",
            "data": accountList
        }
```

```js
})

const addAccount = Mock.mock({
    "id":"@id",
    "url":"@url",
    "username":"@name",
    "email":"@email",
    "groups":"@range(1,2)",
    "groups_names"|1-5:[
        {
            "id":"@id",
            "name":"@name"
        }
    ]
})

Mock.mock(`${process.env.VUE_APP_API_BASE_URL}/account/users/`, 'post', ({body}) => {
    return {
        "code": 0,
        "message": "success",
        "data": addAccount
    }
})

const editAccount = Mock.mock({
    "id":"@id",
    "url":"@url",
    "username":"@name",
    "email":"@email",
    "groups":"@range(1,2)",
    "groups_names"|1-5:[
        {
            "id":"@id",
            "name":"@name"
        }
    ]
})

Mock.mock(`${process.env.VUE_APP_API_BASE_URL}/account/users/`, 'put', ({body}) => {
    return {
        "code": 0,
        "message": "success",
        "data": editAccount
```

```
        }
    })

    Mock.mock(`${process.env.VUE_APP_API_BASE_URL}/account/users/`, 'delete', ({body}) => {
        return {
            "code": 0,
            "message": "success",
            "data": {}
        }
    })
```

在 mock 服务的入口文件 index.js 中引入 account 目录，即完成管理员模块 mock 服务的注册。当管理员页面访问 /account/groups/ 和 /account/users/ 两个路径时，mock 服务层会返回配置的数据。管理员页面访问 mock 服务层的接口需要通过 service 层，所以还需要在 service 层配置管理员模块。

3. service 服务配置

service 服务统一封装了 BiFang 部署平台项目所需要的所有接口路径以及请求方式，并通过 axios 调用下游的 mock 数据，所以 service 服务需要新增管理员模块。在 services 目录下的 api.js 路径集中新增管理员模块的访问路径，代码如下：

```
const API = {
    GROUPLIST: `${BASE_URL}/account/groups/`,
    CREATEGROUP:`${BASE_URL}/account/groups/`,
    DELETEGROUP: `${BASE_URL}/account/groups/{{id}}/`,
    UPDATEGROUP: `${BASE_URL}/account/groups/{{id}}/`,
    USERLIST: `${BASE_URL}/account/users/`,
    CREATEUSER:`${BASE_URL}/account/users/`,
    DELETEUSER: `${BASE_URL}/account/users/{{id}}/`,
    UPDATEUSER: `${BASE_URL}/account/users/{{id}}/`,
}
```

在 services 目录下新增的 account.js 文件，用于配置管理员组件所需服务，代码如下：

```
import {request,METHOD} from '@/utils/request'
import {urlFormat} from '@/utils/util'
import {API} from './api'
/**
 * 获取账户组列表
 * @param {*} params
```

```javascript
 */
async function GroupList(params){
    return request(
        API.GROUPLIST,
        METHOD.GET,
        params
    )
}
/**
 * 新增账户组
 * @param {*} params
 */
async function CreateGroup(params){
    return request(
        API.CREATEGROUP,
        METHOD.POST,
        params
    )
}
/**
 * 删除服务器
 * @param {*} params
 */
async function DeleteGroup(params){
    const url = urlFormat(API.DELETEGROUP, params)
    return request(
        url,
        METHOD.DELETE
    )
}
/**
 * 更新服务器
 * @param {*} params
 */
async function UpdateGroup(params){
    const url = urlFormat(API.UPDATEGROUP, params)
    return request(
        url,
        METHOD.PUT,
        params
    )
}
/**
```

```javascript
 * 获取账户组列表
 * @param {*} params
 */
async function UserList(params){
    return request(
        API.USERLIST,
        METHOD.GET,
        params
    )
}
/**
 * 新增账户组
 * @param {*} params
 */
async function CreateUser(params){
    return request(
        API.CREATEUSER,
        METHOD.POST,
        params
    )
}
/**
 * 删除服务器
 * @param {*} params
 */
async function DeleteUser(params){
    const url = urlFormat(API.DELETEUSER, params)
    return request(
        url,
        METHOD.DELETE
    )
}
/**
 * 更新服务器
 * @param {*} params
 */
async function UpdateUser(params){
    const url = urlFormat(API.UPDATEUSER, params)
    return request(
        url,
        METHOD.PUT,
        params
    )
```

```
}
export default {
    GroupList,
    CreateGroup,
    UpdateGroup,
    DeleteGroup,
    UserList,
    CreateUser,
    UpdateUser,
    DeleteUser
}
```

管理员模块通过 GET/POST/PUT/DELETE 方式请求服务端接口，在用户组管理和用户管理页面中调用，并通过 promise.then(response=>{}) 的回调方式把获取到的数据返回给 afterLogin 处理函数。

至此，管理员组件以及相对应的 mock 服务、service 服务代码都已全部编写完毕，启动项目，单击左侧菜单栏管理下的用户组和用户菜单，管理员模块的两个页面即可呈现在读者面前，操作页面中对用户组和用户的增删查改，即可调用接口返回成功标识。

在 service 服务层中的 api.js 中改变 BASE_UR 和 REAL_URL 即可完成 mock 数据和真实环境的请求切换，方便项目的调试。

16.10 小 结

本章以管理员模块为例，带领读者熟悉了使用 Antd 框架搭建管理平台主界面，封装左侧菜单栏、顶部导航栏和内容组件。以管理员模块为例，介绍业务组件的开发以及如何配置公共服务路由、mock 服务以及 service 服务。

第 17 章将带领大家进入部署项目所需的基础配置、项目与应用模块的搭建、项目与应用的关系以及如何管理应用部署权限。让我们带着这些问题，前进！

成功之前我们要做应该做的事，成功之后我们才能够做喜欢做的事！

第17章　项目应用、服务器模块设计与搭建

> 纸上得来终觉浅，绝知此事要躬行。
>
> ——陆游

本章 GitHub 代码地址：https://github.com/aguncn/bifang-book/tree/main/bifang-ch17。

第16章主要介绍了 BiFang 部署平台管理员模块的设计与搭建、开发主界面布局组件和相关组件，在 mock 服务层、service 服务层以及路由层添加用户组和用户页面专属的模块，熟悉整个流程配置方法，接下来将带领大家进入项目与应用模块的搭建，再次实践开发流程配置。

本章将在前端项目中设计搭建项目应用管理模块以及服务器管理模块，使用 ant-design-vue 提供的组件构建页面组件；新增项目管理和应用管理页面，在 mock 服务、service 服务和路由层添加页面对应的项目与应用模块；新增服务器管理页面，在 mock 服务、service 服务和路由层添加页面对应的服务器管理模块。

17.1　项目与应用

每个公司根据业务应用场景可分为不同的项目，各 IT 项目又包含前端、后端应用集合，前端应用集合包括 H5 应用、后管应用、node 应用等，后端应用集合根据微服务划分为不同的服务应用。

所以项目和应用是一对多的从属关系，一个项目可以包含多个应用，而一个应用只属于一个项目，有相关性的应用归为一类，这样更容易管理一些大型的部署。

在 BiFang 部署平台中的项目及应用模块，主要提供项目和应用的增删查改，及权限管理功能。

17.1.1 功能说明与清单

项目及应用模块功能清单如表 17-1 所列。

表 17-1 项目及应用模块功能表

模块名称	功能名称	所属页面	功能描述
项目及应用	项目管理	项目列表页面	新增项目 编辑项目 删除项目 搜索项目
	应用管理	应用列表页面	新增应用 编辑应用 删除应用 查看应用详情 搜索应用
	权限管理	应用页面	新建发布单权限的用户新增与删除 环境流转权限的用户新增与删除 部署权限的用户新增与删除

项目及应用模块功能说明:

① 由于项目的信息较少,故未提供项目详情查看功能。

② 应用的设置项比较多,为了方便自助操作,相关输入框附有填写说明。

③ 权限在应用一级设置,而不是在项目一级设置。

④ 权限分为三类,新建发布单(包括构建软件包)权限,环境流转及部署权限。

⑤ 应用创建者和管理员(属于 admin 用户组)默认拥有所有权限,其他用户需要授权。

⑥ 所有删除操作,都有二次确认动作,防止误删除。

项目及应用模块截图如图 17-1~图 17-3 所示。

图 17-1　BiFang 项目管理

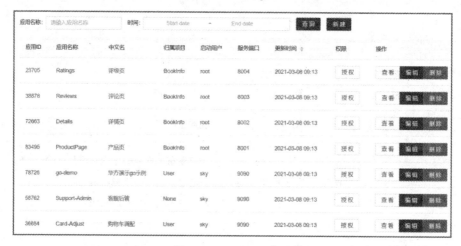

图 17-2　BiFang 应用管理

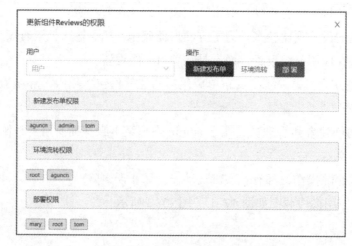

图 17-3　BiFang 应用权限管理

17.1.2 功能技术要点

在项目与应用模块项目管理页面以及应用管理页面搭建中，主要使用 Antd 框架中的组件进行搭建，详细功能点使用的组件如下：
- 查询操作：使用 a-form 表单、a-table 表格、a-pagination 分页组件。
- 新增操作：使用 a-modal 对话框、a-form 表单。
- 编辑操作：使用 a-button 按钮。
- 删除操作：使用 a-button 按钮、a-popconfirm 确认框。
- 权限操作：使用 a-drawer 浮层、a-select 选择框。
- 查看详情操作：封装 DetailList 组件。

上述使用的 Antd 组件常用属性如下，方便读者能清晰地知道各属性的作用。

form 表单是具有数据收集、校验和提交功能的表单，包含复选框、单选框、输入框、下拉选择框等元素，常用的 API 如下：
- layout：表单元素排列方式，水平排列（vertical）、垂直排列（horizontal，默认）、行内排列（inline）。
- labelAlign：label 标签的文本对齐方式，左对齐（left）、右对齐（right）。
- labelCol：label 标签布局，设置 span offset 值，如 {span: 3, offset: 12}。
- wrapperCol：需要为输入控件设置布局样式时，使用该属性，用法同 labelCol。
- getFieldsValue：获取一组输入控件的值，如不传入参数，则获取全部组件的值。
- setFieldsValue：设置一组输入控件的值。
- validateFields：校验并获取一组输入域的值与 error，若 fieldNames 参数为空，则校验全部组件。

table 表格用于当需要对数据进行排序、搜索、分页、自定义操作等复杂行为时展示行列数据，常用的 API 如下：
- dataSource：表格数据源，对应服务端返回的数据。
- columns：表格列的配置，Object 类型数组。
- title：列头显示文字。
- align：内容的对齐方式，'left' | 'right' | 'center'。

- dataIndex：列数据在数据项中对应的 key。
- customRender：生成复杂数据的渲染函数，参数分别为当前行的值、当前行数据、行索引，@return 里面可以设置表格行/列合并。
- scopedSlots：使用 columns 时，可以通过该属性配置支持 slot-scope 的属性，如 scopedSlots: {customRender: 'XXX'}。
- loading：页面是否加载中。
- pagination：分页器，参考配置项或 pagination 文档，设为 false 时不展示和进行分页。
- rowKey：表格行 key 的取值，可以是字符串或一个函数。
- change：分页、排序、筛选变化时触发。

pagination 分页采用分页的形式分隔长列表，每次只加载一个页面，常用的 API 如下：

- current：当前页数。
- pageSize：每页条数。
- total：数据总数。
- showSizeChanger：是否可以改变 pageSize。
- showQuickJumper：是否可以快速跳转至某页。
- showSizeChange：pageSize 变化的回调。

button 按钮标记了一个（或封装一组）操作命令，响应用户单击行为，触发相应的业务逻辑，常用的 API 如下：

- disabled：按钮失效状态。
- icon：设置按钮的图标类型。
- loading：设置按钮载入状态。
- type：设置按钮类型，可选值为 primary dashed danger link。
- click：单击按钮时的回调。

modal 模态对话框在当前页面正中打开一个浮层，承载用户处理事务，又不希望跳转页面以致打断工作流程，常用 API 如下：

- cancelText：取消按钮文字。
- closable：是否显示右上角的关闭按钮。
- destroyOnClose：关闭时销毁 Modal 里的子元素 footer 底部内容，当不需要默认底部按钮时，可以设为 ":footer=\"null\""。

- mask：是否展示遮罩 maskClosable，单击蒙层是否允许关闭。
- okText：确认按钮文字。
- okType：确认按钮类型。
- title：标题。
- width：宽度。

popconfirm 气泡框在目标元素附近弹出浮层提示，询问用户进一步确认操作，常用 API 如下：

- cancelText：取消按钮文字。
- okText：确认按钮文字。
- title：确认框的描述。
- confirm：单击确认的回调。
- cancel：单击取消的回调。

drawer 抽屉从父窗体边缘滑入，覆盖住部分父窗体内容。用户在抽屉内操作时不必离开当前任务，操作完成后，可以平滑地回到原任务，常用 API 如下：

- closable：是否显示右上角的关闭按钮。
- mask：是否展示遮罩。
- title：标题。
- visible：drawer 是否可见。
- bodyStyle：可用于设置 drawer 内容部分的样式。
- close：单击遮罩层或右上角叉或取消按钮的回调。

detailList 组件根据嵌套的 DetailListItem 的数据使用 a-row 和 a-col 组件排列展示数据，开放三个属性供配置：

- title：展示标题。
- col：每一行排列的个数默认为 3 个，可动态传入，不得超过 4 个。
- layout：布局方式可使用横向排列和竖向排列。

在 detailList 组件内部包含 DetailListItem 组件，使用 render 函数布局 label 和内容。

在 components 目录新增 tool 目录，在该目录下新增 DetailList.vue 文件，内容实现代码如下：

```
<template>
    <div :class="['detail-list', layout === 'vertical' ? 'vertical': 'horizontal']">
```

```
            <div v-if="title" class="title">{{title}}</div>
            <a-row>
                <slot></slot>
            </a-row>
        </div>
</template>
<script>
export default {
    name: 'DetailList',
    Item: Item,              //item 为 DetailListItem 组件配置
        props: {
            title: {
                type: String,
                required: false
            },
            col: {
                type: Number,
                required: false,
                default: 3
            },
            size: {
                type: String,
                required: false,
                default: 'large'
            },
            layout: {
                type: String,
                required: false,
                default: 'horizontal'
            }
        },
        provide () {
            return {
                col: this.col > 4 ? 4 : this.col
            }
        }
}
</script>
```

DetailListItem 组件实现代码如下：

```
import ACol from 'ant-design-vue/es/grid/Col'
const Item = {
    name: 'DetailListItem',
```

```
props: {
    term: {
        type: String,
        required: false
    }
},
inject: {
    col: {
        type: Number
    }
},
methods: {
    renderTerm (h, term) {
        return term ? h(
            'div',
            {
                attrs: {
                    class: 'term'
                }
            },
            [term]
        ) : null
    },
    renderContent (h, content) {
        return h(
            'div',
            {
                attrs: {
                    class: 'content'
                }
            },
            [content]
        )
    }
},
render (h) {
    const term = this.renderTerm(h, this.$props.term)
    const content = this.renderContent(h, this.$slots.default)
    return h(
        ACol,
        {
            props: responsive[this.col]
        },
```

```
                [term, content]
            )
        }
    }
}
<style lang="less">
    .detail-list{
        .title {
            font-size: 16px;
            color: @bf-title-color;
            font-weight: bolder;
            margin-bottom: 16px;
        }
        .term {
            line-height: 20px;
            padding-bottom: 16px;
            margin-right: 8px;
            color: @bf-title-color;
            white-space: nowrap;
            display: table-cell;
            &:after {
                content: ':';
                margin: 0 8px 0 2px;
                position: relative;
                top: -0.5px;
            }
        }
        .content{
            line-height: 22px;
            width: 100%;
            padding-bottom: 16px;
            color: @bf-text-color;
            display: table-cell;
        }
        &.vertical{
            .term {
                padding-bottom: 8px;
            }
            .term,.content{
                display: block;
            }
        }
    }
</style>
```

在封装 DetailList 组件时用到了 provide 和 inject，它们是成对出现的，用于父组件向子组件传递数据以实现跨级访问父组件的数据。

以上为搭建项目与应用模块使用的 Antd 组件和封装的 DetailList 组件介绍，接下来进入页面搭建。

17.2 项目管理页面搭建

项目管理页面包含查询项目列表、新增项目、删除项目以及修改项目信息，这里将以实现每个功能的步骤来呈现。

在 src 中的 views 目录下创建 application 目录，并在该目录下创建 project.vue 文件，此文件为项目管理组件。

17.2.1 项目列表查询

项目列表包含项目 ID 列、项目名称列、项目中文名列以及操作列，其中操作列包含编辑和删除两个按钮，单击按钮执行对应的操作。同时列表顶部包含过滤条件，根据用户输入的项目名称和创建时间进行过滤。template 的代码如下：

```
<template>
    <a-card>
        <div class="search">
            <a-form layout="inline" :form="form" @submit="submitHandler">
                <a-form-item label=" 项目名称 ">
                    <a-input
                        placeholder=" 请输入项目名 "
                        v-decorator="['projectName', { rules: [{ required: false }] }]" />
                </a-form-item>
                <a-form-item
                    label=" 时间 "
                >
                    <a-range-picker
                        v-decorator="['timePicker', {rules: [{ type: 'array', required: false }]}]" />
                </a-form-item>
                <a-form-item>
                    <a-button type="primary" html-type="submit">
                        查询
                    </a-button>
                </a-form-item>
```

```html
            <a-form-item>
                <a-button type="primary" @click.prevent="createDialog">
                    新建
                </a-button>
            </a-form-item>
        </a-form>
    </div>
    <div>
        <a-table
            :columns="columns"
            :dataSource="dataSource"
            rowKey="name"
            @clear="onClear"
            @change="onChange"
            :pagination="{
                current: params.currentPage,
                pageSize: params.pageSize,
                total: total,
                showSizeChanger: true,
                showLessItems: true,
                showQuickJumper: true,
                showTotal: (total, range) => `第 ${range[0]}-${range[1]} 条，总计 ${total} 条`
            }"
        >
            <div slot="cn_name" slot-scope="{text, record}">
                <a-tooltip>
                    <template slot="title">
                        {{record.description}}
                    </template>
                    {{text}}
                </a-tooltip>
            </div>

            <div slot="action" slot-scope="{text, record}">
                <a-button-group>
                    <a-button type="primary" @click.prevent="onShowEdit(record)">
                        编辑
                    </a-button>
                    <a-popconfirm
                        title=" 确定执行删除操作么？ "
                        ok-text=" 是 "
                        cancel-text=" 否 "
                        @confirm="onDelete(record)"
```

```
                                >
                                    <a-button type="danger">
                                        删除
                                    </a-button>
                                </a-popconfirm>
                            </a-button-group>
                        </div>
                    </a-table>
                </div>
            </a-card>
        </template>
```

用户组组件以通用的卡片容器 a-card 作为根节点，其中内嵌 a-form 和 a-table 组件，在 a-table 组件中嵌套了 slot 名为 cn_name 和 action 的 template 模板，分别用于渲染项目中文名列以及操作列，其他详细的属性已在技术要点中介绍，在此不再赘述。

在 project.vue 文件中添加 script 标签，项目管理组件中实现的事件包含：
- fetchData：请求接口获取项目列表信息。
- onReset：重置顶部 form 表单数据。
- onChange：用户单击分页或表格个数 pageSize 变更时触发重新请求数据。
- submitHandler：用户单击查询按钮根据 form 表单输入的条件查询项目列表信息。

相关代码如下：

```
<script>
import API from '@/service'
import moment from 'moment' // 引入 moment 时间处理
const columns = [
    {
        title: ' 项目 ID',
        dataIndex: 'project_id'
    },
    {
        title: ' 项目名称 ',
        dataIndex: 'name'
    },
    {
        title: ' 中文名 ',
        dataIndex: 'cn_name',
        scopedSlots: {customRender: 'cn_name'}
```

```js
    },
    {
        title: ' 更新时间 ',
        dataIndex: 'update_date',
        sorter: true,
        customRender: (date) =>{ return
            moment(date).format("YYYY-MM-DD hh:mm")}
    },
    {
        title: ' 操作 ',
        scopedSlots: { customRender: 'action' }
    }
]

export default {
    name: 'project',
    data () {
        return {
            total:0, // 项目总数
            columns: columns,           // 列配置
            dataSource: [],             // 数据源
            visible:false,              // 模态框是否展示
            isEdit:false,               // 是否编辑模式
            params:{                    // 分页配置
                name:"",
                currentPage:1,
                pageSize:20,
                begin_time:"",
                end_time:"",
                sorter:""
            }

        }
    },
    beforeCreate() {
        this.form = this.$form.createForm(this, { name: 'moduleList' });
        this.formDialog = this.$form.createForm(this,
        { name: 'formDialog' });
    },
    created(){
        this.fetchData()
    },
    methods: {
```

```js
submitHandler(e){
    e.preventDefault()
    this.form.validateFields((err, fieldsValue) => {
        if (err) {
            return;
        }
        const rangeValue = fieldsValue['timePicker']
        this.params.name = fieldsValue["projectName"]
        this.params.begin_time =
            rangeValue?rangeValue[0].format("YYYY-MM-DD"):""
        this.params.end_time =
            rangeValue?rangeValue[1].format("YYYY-MM-DD"):""
        this.fetchData()
    });
},
fetchData(){
    API.ProjectList(this.params).then((res)=>{
        let result = res.data
        if(res.status == 200 && result.code == 0){
            this.total = result.data.count
            this.dataSource = result.data.results;
        }
        else{
            this.dataSource = []
        }
    })
},
onChange(pagination, filters, sorter) {
    let {
        current,
        pageSize
    } = pagination
    let {
        order,
        field
    } = sorter
    this.params.currentPage = current
    this.params.pageSize = pageSize
    this.params.sorter = (field?field:"")
    this.fetchData()
},
onReset(){
    this.visible = false
```

```
                this.isEdit = false
                this.formDialog.resetFields()
            },
        }
    }
</script>
```

API.ProjectList 方法为 service 服务配置的请求项目列表的 request 请求，在 service 服务配置中介绍具体的实现。

17.2.2 编辑新增数据

项目的新增和编辑采用模态框的形式实现，其中嵌套 form 表单、项目名称输入框、项目中文名输入框以及隐藏的 id 输入框，用于在编辑时自动提交对应的项目 ID。在 template 中新增代码如下：

```
<a-modal
        :visible="visible"
        :title="isEdit?' 编辑项目 ':' 新增项目 '"
        :okText="isEdit?' 更新 ':' 新建 '"
        @cancel="onReset"
        @ok="isEdit?onUpdateProject():onCreateProject()"
>
    <a-form layout='vertical' :form="formDialog">
        <a-form-item
            label="ID"
            v-show="false"
        >
            <a-input placeholder=" 自动生成 "
            v-decorator="['id']"
            />
        </a-form-item>
        <a-form-item label=' 项目 ID'>
            <a-input
            :disabled="isEdit?true:false"
            v-decorator="[
              'projectId',
              {
                  rules: [{ required: true, message: ' 请输入项目 ID' }],
              }
            ]"
            />
```

```
        </a-form-item>
        <a-form-item label=' 英文名称 '>
          <a-input
            :disabled="isEdit?true:false"
            v-decorator="[
              'name',
              {
                  rules: [{ required: true, message: ' 请输入英文名称 ' }],
              }
            ]"
          />
        </a-form-item>
        <a-form-item label=' 中文名称 '>
          <a-input
            v-decorator="[
              'cnname',
              {
                  rules: [{ required: true, message: ' 请输入中文名称 ' }],
              }
            ]"
          />
        </a-form-item>
        <a-form-item label=' 项目描述 '>
          <a-input
            type='textarea'
            v-decorator="['description']"
          />
        </a-form-item>
    </a-form>
</a-modal>
```

在 script 标签中新增事件包含：

- onUpdateProject：请求接口更新单个项目信息。
- onCreateProject：请求接口新增项目。
- onShowEdit：控制模态框变更为编辑模式并使用行数据初始化表单。
- createDialog：控制模态框变更为新增模式。

新增的事件代码如下：

```
<script>
...// 其他忽略
  methods: {
    ...// 其他忽略
```

```javascript
onShowEdit(record){
    this.visible = true
    this.isEdit = true
        this.$nextTick(()=>{
            this.formDialog.setFieldsValue({
                id:record.id,
                projectId:record.project_id,
                name:record.name,
                cnname:record.cn_name,
                description:record.description
            })
        })
},
onCreateProject () {
    this.formDialog.validateFields((err, fieldsValue) => {
        if (err) {
            return;
        }
        let project_id = fieldsValue["projectId"],
            name = fieldsValue["name"],
            cn_name = fieldsValue["cnname"],
            description = fieldsValue["description"]||""
        let data = {
            project_id,
            name,
            cn_name,
            description
        }
        API.CreateProject(data).then((res)=>{
            let result = res.data
            if(res.status == 200 && result.code == 0){
                this.$message.success(" 项目新增成功 ~")
                this.visible = false
                this.fetchData()
            }
            else{
                this.$message.error(" 项目新增失败 :"+result.message+result.data)
            }
        })
    })
},
onUpdateProject(){
    this.formDialog.validateFields((err, fieldsValue) => {
```

```
                    if (err) {
                        return;
                    }
                    let id = fieldsValue["id"],
                        project_id = fieldsValue["projectId"],
                        name = fieldsValue["name"],
                        cn_name = fieldsValue["cnname"],
                        description = fieldsValue["description"]||""
                    let data = {
                      id,
                      project_id,
                      name,
                      cn_name,
                      description
                    }
                    API.UpdateProject(data).then((res)=>{
                        let result = res.data
                        if(res.status == 200){
                            this.$message.success(" 项目更新成功 ~")
                            this.onReset()
                            this.fetchData()
                        }
                        else{
                            this.$message.error(" 项目更新失败 :"+result.message)
                        }
                    })
                })
            },
            createDialog(){
                this.visible = true
            }
        }
    </script>
```

17.2.3 删除数据

在 a-table 操作列中使用模板渲染了删除按钮，单击按钮可通过 slot-scope 获取当前行数据 record，在单击删除按钮时，可通过该变量获取对应的行数据的 id 进行删除操作。在 script 中添加删除事件 onDelete，相关代码如下：

```
<script>
    ... // 其他忽略
```

```
methods: {
    ... // 其他忽略
    onDelete(record){
        let {id} = record;
        if(id == undefined){
            this.$message.error(" 操作参数非法！ ")
            return false
        }
        API.DeleteProject({id}).then((res)=>{
            let result = res.data
            if(res.status == 200 && result.code == 0){
                this.$message.success(" 删除成功 ~")
                this.fetchData()
            }
            else{
                this.$message.error(" 操作错误 ~:"+result.message)
            }
        })
    },
}
</script>
```

至此项目管理组件已全部实现完毕，method 方法中调用的 API 函数将在 service 服务配置中介绍。

17.3 应用管理页面搭建

应用管理页面包含查询应用列表、新增应用、删除应用以及修改应用信息，这里以实现每个功能的步骤来呈现。

17.3.1 应用列表查询

应用管理列表包含应用名称列、应用中文名列、归属项目列、启动用户列、服务端口列、授权列以及操作列。其中操作列包含编辑和删除两个按钮，单击按钮执行对应的操作。

在 application 目录下创建 app.vue 文件，此文件为应用管理组件，具体的 template 代码如下：

```html
<template>
    <a-card>
        <div class="search">
            <a-form layout="inline" :form="form" @submit="submitHandler">
                <a-form-item label=" 应用名称 " >
                    <a-input
                        placeholder=" 请输入应用名称 "
                        v-decorator="['appName', { rules: [{ required: false }] }]" />
                </a-form-item>
                <a-form-item label=" 时间 " >
                    <a-range-picker
                        v-decorator="['timePicker', {rules: [{ type: 'array', required: false }]}]" />
                </a-form-item>
                <a-form-item>
                    <a-button type="primary" html-type="submit">
                        查询
                    </a-button>
                </a-form-item>
                <a-form-item>
                    <a-button type="primary" @click.prevent="onCreateApp">
                        新建
                    </a-button>
                </a-form-item>
            </a-form>
        </div>
        <div>
            <a-table
                :columns="columns"
                :dataSource="dataSource"
                rowKey="name"
                @clear="onClear"
                @change="onChange"
                :pagination="{
                    current: params.currentPage,
                    pageSize: params.pageSize,
                    total: total,
                    showSizeChanger: true,
                    showLessItems: true,
                    showQuickJumper: true,
                    showTotal: (total, range) => `第 ${range[0]}-${range[1]} 条，总计 ${total} 条`
                }"
            >
                <div slot="permission" slot-scope="text, record">
```

```
                <a-button @click.prevent="showPermission(record)"> 授权 </a-button>
            </div>
            <div slot="action" slot-scope="text, record">
                <a-button-group>
                    <a-button @click.prevent="onShowDetail(record)">
                        查看
                    </a-button>
                    <a-button type="primary" @click.prevent="onShowEdit(record)">
                        编辑
                    </a-button>
                    <a-popconfirm
                        title=" 确定执行删除操作么？ "
                        ok-text=" 是 "
                        cancel-text=" 否 "
                        @confirm="onDelete(record)"
                    >
                        <a-button type="danger">
                            删除
                        </a-button>
                    </a-popconfirm>
                </a-button-group>
            </div>
        </a-table>
        </div>
    </a-card>
</template>
```

应用管理组件使用通用的卡片容器 a-card 作为根节点，其中内嵌 a-form 和 a-table 组件，在 a-table 组件中嵌套了 slot 名为 permission 和 action 的 template 模板，分别用于权限列以及操作列，使用的 Antd 组件的详细属性已在技术要点中介绍，在此不再赘述。

在 app.vue 文件中添加 script 标签，用户组件中实现的事件如下：

● fetchData：请求接口获取应用列表信息。

● onChange：用户单击分页或表格个数 pageSize 变更时触发重新请求数据。

相关代码如下：

```
<script>
import moment from 'moment'
import DetailList from '@/components/tool/DetailList'
import API from '@/service'
```

```
const DetailListItem = DetailList.Item
const columns = [
    {
        title: ' 应用 ID',
        dataIndex: 'app_id'
    },
    {
        title: ' 应用名称 ',
        dataIndex: 'name'
    },
    {
        title: ' 中文名 ',
        dataIndex: 'cn_name'
    },
    {
        title: ' 归属项目 ',
        dataIndex: 'project_name',
    },
    {
        title: ' 启动用户 ',
        dataIndex: 'service_username'
    },
    {
        title: ' 服务端口 ',
        dataIndex: 'service_port'
    },
    {
        title: ' 更新时间 ',
        dataIndex: 'update_date',
        sorter: true,
        customRender: (date) =>{ return moment(date).format("YYYY-MM-DD hh:mm")}
    },
    {
        title: ' 权限 ',
        scopedSlots: { customRender: 'permission' }
    },
    {
        title: ' 操作 ',
        scopedSlots: { customRender: 'action' }
    }
]

export default {
```

```
name: 'app',
components: {DetailList,DetailListItem},
data () {
    return {
        total:0,                              // 应用总数
        visible: false,                       // 模态框开关
        columns: columns,                     // 列数据设置
        dataSource: [],                       // 数据源
        userOptions: [],                      // 选择用户源
        selectUser: "",                       // 选择的用户
        selectApp: "",                        // 选择的应用
        permissionDataSource: [],             // 所有权限数据源
        createPermissionUser: [],             // 已有 create 权限用户
        envPermissionUser: [],                // 已有流转权限用户
        deployPermissionUser: [],             // 已有部署权限用户
        detailVisible:false,                  // 应用详情模态框
        detailRecord:{                        // 单应用基础信息
            app_id:null,
            name:null,
            cn_name:null,
            project_name:null,
            description:null,
            git_name:null,
            git_app_id:null,
            git_trigger_token:null,
            build_script:null,
            deploy_script:null,
            zip_package_name:null,
            service_username:null,
            service_port:null,
            update_date:null
        },
        params:{                              // 分页参数
            name:"",
            currentPage:1,
            pageSize:20,
            begin_time:"",
            end_time:"",
            sorter:""
        }
    }
},
beforeCreate() {
```

```js
            this.form = this.$form.createForm(this, { name: 'moduleList' });
            this.formPermission = this.$form.createForm(this, { name: 'permissionModuleList' });
    },
    created(){
            this.fetchData()
    },
    methods: {
            submitHandler(e){
                    e.preventDefault()
                    this.form.validateFields((err, fieldsValue) => {
                            if (err) {
                                    return;
                            }
                            const rangeValue = fieldsValue['timePicker']
                            this.params.name = fieldsValue["appName"]
                            this.params.begin_time =
                                    rangeValue?rangeValue[0].format("YYYY-MM-DD"):""
                            this.params.end_time =
                                    rangeValue?rangeValue[1].format("YYYY-MM-DD"):""
                            this.fetchData()
                    });
            },
            fetchData(){
                    API.AppList(this.params).then((res)=>{
                            let result = res.data
                            if(res.status == 200 && result.code == 0){
                                    this.total = result.data.count
                                    this.dataSource = result.data.results;
                            }
                            else{
                                    this.dataSource = []
                            }
                    })
            },
            onChange(pagination, filters, sorter) {
                    let {
                            current,
                            pageSize
                    } = pagination
                    let {
                            order,
                            field
                    } = sorter
```

```
                this.params.currentPage = current
                this.params.pageSize = pageSize
                this.params.sorter = (field?field:"")
                this.fetchData()
            }
        }
    }
</script>
```

在应用管理组件创建前使用 this.$form.createForm 收集 form 表单对象供后续做表单校验、获取表单内组件数据以及表单内部组件的赋值操作。

17.3.2 新增编辑应用

项目归属的应用新增和编辑由于应用信息较多，采用新的页面去承接。

在 application 目录下新增 appdetail.vue 文件，该文件即为应用详情组件，使用 form 表单收集应用中文名、英文名、归属项目、应用描述、应用 GIT 项目 ID、GIT 服务地址、GIT Token、编译脚本、部署脚本以及应用压缩包路径信息提交后台进行操作。

在 template 中新增代码如下：

```
<template>
    <a-card :body-style="{padding: '24px 32px'}" :bordered="false">
        <a-form :form="form"
            @submit="isEdit?onUpdateApplication():onCreateApplication()">
            <a-form-item
                label="ID"
                v-show="false"
            >
                <a-input placeholder=" 自动生成 "
                    v-decorator="['id']"
                />
            </a-form-item>
            <a-form-item
                label=' 应用 ID'
                :labelCol="{span: 7}"
                :wrapperCol="{span: 10}"
            >
                <a-input
                    :disabled="isEdit?true:false"
                    v-decorator="[
```

```
                'app_id',
                {
                    rules: [{ required: true, message: ' 请输入应用 ID' }],
                }
            ]"
        />
    </a-form-item>
    <a-form-item
        label=' 英文名称 '
        :labelCol="{span: 7}"
        :wrapperCol="{span: 10}"
    >
        <a-input
        :disabled="isEdit?true:false"
        v-decorator="[
            'name',
            {
                rules: [{ required: true, message: ' 请输入英文名称 ' }],
            }
        ]"
        />
    </a-form-item>
    <a-form-item
        label=' 中文名称 '
        :labelCol="{span: 7}"
        :wrapperCol="{span: 10}">
        <a-input
        v-decorator="[
            'cn_name',
            {
                rules: [{ required: true, message: ' 请输入中文名称 ' }],
            }
        ]"
        />
    </a-form-item>
    <a-form-item
        label=" 归属项目 "
        :labelCol="{span: 7}"
        :wrapperCol="{span: 10}"
    >
        <a-select
            show-search
            placeholder=" 请选择项目 "
```

```
                    option-label-prop="children"
                    style="width: 100%"
                    :options="projectOptions"
                    :filter-option="filterOption"
                    v-decorator="['project_id', { rules: [{ required: true, message: ' 请选择归属项目 !' }] }]"
                >
                </a-select>
            </a-form-item>
            <a-form-item
                label=" 应用描述 "
                :labelCol="{span: 7}"
                :wrapperCol="{span: 10}"
            >
                <a-textarea rows="4"
                    placeholder=" 请输入应用描述 "
                    v-decorator="['description']"
                />
            </a-form-item>
            <a-form-item
                label="GIT 服务器 "
                :labelCol="{span: 7}"
                :wrapperCol="{span: 10}"
            >
                <a-select
                    show-search
                    placeholder=" 请选择 Git 服务器 "
                    option-filter-prop="children"
                    style="width: 100%"
                    :filter-option="filterOption"
                    v-decorator="['git_id', { rules: [{ required: true, message: ' 请选择 Git 服务器 !' }] }]"
                >
                    <a-select-option v-for="d in gitOptions" :key="d.value">
                        {{ d.label }}
                    </a-select-option>
                </a-select>
            </a-form-item>
            <a-form-item
                label="GIT 项目 ID"
                :labelCol="{span: 7}"
                :wrapperCol="{span: 10}"
            >
                <a-input
                    placeholder=" 请输入 GIT 项目 ID, 此 ID 在每个 GIT 服务器上是唯一的。"
```

```
                v-decorator="['git_app_id',
                    { rules: [{ required: true, message: ' 请输入 git appId!' }] }]"
            />
        </a-form-item>
        <a-form-item
            label="Git Trigger Token"
            :labelCol="{span: 7}"
            :wrapperCol="{span: 10}"
        >
            <a-input
                placeholder=" 请输入 Git Trigger Token, 此 token 用于触发 gitlab 的 ci/cd 的 pipeline"
                v-decorator="['git_trigger_token', { rules: [{ required: true,
                    message: ' 请输入 git token!' }] }]"
            />
        </a-form-item>
        <a-form-item
            label=" 编译脚本 "
            :labelCol="{span: 7}"
            :wrapperCol="{span: 10}"
        >
            <a-input
                placeholder=" 请输入编译脚本路径, 是相对于 gitalb 项目的根目录, 如 script/build.sh"
                v-decorator="['build_script', { rules: [{ required: true,
                    message: ' 请输入编译脚本路径!' }] }]"
            />
        </a-form-item>
        <a-form-item
            label=" 部署脚本 "
            :labelCol="{span: 7}"
            :wrapperCol="{span: 10}"
        >
            <a-input
                placeholder=" 请输入部署脚本路径, 是相对于 gitalb 项目的根目录, 如
                    script/deploy.sh"
                v-decorator="['deploy_script', { rules: [{ required: true,
                    message: ' 请输入部署脚本路径!' }] }]"
            />
        </a-form-item>
        <a-form-item
            label=" 应用压缩包 "
            :labelCol="{span: 7}"
            :wrapperCol="{span: 10}"
        >
```

```
            <a-input
                placeholder=" 请输入应用压缩包路径，目前是 tar.gz 格式压缩包，如 go-demo.tar.gz"
                v-decorator="['zip_package_name', { rules: [{ required: true,
                    message: ' 请输入应用压缩包路径 !' }] }]"
            />
        </a-form-item>
        <a-form-item
            label=" 服务端口 "
            :labelCol="{span: 7}"
            :wrapperCol="{span: 10}"
        >
            <a-input
                placeholder=" 请输入服务端口 "
                v-decorator="['service_port', { rules: [{ required: true, message: ' 请输入服务端口 !' }] }]"
            />
        </a-form-item>
        <a-form-item style="margin-top: 24px" :wrapperCol="{span: 10, offset: 7}">
            <a-button type="primary" html-type="submit">{{btnDesc}}</a-button>
        </a-form-item>
    </a-form>
</a-card>
</template>
```

使用 Antd 框架 a-form 组件嵌套 input 输入框、select 选择框组件，详细的属性已在技术要点中介绍，在此不再赘述。

在应用详情组件中需要实现事件包含：

- fetchProjectList：请求接口查询项目列表。
- fetchGitList：请求接口查询 Git 服务器列表。
- onCreateApplication：根据录入的信息创建新应用。
- onUpdateApplication：根据修改的应用信息提交服务器进行更新。
- filterOption：select 选择框根据输入条件本地过滤函数。

应用详情 script 代码如下：

```
<script>
import API from '@/service'
export default {
    name: 'appDetail',
    data () {
        return {
            isEdit:false,
            type:"build",
```

```
            value: 1,
            fetching:false,
            projectOptions:[],
            projectList:[],
            gitOptions:[],
            gitList:[],
            btnDesc:" 创建 "
        }
},
beforeCreate() {
    this.form = this.$form.createForm(this, { name: 'applicaitonDetail' });
},
created(){
    this.fetchProjectList()
    this.fetchGitList()
    // 根据传入的参数判断编辑模式还是新增模式
    if(this.$route.query.id){
            this.isEdit = true
            this.btnDesc = " 更新 "
    }

},
mounted(){
    if(this.isEdit){
            this.$nextTick(()=>{
                    this.form.setFieldsValue({
                            ...this.$route.query,
                            project_id: this.$route.query["project"],
                            git_id: this.$route.query["git"]
                    })
            })
    }
},
methods: {
    fetchProjectList(){
        API.ProjectList({}).then((res)=>{
            let result = res.data
            if(res.status == 200 && result.code == 0){
                this.projectList = result.data.results
                result.data.results.forEach(item=>{
                    this.projectOptions.push({
                        label:item.cn_name,
                        key:item.id.toString()
```

```javascript
                })
            })
        }
        else{
            this.$message.error(" 无法获取应用列表 ~")
        }
    })
},
fetchGitList(){
    API.GitList({}).then((res)=>{
        let result = res.data
        if(res.status == 200 && result.code == 0){
            this.gitList = result.data.results
            result.data.results.forEach(item=>{
                this.gitOptions.push({
                    label:item.name,
                    value:item.id.toString()
                })
            })
        }
        else{
            this.$message.error(" 无法获取应用列表 ~")
        }
    })
},
onCreateApplication () {
    let self = this
    this.form.validateFields((err, fieldsValue) => {
        if (err) {
            return;
        }
        let project = self.projectList.find(item=>item.id == fieldsValue["project_id"])
        let data = {
            ...fieldsValue,
            project_name: project.name
        }
        API.CreateApplication(data).then((res)=>{
            let result = res.data
            if(res.status == 200 && result.code == 0){
                this.$message.success(" 应用创建成功 ~")
                this.$router.push("app")
            }
            else{
```

```
                        this.$message.error(" 应用创建失败 ~:"+result.message)
                    }
                })
            });
        },
        onUpdateApplication(){
            let self = this
            this.form.validateFields((err, fieldsValue) => {
                if (err) {
                    return;
                }
                let project = self.projectList.find(item=>tem.id == fieldsValue["project_id"])
                let data = {
                    ...fieldsValue,
                    project_name:project.name
                }
                API.UpdateApplication(data).then((res)=>{
                    let result = res.data
                    if(res.status == 200){
                        this.$message.success(" 应用更新成功 ~")
                        this.$router.push('/application/app')
                    }
                    else{
                        this.$message.error(" 应用更新失败 :"+result.message)
                    }
                })
            })
        },
        filterOption(input, option) {
            return (
                option.componentOptions.children[0].text
                    .toLowerCase().indexOf(input.toLowerCase()) >= 0
            );
        },
    }
}
</script>
```

17.3.3 删除应用

在 a-table 表单操作列中使用 action 模板渲染了删除按钮，单击按钮可通过 slot-scope 获取当前行数据 record，在单击删除按钮时，可通过该变量获取对应行数据的 id 进行删除操作。

在 script 中添加删除事件 onDelete，代码如下：

```
<script>
    ... // 其他忽略
    methods: {
        ... // 其他忽略
        onDelete(record){
            let {id} = record;
            if(id == undefined){
                this.$message.error(" 操作参数非法！ ")
                return false
            }
            API.DeleteApplication({id}).then((res)=>{
                let result = res.data
                if(res.status == 200 && result.code == 0){
                    this.$message.success(" 删除成功 ~")
                    this.fetchData()
                }
                else{
                    this.$message.error(" 操作错误 ~:"+result.message)
                }
            })
        },
    }
}
</script>
```

17.3.4　查看应用详情

在 a-table 表单操作列中使用 action 模板渲染了查看按钮，单击查看按钮可通过 slot-scope 获取当前行数据 record。使用 a-modal 模态框内嵌自定义 DetailList 组件渲染应用基础信息和部署相关信息。在 template 中新增代码如下：

```
<template>
    <!-- 详情弹出框 -->
    <a-modal
        :visible="detailVisible"
        :title="' 应用详情 '"
        :dialog-style="{ top: '30px' }"
        width="50%"
```

```
    okText=" 确定 "
    :footer="null"
    @cancel="onDetailModalClose"
>

<a-card :border="false">
    <detail-list title=" 基础信息 ">
        <detail-list-item term=" 应用 ID">{{detailRecord.app_id}}</detail-list-item>
        <detail-list-item term=" 应用名称 ">{{detailRecord.name}}</detail-list-item>
        <detail-list-item term=" 中文名 ">{{detailRecord.cn_name}}</detail-list-item>
        <detail-list-item term=" 归属项目 ">{{detailRecord.project_name}}</detail-list-item>
        <detail-list-item term=" 更新时间 ">{{dateTransformer(detailRecord.update_date)}}
        </detail-list-item>
        <a-row>
            <a-col :span="24">
                <detail-list-item term=" 应用描述 ">{{detailRecord.description}}</detail-list-item>
            </a-col>
        </a-row>
    </detail-list>
    <detail-list title=" 部署相关 " >
        <detail-list-item term="GIT 服务器 ">
            {{detailRecord.git_name}}</detail-list-item>
        <detail-list-item term="GIT 项目 ID">
            {{detailRecord.git_app_id}}</detail-list-item>
        <detail-list-item term=" 启动用户 ">
            {{detailRecord.service_username}}</detail-list-item>
        <detail-list-item term=" 服务端口 ">
            {{detailRecord.service_port}}</detail-list-item>
        <a-row>
            <a-col :span="24">
                <detail-list-item term=" 应用描述 ">
                    {{detailRecord.description}}</detail-list-item>
            </a-col>
            <a-col :span="24">
                <detail-list-item term="Git Trigger Token">
                    {{detailRecord.git_trigger_token}}</detail-list-item>
            </a-col>
            <a-col :span="24">
                <detail-list-item term=" 编译脚本 ">
                    {{detailRecord.build_script}}</detail-list-item>
            </a-col>
            <a-col :span="24">
                <detail-list-item term=" 部署脚本 ">
                    {{detailRecord.deploy_script}}</detail-list-item>
```

```
                    </a-col>
                    <a-col :span="24">
                        <detail-list-item term=" 应用压缩包 ">
                            {{detailRecord.zip_package_name}}</detail-list-item>
                    </a-col>
                </a-row>
            </detail-list>
        </a-card>
    </a-modal>
    <!---->
</template>
```

单击按钮把 table 中的行数据 record 合并给 data 中的 detailRecord 变量，实现应用详情数据的展示。在 script 中添加 onShowDetail 事件的实现，代码如下：

```
<script>
    ... // 其他忽略
    methods: {
        ... // 其他忽略
        onShowDetail(record){
            Object.assign(this.detailRecord,record)
            this.detailVisible = true
        }
    }
}
</script>
```

17.3.5　应用权限授权

在 a-table 表单操作列中使用 permission 模板渲染授权按钮，单击授权按钮通过 slot-scope 获取当前行数据 record。使用 a-drawer 抽屉内嵌 a-form 表单、a-tag 标签渲染新建发布单权限用户、环境流转权限用户以及部署权限用户，在 a-select 用户列表中选择用户，单击对应权限按钮即可对用户添加权限。在 template 中新增代码如下：

```
<template>
    <!---->
    <div>
        <a-drawer
            :title="` 更新组件 ${selectApp} 的权限 `"
            :width="720"
            :visible="visible"
```

```
            :body-style="{ paddingBottom: '80px' }"
            @close="onClose"
        >
            <a-form :form="formPermission"
                    layout="vertical" hide-required-mark>
                <a-row :gutter="16">
                    <a-col :span="12">
                        <a-form-item label=" 用户 ">
                            <a-select
                                show-search
                                placeholder=" 用户 "
                                option-filter-prop="children"
                                style="width: 320px"
                                @change="handleChange"
                                v-decorator="['selectUser', { rules: [{ required: true,
                                    message: ' 请选择授权用户！' }] }]"
                            >
                                <a-select-option
v-for="d in userOptions" :key="d.value">
                                    {{ d.label }}
                                </a-select-option>
                            </a-select>
                        </a-form-item>
                    </a-col>
                    <a-col :span="12">
                        <a-form-item label=" 操作 ">
                            <a-button type="primary"
                                        @click.prevent="onCreatePermission('Create')" >
                                新建发布单
                            </a-button>
                            <a-button @click.prevent="onCreatePermission('Env')">
                                            环境流转 </a-button>
                            <a-button type="danger"
                                        @click.prevent="onCreatePermission('Deploy')">
                                部署
                            </a-button>
                        </a-form-item>
                    </a-col>
                </a-row>
                <a-row :gutter="16">
                    <a-col :span="24">
                        <a-alert message=" 新建发布单权限 "
type="success" /><br/>
```

```html
                    <a-tag color="blue" v-for="item in createPermissionUser">
                        <a-popconfirm
                                title=" 删除此用户权限 ?"
                                ok-text=" 确定 "
                                cancel-text=" 否 "
                                @confirm="confirmDeleteUser(item.id)"
                                @cancel="cancelDeleteUser"
                        >
                                {{item.name}}
                        </a-popconfirm>
                    </a-tag>
                    <br/>
                    <br/>
            </a-col>
            <a-col :span="24">
                    <a-alert message=" 环境流转权限 " type="success" /><br/>
                    <a-tag color="blue" v-for="item in envPermissionUser">
                        <a-popconfirm
                                title=" 删除此用户权限 ?"
                                ok-text=" 确定 "
                                cancel-text=" 否 "
                                @confirm="confirmDeleteUser(item.id)"
                                @cancel="cancelDeleteUser"
                        >
                                {{item.name}}
                        </a-popconfirm>
                    </a-tag>
                    <br/>
                    <br/>
            </a-col>
            <a-col :span="24">
                    <a-alert message=" 部署权限 " type="success" /><br/>
                    <a-tag color="blue"
                            v-for="item in deployPermissionUser">
                        <a-popconfirm
                                title=" 删除此用户权限 ?"
                                ok-text=" 确定 "
                                cancel-text=" 否 "
                                @confirm="confirmDeleteUser(item.id)"
                                @cancel="cancelDeleteUser"
                        >
                                {{item.name}}
                        </a-popconfirm>
```

```
                        </a-tag>
                        <br/>
                        <br/>
                    </a-col>
                </a-row>
            </a-form>
            <div
                :style="{
                    position: 'absolute',
                    right: 0,
                    bottom: 0,
                    width: '100%',
                    borderTop: '1px solid #e9e9e9',
                    padding: '10px 16px',
                    background: '#fff',
                    textAlign: 'right',
                    zIndex: 1,
                }"
            >
                <a-button type="primary" @click="onClose">
                    关闭
                </a-button>
            </div>
        </a-drawer>
    </div>
    <!---->
</template>
```

顶部导航组件中的其他实现方法可参见代码以及备注，不再详细解释，接下来介绍内容区域组件 PageView。

17.4 项目与应用基础服务配置

项目与应用管理页面开发完成后，就需要在浏览器中测试开发的页面展示以及功能是否有问题，即进入开发本地测试阶段，需要用到第 14 章介绍的路由、service 服务以及 mock 服务配置。

17.4.1 路由配置

若要在浏览器能够展示登录页面，只需要把登录页面的组件配置到 router

目录下的 config.js 文件中，代码如下：

```
const routeConfig = [
        const routeConfig = [
    {
                    path: '/application',
                    name: ' 项目应用 ',
                    meta: {
                        icon: 'appstore'
                    },
                    children:[
                        {
                            path: '/project',
                            name: ' 项目 ',
                            component: () => import('@/views/application/project')
                        },
                        {
                            path: '/app',
                            name: ' 应用 ',
                            component: () => import('@/views/application/app')
                        },
                        {
                            path: '/appDetail',
                            name: ' 应用详情 ',
                            meta: {
                                isShowMenu: false
                            },
                            component: () => import('@/views/application/appdetail')
                        },
                    ]
                },
        ]
]
```

通过在浏览器地址中添加 hash 值为 /application/project，即访问 http://localhost:8080/#/application/project，便可在浏览器中看到开发完的项目管理页面，同理添加 /application/app 和 /application/appDetail，可查看应用管理和应用详情页面。

17.4.2 mock 服务配置

项目中的本地 mock 服务用于模拟服务端请求，并返回自定义的模拟数据。

项目与应用页面请求服务端项目与应用模块增删查改接口，所以需要在 mock 目录下新增 application 目录，目录中新增 index.js 文件，文件内容如下：

```js
import Mock from 'mockjs'
const projectList = Mock.mock({
    "count": "@integer(1, 100)",
    "results|100": [
        {
            "id": "@id",
            "create_user_name": "@name",
            "name": "@name",
            "description": "@@sentence(4)",
            "update_date": "@date",
            "create_date": "@date",
            "base_status": "@boolean",
            "cn_name": "@name",
            "project_id": "@id",
            "create_user": "@id"
        }
    ]
})

Mock.mock(`${process.env.VUE_APP_API_BASE_URL}/project/list/`, 'get', ({body}) => {
    return {
        "code": 0,
        "message": "success",
        "data": projectList
    }
})

const addProject = Mock.mock({
    "create_user_name": "@name",
    "name": "@name",
    "description": "@@sentence(5)",
    "cn_name": "@name",
    "project_id": "@id",
    "create_user": "@id"
})

Mock.mock(`${process.env.VUE_APP_API_BASE_URL}/project/create/`, 'post', ({body}) => {
    return {
        "code": 0,
        "message": "success",
```

```
                "data": addProject
        }
})

Mock.mock(`${process.env.VUE_APP_API_BASE_URL}/project/update/`, 'put', ({body}) => {
        return {
                "code": 0,
                "message": "success",
                "data": {}
        }
})

Mock.mock(`${process.env.VUE_APP_API_BASE_URL}/project/delete/`, 'delete', ({body}) => {
        return {
                "code": 0,
                "message": "success",
                "data": {}
        }
})

const appList = Mock.mock({
        "count": "@integer(1, 100)",
        "results|100": [
                {
                        "id": "@id",
                        "project_name": "@name",
                        "git_name": "@name",
                        "name": "@name",
                        "description": "@sentence(5)",
                        "update_date": "@date",
                        "create_date": "@date",
                        "base_status": "boolean",
                        "cn_name": "@name",
                        "app_id": "@id",
                        "git_app_id": "@id",
                        "git_trigger_token": "c14b9eead3021965918e187eca16a0",
                        "build_script": "script/build.sh",
                        "deploy_script": "script/deploy.sh",
                        "zip_package_name": "Ratings.tar.gz",
                        "service_port": "@integer(1000,30000)",
                        "service_username": "@name",
                        "service_group": "@name",
                        "create_user": "@id",
```

```js
                    "git": "@id",
                    "project": "@id"
                }
            ]
        })

Mock.mock(`${process.env.VUE_APP_API_BASE_URL}/app/list/`, 'get', ({body}) => {
    return {
        "code": 0,
        "message": "success",
        "data": appList
    }
})

const addApp = Mock.mock({
    "id": "@id",
    "project_name": "@name",
    "git_name": "@name",
    "name": "@name",
    "description": "@sentence(5)",
    "update_date": "@date",
    "create_date": "@date",
    "base_status": "boolean",
    "cn_name": "@name",
    "app_id": "@id",
    "git_app_id": "@id",
    "git_trigger_token": "c14b9eead3021965918e187eca16a0",
    "build_script": "script/build.sh",
    "deploy_script": "script/deploy.sh",
    "zip_package_name": "Ratings.tar.gz",
    "service_port": "@integer(1000,30000)",
    "service_username": "@name",
    "service_group": "@name",
    "create_user": "@id",
    "git": "@id",
    "project": "@id"
})

Mock.mock(`${process.env.VUE_APP_API_BASE_URL}/app/create/`, 'post', ({body}) => {
    return {
        "code": 0,
        "message": "success",
        "data": addApp
```

```
        }
})

const editApp = Mock.mock({
        "id": "@id",
        "project_name": "@name",
        "git_name": "@name",
        "name": "@name",
        "description": "@sentence(5)",
        "update_date": "@date",
        "create_date": "@date",
        "base_status": "boolean",
        "cn_name": "@name",
        "app_id": "@id",
        "git_app_id": "@id",
        "git_trigger_token": "c14b9eead3021965918e187eca16a0",
        "build_script": "script/build.sh",
        "deploy_script": "script/deploy.sh",
        "zip_package_name": "Ratings.tar.gz",
        "service_port": "@integer(1000,30000)",
        "service_username": "@name",
        "service_group": "@name",
        "create_user": "@id",
        "git": "@id",
        "project": "@id"
})

Mock.mock(`${process.env.VUE_APP_API_BASE_URL}/app/update/`, 'put', ({body}) => {
        return {
                "code": 0,
                "message": "success",
                "data": editApp
        }
})

Mock.mock(`${process.env.VUE_APP_API_BASE_URL}/app/delete/`, 'delete', ({body}) => {
        return {
                "code": 0,
                "message": "success",
                "data": {}
        }
})
```

```js
const permissionList = Mock.mock({
    "count": "@integer(1, 100)",
    "results|100": [
        {
            "id": "@id",
            "app_name": "@name",
            "action_name": "@name",
            "create_username": "@name",
            "pm_username": "@name",
            "name": "@name",
            "description": "@sentence(5)",
            "update_date": "@date",
            "create_date": "@date",
            "base_status": "@boolean",
            "create_user": "@id",
            "app": "@id",
            "action": "@id",
            "pm_user": "@id"
        }
    ]
})

Mock.mock(`${process.env.VUE_APP_API_BASE_URL}/permission/list/`, 'get', ({body}) => {
    return {
        "code": 0,
        "message": "success",
        "data": permissionList
    }
})

const addPermisson = Mock.mock({
    "app_name": "@name",
    "action_name": "@name",
    "create_username": "@name",
    "pm_username": "@name",
    "name": "@name",
    "description": "@sentence(5)",
    "create_user": "@id",
    "app": "@id",
    "action": "@id",
    "pm_user": "@id"
})
```

```javascript
Mock.mock(`${process.env.VUE_APP_API_BASE_URL}/permission/create/`, 'post', ({body}) => {
    return {
        "code": 0,
        "message": "success",
        "data": addPermisson
    }
})

Mock.mock(`${process.env.VUE_APP_API_BASE_URL}/permission/delete/`, 'delete', ({body}) => {
    return {
        "code": 0,
        "message": "success",
        "data": {}
    }
})
```

在 mock 服务的入口文件 index.js 中引入 application 目录，即完成了项目与应用模块 mock 服务的注册，当项目应用组件通过 GET/POST/PUT/DELETE 方式访问 /project/*、/app/* 和 /permisson/* 路径时，mock 服务层会返回配置的数据。项目应用组件访问 Mock 服务层的接口获取数据需要通过 service 层，所以还需要在 service 层配置项目应用模块。

17.4.3 service服务配置

service 服务统一封装了 BiFang 部署平台项目所需要的所有接口路径以及请求方式，并通过 axios 调用下游的 mock 数据，所以 service 服务需要新增登录模块。在 services 目录下的 api.js 路径集中新增登录和注册的访问路径，代码如下：

```javascript
const API = {
    // 查询项目列表
    PROJECTLIST: `${REAL_URL}/project/list/`,
    // 创建新项目
    CREATEPROJECT: `${REAL_URL}/project/create/`,
    // 删除项目
    DELETEPROJECT: `${REAL_URL}/project/delete/{{id}}/`,
    // 更新项目
    UPDATEPROJECT: `${REAL_URL}/project/update/{{id}}/`,
    // 查询应用列表
    APPLICATIONLIST: `${REAL_URL}/app/list/`,
    // 删除应用
```

```
        DELETEAPPLICATION: `${REAL_URL}/app/delete/{{id}}/`,
        // 创建新应用
        CREATEAPPLICATION: `${REAL_URL}/app/create/`,
        // 更新应用
        UPDATEAPPLICATION: `${REAL_URL}/app/update/{{id}}/`,
}
```

在 services 目录下的新增 application.js 文件，用于配置项目与应用模块所需服务，代码如下：

```
import {request,METHOD } from '@/utils/request'
import {urlFormat} from '@/utils/util'
import {API} from './api'
/**
 * 获取项目列表
 * @param {*} params
 */
async function ProjectList(params){
        return request(
                API.PROJECTLIST,
                METHOD.GET,
                params
        )
}
/**
 * 新增项目
 * @param {*} params
 */
async function CreateProject(params){
        return request(
                API.CREATEPROJECT,
                METHOD.POST,
                params
        )
}
/**
 * 删除项目
 * @param {*} params
 */
async function DeleteProject(params){
        const url = urlFormat(API.DELETEPROJECT, params)
        return request(
                url,
```

```
                METHOD.DELETE
        )
}
/**
 * 更新项目
 * @param {*} params
 */
async function UpdateProject(params){
        const url = urlFormat(API.UPDATEPROJECT, params)
        return request(
                url,
                METHOD.PUT,
                params
        )
}
/**
 * 获取应用列表
 * @param {*} params
 */
async function AppList(params){
        return request(
                API.APPLICATIONLIST,
                METHOD.GET,
                params
        )
}
/**
 * 删除应用
 * @param {*} params
 */
async function DeleteApplication(params){
        const url = urlFormat(API.DELETEAPPLICATION, params)
        return request(
                url,
                METHOD.DELETE
        )
}
/**
 * 新增应用
 * @param {*} params
 */
async function CreateApplication(params){
        return request(
```

```
                    API.CREATEAPPLICATION,
                    METHOD.POST,
                    params
            )
    }
    /**
     * 更新应用
     * @param {*} params
     */
    async function UpdateApplication(params){
            const url = urlFormat(API.UPDATEAPPLICATION, params)
            return request(
                    url,
                    METHOD.PUT,
                    params
            )
    }
    export default {
            AppList,
            DeleteApplication,
            CreateApplication,
            UpdateApplication,
            ProjectList,
            CreateProject,
            DeleteProject,
            UpdateProject
    }
```

在项目与应用模块中调用 service 层配置的接口名称，并通过 promise.then(response=>{}) 的回调方式把获取的数据返回给处理函数。

至此，项目与应用模块以及相对应的路由、mock 服务、service 服务代码都已全部编写完毕，启动项目，点击菜单栏中的项目应用菜单，项目管理和应用管理页面即可呈现在读者面前，页面调用 service 服务提供的项目应用增删查改接口返回成功标识。

在 service 服务层中的 api.js 中改变 BASE_UR 和 BASE_URL 即可完成 mock 数据和真实环境的请求切换，方便项目的调试。

17.5 服务器管理模块

传统 IT 项目部署通常是运维人员（也可能是开发人员）通过 SHH 连接到指定服务器进行应用部署，毕方部署平台服务器模块管理的是部署服务器的相关信息，主要提供服务器的增删查改等功能。

服务器管理模块主要提供服务器的增删查改等功能。

1. 服务器管理模块功能清单

（1）服务器管理模块功能清单如表 17-2 所列。

表 17-2 服务器管理模块功能表

模块名称	功能名称	所属页面	功能描述
服务器	服务器管理	服务器列表页面	新增服务器 编辑服务器 删除服务器 搜索服务器

（2）服务器管理模块功能说明。

①服务器和端口组合，形成一条唯一记录。

②同一个服务器的不同端口，可以部署不同的应用。

③服务器和端口组合必须属于一个环境（测试环境，线上环境等）。

④服务器和端口组合必须属于一个应用。

（3）服务器管理模块截图如图 17-4 所示。

图 17-4 BiFang 服务器管理

2. 功能技术要点

在服务器管理模块相关页面的搭建中，主要使用 Antd 框架中的组件进行搭建，详细功能点使用的组件如下：
- 查询操作：使用 a-form 表单、a-table 表格、a-pagination 分页组件。
- 新增操作：使用 a-modal 对话框、a-form 表单。
- 编辑操作：使用 a-button 按钮。
- 删除操作：使用 a-button 按钮、a-popconfirm 确认框。

上述使用的 Antd 组件常用属性，方便读者能清晰地知道各属性的作用。
- Form 表单是具有数据收集、校验和提交功能的表单，包含复选框、单选框、输入框、下拉选择框等元素。
- Table 表格用于当需要对数据进行排序、搜索、分页、自定义操作等复杂行为时展示行列数据。
- Pagination 分页采用分页的形式分隔长列表，每次只加载一个页面。
- Button 按钮标记了一个（或封装一组）操作命令，响应用户单击行为，触发相应的业务逻辑。
- Modal 模态对话框在当前页面正中打开一个浮层，承载用户处理事务，又不希望跳转页面以致打断工作流程。

17.6 服务器管理页面搭建

服务器管理页面包含查询服务器列表、新增服务器、删除服务器以及修改服务器信息，这里将以实现每个功能的步骤来呈现。

在 src 中的 views 目录下创建 server 目录，并在该目录下创建 list.vue 文件，此文件为服务器管理组件。

1. 查询服务器列表

服务器列表包含服务器 IP 列、端口列、系统类型列、对应环境列、更新时间列以及操作列，其中操作列包含编辑和删除两个按钮，单击按钮执行对应的操作。同时列表顶部包含过滤条件，根据用户输入的服务器 IP 和创建时间进行过滤。template 的代码如下：

```
<template>
    <a-card>
```

```html
<div class="search">
    <a-form layout="inline" :form="form" @submit="submitHandler">
        <a-form-item
            label=" 服务器地址 "
        >
            <a-input
                placeholder=" 请输入服务器 IP"
                v-decorator="['IP', { rules: [{ required: false, message: ' 请输入服务器 IP' }] }]" />
        </a-form-item>
        <a-form-item
            label=" 时间 "
        >
            <a-range-picker
                v-decorator="['timePicker', {rules: [{ type: 'array', required: false }]}]" />
        </a-form-item>
        <a-form-item>
            <a-button type="primary" html-type="submit">
                查询
            </a-button>
        </a-form-item>
        <a-form-item>
            <a-button type="primary" @click.prevent="createDialog">
                新建
            </a-button>
        </a-form-item>
    </a-form>
</div>
<div>
    <a-table
        :columns="columns"
        :dataSource="dataSource"
        rowKey="name"
        @clear="onClear"
        @change="onChange"
        :pagination="{
            current: params.currentPage,
            pageSize: params.pageSize,
            total: total,
            showSizeChanger: true,
            showLessItems: true,
            showQuickJumper: true,
            showTotal: (total, range) => `第 ${range[0]}-${range[1]} 条，总计 ${total} 条`
        }"
```

```
            >
                <template slot="env_name" slot-scope="text,record">
                    <a-tag color='blue'>{{text}}</a-tag>
                </template>
                <div slot="action" slot-scope="text, record">
                    <a-button-group>
                        <a-button type="primary" @click.prevent="onShowEdit(record)">
                            编辑
                        </a-button>
                        <a-popconfirm
                            title=" 确定执行删除操作么？ "
                            ok-text=" 是 "
                            cancel-text=" 否 "
                            @confirm="onDelete(record)"
                        >
                            <a-button type="danger">
                                删除
                            </a-button>
                        </a-popconfirm>
                    </a-button-group>
                </div>
            </a-table>
        </div>
    </a-card>
</template>
```

服务器管理组件以通用的卡片容器 a-card 作为根节点，其中内嵌 a-form 和 a-table 组件，在 a-table 组件中嵌套了 slot 名为 env_name 和 action 的 template 模板，分别用于渲染服务器归属环境列以及操作列，其他详细的属性已在技术要点中介绍，在此不再赘述。

在 list.vue 文件中添加 script 标签，项目管理组件中实现的事件包含：

- fetchData：请求接口获取服务器列表信息。
- onReset：重置顶部 form 表单数据。
- onChange：用户单击分页或表格个数 pageSize 变更时触发重新请求数据。
- submitHandler：用户单击查询按钮根据 form 表单输入的条件查询服务器列表信息。

相关代码如下：

```
<script>
import API from '@/service'
```

```js
import moment from 'moment'
const columns = [
    {
        title: ' 服务器 IP',
        dataIndex: 'name'
    },
    {
        title: ' 端口 ',
        dataIndex: 'port'
    },
    {
        title: ' 系统类型 ',
        dataIndex: 'system_type'
    },
    {
        title: ' 所属应用 ',
        dataIndex: 'app_name'
    },
    {
        title: ' 环境 ',
        dataIndex: 'env_name',
        scopedSlots: { customRender: 'env_name' }
    },
    {
        title: ' 更新时间 ',
        dataIndex: 'update_date',
        sorter: true,
        customRender: (date) =>{ return moment(date).format("YYYY-MM-DD hh:mm")}
    },
    {
        title: ' 操作 ',
        scopedSlots: { customRender: 'action' }
    }
]

export default {
    name: 'serverList',
    data () {
        return {
            total:0,                          // 总条数
            columns: columns,                 // 列配置
            dataSource: [],                   // 数据源
```

```
            visible:false,            // 模态框开关
            options:[],               // 组件选择框数据源
            envOptions:[],            // 环境选择数据源
            isEdit:false,             // 是否编辑模式
            params:{                  // 分页参数
                name:"",
                currentPage:1,
                pageSize:20,
                begin_time:"",
                end_time:"",
                sorter:""
            }

        }
    },
    beforeCreate() {
        this.form = this.$form.createForm(this, { name: 'serverList' });
        this.formDialog = this.$form.createForm(this, { name: 'formDialog' });
    },
    created(){
        this.fetchData()                // 获取列表数据
        this.fetchComponentList()
        this.fetchEnv()
    },
    methods: {
        submitHandler(e){
            e.preventDefault()
            this.form.validateFields((err, fieldsValue) => {
                if (err) {
                    return;
                }
                const rangeValue = fieldsValue['timePicker']
                this.params.name = fieldsValue["IP"]
                this.params.begin_time = rangeValue?rangeValue[0].format("YYYY-MM-DD"):""
                this.params.end_time = rangeValue?rangeValue[1].format("YYYY-MM-DD"):""
                this.fetchData()
            });
        },
        fetchData(){
            API.ServerList(this.params).then((res)=>{
                let result = res.data
```

```javascript
                if(res.status == 200 && result.code == 0){
                    this.total = result.data.count
                    this.dataSource = result.data.results;
                }
                else{
                    this.dataSource = []
                }
            })
        },
        onReset(){
            this.visible = false
            this.isEdit = false
            this.formDialog.resetFields()
        },
        onChange(pagination, filters, sorter) {
            let {
                current,
                pageSize
            } = pagination
            let {
                order,
                field
            } = sorter
            this.params.currentPage = current
            this.params.pageSize = pageSize
            this.params.sorter = (field?field:"")
            this.fetchData()
        },
    }
}
</script>
```

API.ServerList 方法为 service 服务配置的请求服务器列表的 request 请求，在 service 服务配置中介绍具体的实现。

2. 编辑新增服务器

服务器信息的新增和编辑采用模态框的形式实现，其中嵌套 form 表单、服务器 IP 输入框、服务器端口输入框、组件选择框、归属环境选择框以及隐藏的 id 输入框（用于在编辑时自动提交对应的项目 id）。在 template 中新增代码如下：

```
<a-modal
    :visible="visible"
    :title="isEdit?' 编辑服务器 ':' 新增服务器 '"
```

```
    :okText="isEdit?' 更新 ':' 新建 '"
    @cancel="onReset"
    @ok="isEdit?onUpdateServer():onCreateServer()"
>
    <a-form layout='vertical' :form="formDialog">
        <a-form-item
            label="id"
            v-show="false"
        >
            <a-input placeholder=" 自动生成 "
                v-decorator="['id']"
            />
        </a-form-item>
        <a-form-item label=' 服务器 IP'>
            <a-input
                :disabled="isEdit?true:false"
                v-decorator="[
                    'ip',
                    {
                        rules: [{ required: true, message: ' 请输入服务器 IP' }],
                    }
                ]"
            />
        </a-form-item>
        <a-form-item label=' 服务器端口 '>
            <a-input
                :disabled="isEdit?true:false"
                v-decorator="[
                    'port',
                    {
                        rules: [{ required: true, message: ' 请输入服务器端口 ' }],
                    }
                ]"
            />
        </a-form-item>
        <a-form-item
            label=" 组件选择 "
        >
            <a-select
                show-search
                placeholder=" 请选择组件 "
                option-filter-prop="children"
                style="min-width: 200px;width:100%"
```

```
                :filter-option="filterOption"
                v-decorator="['appId', { rules: [{ required: true, message: ' 请选择所属的组件 !' }] }]"
            >
                <a-select-option v-for="d in options" :key="d.value">
                    {{ d.label }}
                </a-select-option>
            </a-select>
        </a-form-item>
        <a-form-item
            label=" 环境选择 "
        >
            <a-select
                show-search
                placeholder=" 请选择组件 "
                option-filter-prop="children"
                style="min-width: 200px;width:100%"
                :filter-option="filterOption"
                v-decorator="['envId', { rules: [{ required: true,
                            message: ' 请选择所属的环境 !' }] }]"
            >
                <a-select-option v-for="d in envOptions" :key="d.value">
                    {{ d.label }}
                </a-select-option>
            </a-select>
        </a-form-item>
        <a-form-item
            label=" 操作系统 "
        >
            <a-select
                v-decorator="[ 'system', { rules: [{ required: true, message: ' 请选择操作系统 ' }] }]"
            >
                <a-select-option value="WINDOWS">
                    WINDOWS
                </a-select-option>
                <a-select-option value="LINUX">
                    LINUX
                </a-select-option>
            </a-select>
        </a-form-item>
        <a-form-item label=' 描述 '>
            <a-input
                type='textarea'
                v-decorator="['description']"
```

```
            />
        </a-form-item>
    </a-form>
</a-modal>
```

在 script 标签中新增事件包含：

- onUpdateServer：请求接口更新单个服务器信息。
- onCreateServer：请求接口新增服务器信息。
- fetchEnv：请求环境配置。
- fetchComponentList：请求应用列表信息，用于选择服务器归属应用。
- onShowEdit：控制模态框变更为编辑模式并使用行数据初始化表单。
- createDialog：控制模态框变更为新增模式。

新增的事件代码如下：

```
<script>
... // 其他忽略
    methods: {
    ...// 其他忽略
    onShowEdit(record){
        this.visible = true
        this.isEdit = true
        this.$nextTick(()=>{
            this.formDialog.setFieldsValue({
                id:record.id,
                ip:record.ip,
                port:record.port,
                system:record.system_type,
                appId:record.app,
                envId:record.env,
                description:record.description
            })
        })
    },
    onUpdateServer(){
        this.formDialog.validateFields((err, fieldsValue) => {
            if (err) {
                return;
            }
            let id = fieldsValue["id"],
                ip = fieldsValue["ip"],
                port = fieldsValue["port"],
```

```javascript
                    app_id = fieldsValue["appId"],
                    env_id = fieldsValue["envId"],
                    system_type = fieldsValue["system"],
                    description = fieldsValue["description"]||""
                let data = {
                    id,
                    ip,
                    port,
                    app_id,
                    env_id,
                    system_type,
                    description
                }
                API.UpdateServer(data).then((res)=>{
                    let result = res.data
                    if(res.status == 200){
                        this.$message.success(" 服务器更新成功 ~")
                        this.onReset()
                        this.fetchData()
                    }
                    else{
                        this.$message.error(" 服务器更新失败 :"+result.message)
                    }
                })
            })
        },
        fetchComponentList(){
            API.AppList({}).then((res)=>{
                let result = res.data
                if(res.status == 200 && result.code == 0){
                    result.data.results.forEach(item=>{
                        this.options.push({
                            label:item.name,
                            value:item.id
                        })
                    })
                }
                else{
                    this.$message.error(" 无法获取应用列表 ~")
                }
            })
        },
        onCreateServer () {
```

```javascript
                this.formDialog.validateFields((err, fieldsValue) => {
                    if (err) {
                        return;
                    }
                    let ip = fieldsValue["ip"],
                        port = fieldsValue["port"],
                        app_id = fieldsValue["appId"],
                        env_id = fieldsValue["envId"],
                        system_type = fieldsValue["system"],
                        description = fieldsValue["description"]||""
                    let data = {
                        ip,
                        port,
                        app_id,
                        env_id,
                        system_type,
                        description
                    }
                    API.CreateServer(data).then((res)=>{
                        let result = res.data
                        if(res.status == 200 && result.code == 0){
                            this.$message.success(" 服务器新增成功 ~")
                            this.visible = false
                            this.fetchData()
                        }
                        else{
                            this.$message.error(" 服务器新增失败 :"+result.message)
                        }
                    })
                })
        },
        createDialog(){
            this.visible = true
        }
    }
</script>
```

3. 删除服务器

在 a-table 操作列中使用模板渲染了删除按钮，单击按钮可通过 slot-scope 获取当前行数据 record，在单击删除按钮时，可通过该变量获取对应的行数据的 id 进行删除操作。

在 script 中添加删除事件 onDelete，相关代码如下：

```
<script>
    ... // 其他忽略
    methods: {
        ... // 其他忽略
        onDelete(record){
            let {id} = record;
            if(id == undefined){
                this.$message.error(" 操作参数非法！ ")
                return false
            }
            API.DeleteServer({id}).then((res)=>{
                let result = res.data
                if(res.status == 200 && result.code == 0){
                    this.$message.success(" 删除成功 ~")
                    this.fetchData()
                }
                else{
                    this.$message.error(" 操作错误 ~:"+result.message)
                }
            })
        },
    }
}
</script>
```

至此，服务器管理组件已全部实现完毕，组件通过 import 引入 service API，调用的服务器用增删查改方法，为 service 服务配置的请求发布单信息的 request 请求，在 service 服务配置中介绍具体的实现。

17.7 服务器基础服务配置

服务器管理页面开发完成后，若要在浏览器中测试开发的页面展示以及功能是否有问题，需要配置路由、service 以及 mock 服务。

17.7.1 路由配置

若要在浏览器能够展示服务器管理页面，只需把组件配置到 router 目录下的 config.js 文件中，代码如下：

```
const routeConfig = [
    {
```

```
            path: '/server',
            name: '服务器',
            meta: {
                    icon: 'cluster'
            },
            children:[
                    {
                            path: '/serverlist',
                            name: '列表',
                            component: () => import('@/views/server/list')
                    }
            ]
    },
]
```

在浏览器地址中添加 hash 值为 server/list，即访问 http://localhost:8080/#/server/list，便可以在浏览器中看到开发完的服务器管理页面。

17.7.2　mock服务配置

项目中的本地 mock 服务用于模拟服务端请求，并返回自定义的模拟数据。服务器管理页面请求服务端针对服务器的增删查改接口，所以需要在 mock 目录下新增 server 目录，目录中新增 index.js 文件，文件内容如下。在 template 中新增代码如下：

```
import Mock from 'mockjs'
const serverList = Mock.mock({
        "count":"@integer(1, 100)",
        "results|100":[
                {
                        "id": "@id",
                        "app_name": "@name",
                        "project_name": "@name",
                        "env_name": "@name",
                        "create_username": "@name",
                        "name": "@ip",
                        "description": "@csentence()",
                        "update_date": "@date",
                        "create_date": "@date",
                        "base_status": "@boolean",
                        "ip": "@ip",
                        "port": "@integer(1000,30000)",
```

```js
                    "system_type"|1: ["WINDOWS","LINUX"],
                    "deploy_no": "@integer",
                    "create_user": "@id",
                    "app": "@id",
                    "env": "@id",
                    "deploy_status": "@id"
                }
            ]
})

Mock.mock(`${process.env.VUE_APP_API_BASE_URL}/server/list/`, 'get', ({body}) => {
        return {
                "code": 0,
                "message": "success",
                "data": serverList
        }
})

const addServer = Mock.mock({
        "id": "@id",
        "app_name": "@name",
        "project_name": "@name",
        "env_name": "@name",
        "create_username": "@name",
        "name": "@ip",
        "description": "@csentence()",
        "update_date": "@date",
        "create_date": "@date",
        "base_status": "@boolean",
        "ip": "@ip",
        "port": "@integer(1000,30000)",
        "system_type"|1: ["WINDOWS","LINUX"],
        "deploy_no": "@integer",
        "create_user": "@id",
        "app": "@id",
        "env": "@id",
        "deploy_status": "@id"
})

Mock.mock(`${process.env.VUE_APP_API_BASE_URL}/server/create/`, 'post', ({body}) => {
        return {
                "code": 0,
                "message": "success",
```

```
                "data": addServer
        }
})

Mock.mock(`${process.env.VUE_APP_API_BASE_URL}/server/update/`, 'put', ({body}) => {
        return {
                "code": 0,
                "message": "success",
                "data": {}
        }
})

Mock.mock(`${process.env.VUE_APP_API_BASE_URL}/server/delete/`, 'delete', ({body}) => {
        return {
                "code": 0,
                "message": "success",
                "data": {}
        }
})
```

在mock服务的入口文件index.js中引入server目录，即完成了服务器mock服务的注册，当服务器组件通过GET/POST/PUT/DELETE方式访问/server/*路径时，mock服务层会返回配置的数据。服务器管理模块访问mock服务层的接口数据需要通过service层，所以还需要在service层配置服务器管理模块。

17.7.3　service服务配置

service服务统一封装了BiFang部署平台项目所需要的所有接口路径以及请求方式，并通过axios调用下游的mock数据，所以service服务需要新增服务器管理模块。

在services目录下的api.js路径集中新增服务器管理所需的访问路径，代码如下：

```
const API = {
        SERVERLIST: `${REAL_URL}/server/list/`,                 // 服务器列表
        CREATESERVER: `${REAL_URL}/server/create/`,             // 创建服务器列
        // 删除服务器
        DELETESERVER: `${REAL_URL}/server/delete/{{id}}/`,
        // 更新服务器
```

```
            UPDATESERVER: `${REAL_URL}/server/update/{{id}}/`,
}
```

在 services 目录下新增 server.js 文件，用于配置服务器管理组件所需服务，代码如下：

```
import {request,METHOD} from '@/utils/request'
import {urlFormat} from '@/utils/util'
import {API} from './api'

/**
* 获取服务器列表
* @param {*} params
*/
async function ServerList(params){
        return request(
                API.SERVERLIST,
                METHOD.GET,
                params
        )
}

/**
* 新增服务器
* @param {*} params
*/
async function CreateServer(params){
        return request(
                API.CREATESERVER,
                METHOD.POST,
                params
        )
}

/**
* 删除服务器
* @param {*} params
*/
async function DeleteServer(params){
        const url = urlFormat(API.DELETESERVER, params)
        return request(
                url,
                METHOD.DELETE
        )
```

```
}

/**
 * 更新服务器
 * @param {*} params
 */
async function UpdateServer(params){
        const url = urlFormat(API.UPDATESERVER, params)
        return request(
                url,
                METHOD.PATCH,
                params
        )
}

export default {
        ServerList,
        CreateServer,
        DeleteServer,
        UpdateServer
}
```

在 services 目录下新增 permission.js 文件,用于配置项目与应用组件中权限所需服务,代码如下:

```
import {request,METHOD} from '@/utils/request'
import {urlFormat} from '@/utils/util'
import {API} from './api'

/**
 * 获取权限列表
 * @param {*} params
 */
async function PermissionList(params){

        return request(
                API.PERMISSIONLIST,
                METHOD.GET,
                params
        )
}

/**
```

```
/**
 * 新增权限
 * @param {*} params
 */
async function CreatePermission(params){
        return request(
                API.CREATEPERMISSION,
                METHOD.POST,
                params
        )
}
/**
 * 删除权限
 * @param {*} params
 */
async function DeletePermission(params){
        const url = urlFormat(API.DELETEPERMISSION, params)
        return request(
                url,
                METHOD.DELETE
        )
}
export default {
        PermissionList,
        CreatePermission,
        DeletePermission
}
```

service 目录下主入口文件 index.js 中引入项目与应用模块和权限模块，代码如下：

```
import application from './application'
import permission from './permission'
export default {
        ...application,
        ...permission,
}
```

服务器管理组件中调用 service 层配置的接口，axios 发起 request 请求并通过 promise.then(response=>{}) 的回调方式把获取到的数据返回给处理函数。

到此为止，服务器管理页面相对应的 mock 服务、service 服务代码都已全部编写完毕，启动项目，单击左侧菜单导航中的服务器管理菜单，页面即可呈现在读者面前，增删查改操作调用接口返回成功标识。

在 service 服务层中的 api.js 中改变 BASE_UR 和 REAL_URL 即可完成 mock 数据和真实环境的请求切换，方便项目的调试。

17.8 小　　结

本章带领读者熟悉了 BiFang 部署平台中项目与应用模块以及服务器模块的页面设计、页面搭建到功能实现的流程，掌握了如何在公共服务路由、mock 服务以及 service 服务中添加项目应用模块和服务器模块，中间也穿插了用到的 Ant-design-vue 框架相关的组件知识点，相信没有了解过相关知识点的读者能对使用框架组件布局页面有全面的认识。

第 18 章将带领大家进入 BiFang 部署平台发布单的管理和环境流转，发布单是什么？如何流转发布单状态以及它跟项目应用以及服务器关系是怎样的？让我们带着这些问题，前进！

每一个闪闪发光的人都在背后熬过了一个又一个不为人知的黑夜！

第18章 发布单生成、流转模块设计与搭建

> 有志者，事竟成，破釜沉舟，百二秦关终属楚。
>
> ——蒲松龄

本章 GitHub 代码地址：https://github.com/aguncn/bifang-book/tree/main/bifang-ch18。

第17章主要介绍了 BiFang 部署平台基础配置中项目与应用模块以及服务器模块的设计与搭建，开发项目管理、应用管理以及服务器管理组件，调用对应模块的增删查改接口。在 mock 服务层、service 服务层以及路由层添加项目与应用模块和服务器管理专属的模块，熟悉 Web 页面流程配置方法。接下来将带领大家进入发布单生成模块和环境流转模块的搭建，再次实践开发流程配置。

本章将在前端项目中设计搭建发布单生成模块以及环境流转模块，使用 ant-design-vue 提供的组件构建页面组件。新增发布单生成模块中的新增发布单页面、发布单列表以及发布单详情页面，新增环境流转模块待流转发布单页面。在 mock 服务、service 服务和路由层添加页面对应的发布单生成和环境流转模块实现接口服务请求和页面展示。

18.1 发布单生成模块

发布单是应用发布时间移交的应用代码集合，它和应用是多对一的关系，一个应用在不同时间有不同的版本。发布单是贯穿在整个 BiFang 部署平台中，应用的发布就是执行对应操作，更新发布单状态，直至发布成功。

发布单状态分为 create（新建发布单）、building（应用代码构建中）、build（应用代码构建成功）、buildfailed（应用代码构建失败）、ready（发布单扭转）、ongoing（发布单发布中）、success（发布单发布成功）和 failed（发布单发布失败）。发布单生成模块主要有发布单新建、列表、详细及软件包构建功能。

第 18 章　发布单生成、流转模块设计与搭建

1. 功能说明与清单

（1）发布单生成模块功能清单如表 18-1 所列。

表 18-1　发布单生成模块功能表

模块名称	功能名称	所属页面	功能描述
发布单	新建发布单	新建发布单页面	发布单号自动生成 选择相应项目下的应用 输入 git 代码分支 输入发布单描述
	发布单列表	发布单列表页面	所有发布单按时间排序 可按项目应用搜索发布单
	构建发布单	发布单列表页面	显示发布单具体信息 构建此发布单软件包 失败后可重复构建 构建完成后可快速进入环境流转环节
	发布单详情	发布单详情页面	显示发布单所有信息 显示发布单操作历史

（2）发布单生成模块功能说明。

① 发布单名称以时间戳和两位随机字母构成，预防冲突。

② 输入的 git 分支会传到 GitLab 的 CI/CD 流水线中。

③ 构建时，会通过 trigger token 触发 GitLab 的 CI/CD 流水线。

④ 在发布单详情中，同时展示发布单的历史操作记录。

发布单生成模块截图如图 18-1 ～图 18-4 所示。

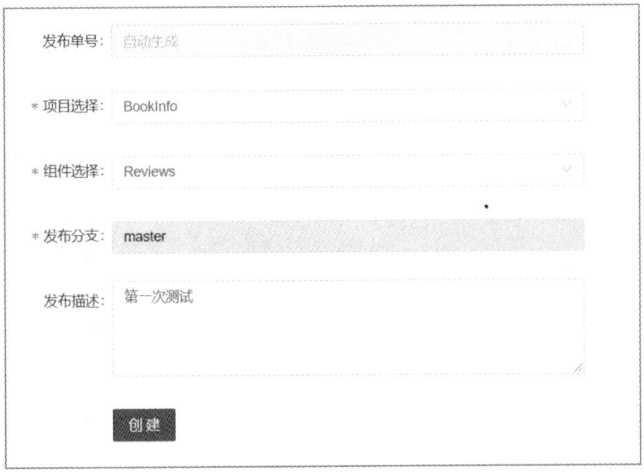

图 18-1　BiFang 新建发布单

图 18-2　BiFang 发布单列表

图 18-3　BiFang 构建发布单

图 18-4　BiFang 发布单详情

2. 功能技术要点

在发布单生成模块发布单列表、发布单详情以及发布单历史页面搭建中，主要使用 Antd 框架中的组件进行搭建，详细功能点使用的组件如下：

- 查询操作：使用 a-form 表单、a-table 表格、a-pagination 分页组件。
- 新建操作：使用 a-form 表单、a-select 选择框、a-input 输入框组件。
- 构建操作：使用 a-modal 模态框、a-button 按钮组件组件。
- 查看详情操作：使用 a-list 列表、a-timeline 时间轴组件。

上述使用的 Antd 组件常用属性，方便读者能清晰地知道各属性的作用。

Form 表单是具有数据收集、校验和提交功能的表单，包含复选框、单选框、输入框、下拉选择框等元素，常用的 API 如下：

- layout：表单元素排列方式，水平排列（vertical）、垂直排列（horizontal，默认）、行内排列（inline）。
- labelAlign：label 标签的文本对齐方式，左对齐（left）、右对齐（right）。
- labelCol：label 标签布局，设置 span offset 值，如 {span: 3, offset: 12}。
- wrapperCol：需要为输入控件设置布局样式时，使用该属性，用法同 labelCol。
- getFieldsValue：获取一组输入控件的值，如不传入参数，则获取全部组件的值。
- setFieldsValue：设置一组输入控件的值。
- validateFields：校验并获取一组输入域的值与 error，若 fieldNames 参数为空，则校验全部组件。

Table 表格用于当需要对数据进行排序、搜索、分页、自定义操作等复杂行为时展示行列数据，常用的 API 如下：

- dataSource：表格数据源，对应服务端返回的数据。
- columns：表格列的配置，Object 类型数组。
- title：列头显示文字。
- align：内容的对齐方式，'left' | 'right' | 'center'。
- dataIndex：列数据在数据项中对应的 key。
- customRender：生成复杂数据的渲染函数，参数分别为当前行的值、当前行数据、行索引，@return 里面可以设置表格行/列合并。
- scopedSlots：使用 columns 时，可以通过该属性配置支持 slot-scope 的属性，

如 scopedSlots: {customRender: 'XXX'}。
- loading：页面是否加载中。
- pagination：分页器，参考配置项或 pagination 文档，设为 false 时不展示和进行分页。
- rowKey：表格行 key 的取值，可以是字符串或一个函数。
- change：分页、排序、筛选变化时触发。

Pagination 分页采用分页的形式分隔长列表，每次只加载一个页面，常用的 API 如下：
- current：当前页数。
- pageSize：每页条数。
- total：数据总数。
- showSizeChanger：是否可以改变 pageSize。
- showQuickJumper：是否可以快速跳转至某页。
- showSizeChange：pageSize 变化的回调。

Button 按钮标记了一个（或封装一组）操作命令，响应用户单击行为，触发相应的业务逻辑，常用的 API 如下：
- disabled：按钮失效状态。
- icon：设置按钮的图标类型。
- loading：设置按钮载入状态。
- type：设置按钮类型，可选值为 primary dashed danger link。
- click：单击按钮时的回调。

Modal 模态对话框在当前页面正中打开一个浮层，承载用户处理事务，又不希望跳转页面以致打断工作流程，常用 API 如下：
- cancelText：取消按钮文字。
- closable：是否显示右上角的关闭按钮。
- destroyOnClose：关闭时销毁 Modal 里的子元素 footer 底部内容，当不需要默认底部按钮时，可以设为 ":footer="null""。
- mask：是否展示遮罩 maskClosable 单击蒙层是否允许关闭。
- okText：确认按钮文字。
- okType：确认按钮类型。
- title：标题。

- width：宽度。

List 通用列表可承载文字、列表、图片、段落，常用于后台数据展示页面，常用 API 如下：
- footer：列表底部。
- grid：列表栅格配置。
- itemLayout：设置 List.Item 布局，设置成 vertical 则竖直样式显示，默认横排。
- loadMore：加载更多。
- Split：是否展示分割线。
- dataSource：列表数据源。

Timeline 垂直展示的时间流信息，当有一系列信息需按时间排列时，可正序和倒序，常用 API 如下：
- pending：指定最后一个幽灵节点是否存在内容。
- reverse：节点排序。
- pendingDot：当最后一个幽灵节点存在时，指定其时间图点。
- mode：通过设置 mode 可以改变时间轴和内容的相对位置。

以上为搭建项目与应用模块使用的 Antd 组件介绍，接下来进入页面搭建。

18.2　发布单列表页面搭建

发布单列表页面包含查询发布单列表、新增发布单按钮以及发布单构建，其中单击新增发布单将跳转创建发布单页面以创建发布单，这里将以实现每个功能的步骤来呈现。

在 src 中的 views 目录下创建 release 目录，并在该目录下创建 list.vue 文件，此文件为发布单列表组件。

1. 发布单列表查询

发布单列表包含发布单编号列、归属项目列、归属应用列、编译分支名称列、创建用户列以及构建按钮列，其中构建按钮分为已构建和未构建，已构建按钮无法单击。同时列表顶部包含过滤条件，根据用户选择的项目名称、应用名称和发布单编号进行过滤。具体的 template 代码如下：

```html
<template>
    <a-card>
        <div class="search">
            <a-form layout="inline" :form="form" @submit="submitHandler">
                <a-form-item
                    label=" 项目选择 "
                >
                    <a-select
                        show-search
                        placeholder=" 请选择项目 "
                        option-filter-prop="children"
                        :filter-option="filterOption"
                        @change="handleChange"
                        style="width:200px"
                        v-decorator="['projectName', { rules: [{ required: false,
                            message: ' 请选择项目 !' }] }]"
                    >
                        <a-select-option v-for="d in projectOption" :key="d">
                            {{ d }}
                        </a-select-option>
                    </a-select>
                </a-form-item>
                <a-form-item
                    label=" 组件选择 "
                >
                    <a-select
                        show-search
                        placeholder=" 请选择组件 "
                        option-filter-prop="children"
                        :filter-option="filterOption"
                        style="width:200px"
                        v-decorator="['appName', { rules: [{ required: false,
                            message: ' 请选择发布的组件 !' }] }]"
                    >
                        <a-select-option v-for="d in options" :key="d.label">
                            {{ d.label }}
                        </a-select-option>
                    </a-select>
                </a-form-item>
                <a-form-item
                    label=" 发布单号 "
                >
                    <a-input
```

```html
                    placeholder=" 请输入发布单 "
                    v-decorator="['releaseNo', { rules: [{ required: false,
                        message: ' 请输入发布单 ' }] }]" />
            </a-form-item>
            <a-form-item>
                <a-button type="primary" html-type="submit">
                    查询
                </a-button>
            </a-form-item>
            <a-form-item>
                <a-button type="primary" @click.prevent="onCreateRelease">
                    新建
                </a-button>
            </a-form-item>
        </a-form>
</div>
<div>
    <a-table
        :columns="columns"
        :dataSource="dataSource"
        rowKey="name"
        @change="onChange"
        :pagination="{
            current: params.currentPage,
            pageSize: params.pageSize,
            total: total,
            showSizeChanger: true,
            showLessItems: true,
            showQuickJumper: true,
            showTotal: (total, range) => `第 ${range[0]}-${range[1]} 条，总计 ${total} 条`
        }"
    >
        <div slot="name" slot-scope="text, record">
            <a-tooltip>
                <template slot="title">
                    {{record.description}}
                </template>
                <a @click="showReleaseHistory(record)">
                    {{text}}
                </a>
            </a-tooltip>
        </div>
        <div slot="action" slot-scope="text, record">
```

```html
                    <a-button type="primary"
                        v-if="record.release_status_name == 'Create'"
                        @click="buildShow(record)">
                        构建
                    </a-button>
                    <a-button type="primary"
                        v-else-if="record.release_status_name == 'BuildFailed'"
                        @click="buildShow(record)">
                        构建
                    </a-button>
                    <a-button type="default" v-else     disabled>
                        已构建
                    </a-button>
                </div>
                <template slot="git_branch" slot-scope="text,record">
                    <a-tooltip>
                        <template slot="title">
                            {{record.description}}
                        </template>
                        <a-tag color='blue'>{{text}}</a-tag>
                    </a-tooltip>
                </template>
                <template slot="statusTitle">
                    <a-icon @click.native="onStatusTitleClick" type="info-circle" />
                </template>
            </a-table>
        </div>
    </a-card>
</template>
```

发布单列表组件以通用的卡片容器 a-card 作为根节点，其中内嵌 a-form 和 a-table 组件，在 a-table 组件中嵌套了 slot 名为 name、git_branch 和 action 的 template 模板。name 模板用于渲染组件名称，单击发布单编号可以进入组件部署历史详情页面；git_branch 模板用于渲染编译分支，使用 a-tag 组件显示分支名称；action 模板用于渲染操作列，若该发布单已构建，则显示"已构建"，用户无法单击。其他详细的属性已在技术要点中介绍，在此不再赘述。

在 list.vue 文件中添加 script 标签，发布单列表组件中实现的事件包含：

- fetchData：请求接口获取已生成的发布单列表信息。
- fetchAppList：请求接口获取应用列表，供过滤条件中的组件列表使用。
- handleChange：当过滤条件中项目列表被选择时，更新组件列表数据。

- filterOption：select 过滤函数。
- showReleaseHistory：跳转发布单部署历史页面。
- onCreateRelease：跳转新建发布单页面。
- onChange：用户单击分页或表格个数 pageSize 变更时触发重新请求数据。
- submitHandler：用户单击查询按钮根据 form 表单输入的条件查询项目列表信息。

相关代码如下：

```
<script>
import API from '@/service'
import moment from 'moment'
const columns = [
    {
        title: ' 发布单编号 ',
        dataIndex: 'name',
        scopedSlots: {customRender: 'name'}
    },
    {
        title: ' 项目 ',
        dataIndex: 'project_name'
    },
    {
        title: ' 组件 ',
        dataIndex: 'app_name'
    },
    {
        title: ' 编译分支 ',
        dataIndex: 'git_branch',
        scopedSlots: {customRender: 'git_branch'}
    },
    {
        title: ' 用户 ',
        dataIndex: 'create_user_name'
    },
    {
        title: ' 更新时间 ',
        dataIndex: 'update_date',
        sorter: true,
        defaultSortOrder: 'descend',
        sortDirections: ['descend', 'ascend'],
        customRender: (date) =>{ return
```

```js
            moment(date).format("YYYY-MM-DD hh:mm")}
    },
    {
        title: ' 操作 ',
        scopedSlots: { customRender: 'action' }
    }
]

export default {
    name: 'releaseList',
    data () {
        return {
            total:0,                    // 发布单总数量
            visiable:false,             // 构建模态框
            modelData: {},              // 单发布单信息
            buildTimer: null,           // 定时器
            buildStatus: "notBegin",    // 初始构建状态
            columns: columns,           // 列配置项
            dataSource: [],             // 数据源
            projects:{},                // 项目列表
            projectOption: [],          // 项目选择数据源
            options:[],                 // 应用选择数据源
            params:{                    // 分页配置
                name:"",
                currentPage:1,
                pageSize:20,
                sort:""
            }
        }
    },
    beforeCreate() {
        this.form = this.$form.createForm(this, { name: 'releaseList' });
    },
    created(){
        this.fetchAppList()
        this.fetchData()
    },
    methods: {
        submitHandler(e){
            e.preventDefault()
            this.form.validateFields((err, fieldsValue) => {
                if (err) {
                    return;
                }
```

```js
                    const rangeValue = fieldsValue['timePicker']
                    this.params.name = fieldsValue["releaseNo"]
                    this.params.project_name = fieldsValue["projectName"]
                    this.params.app_name = fieldsValue["appName"]
                    this.fetchData()
                });
        },
        fetchData(){
            API.ReleaseList(this.params).then((res)=>{
                let result = res.data
                if(res.status == 200 && result.code == 0){
                    this.total = result.data.count
                    this.dataSource = result.data.results;
                    console.log(this.dataSource)
                } else {
                    this.dataSource = []
                }
            })
        },
        fetchAppList(){
            API.AppList({}).then((res)=>{
                let result = res.data
                if(res.status == 200 && result.code == 0){
                    result.data.results.forEach(item=>{
                        this.projects[item.project_name]?
                            this.projects[item.project_name].push({
                            label:item.name,
                            value:item.id
                        }):this.projects[item.project_name] = [{
                            label:item.name,
                            value:item.id
                        }]
                    })
                    this.projectOption = Object.keys(this.projects)
                }
                else{
                    this.$message,error(" 无法获取应用列表 ~")
                }
            })
        },
        handleChange(value) {
            this.form.setFieldsValue({
                appName:""
```

```
                })
                this.options = this.projects[value]
        },
        filterOption(input, option) {
            return (
                option.componentOptions.children[0].text.toLowerCase().indexOf(input.toLowerCase()) >= 0
            );
        },
        showReleaseHistory(data) {
            console.log(data)
            // 链接到发布单历史部署详细页面
            this.$router.push({ name: '发布单部署历史', params: { releaseId: data.id }});
        },
        onChange(pagination, filters, sorter) {
            let {
                current,
                pageSize
            } = pagination
            let {
                order,
                field
            } = sorter
            this.params.currentPage = current
            this.params.pageSize = pageSize
            console.log("sort order",order)
            if(!order){
                field = ""
            }
            else if(order == 'descend'){
                field = '-'+field
            }
            else{}
            this.params.sort = field
            this.fetchData()
        },
        onCreateRelease(){
            this.$router.push("createRelease")
        }
    }
}
</script>
```

通过 import 引入 service API，调用的方法为 service 服务配置的请求发布单列表的 request 请求，在 service 服务配置中介绍具体的实现。

2. 新增发布单

新增发布单在新的页面中实现，使用 form 表单内嵌不可编辑的 id 输入框、项目选择框、应用组件选择框、发布单构建分支以及发布单描述输入框。

在 views 下的 release 目录创建 create.vue 文件，此文件为新建发布单组件。在 template 中新增代码如下：

```
<template>
    <a-card :body-style="{padding: '24px 32px'}" :bordered="false">
        <a-form :form="form" @submit="submitHandler">
            <a-form-item
                label=" 发布单号 "
                :labelCol="{span: 7}"
                :wrapperCol="{span: 10}"
            >
                <a-input disabled value=" 自动生成 " placeholder=" 自动生成 " />
            </a-form-item>
            <a-form-item
                label=" 项目选择 "
                :labelCol="{span: 7}"
                :wrapperCol="{span: 10}"
            >
                <a-select
                    show-search
                    placeholder=" 请选择项目 "
                    option-filter-prop="children"
                    :filter-option="filterOption"
                    @focus="handleFocus"
                    @blur="handleBlur"
                    @change="handleChange"
                    v-decorator="['projectId', { rules: [{ required: true, message: ' 请选择项目 !' }] }]"
                >
                    <a-select-option v-for="d in projectOption" :key="d">
                        {{ d }}
                    </a-select-option>
                </a-select>
            </a-form-item>
            <a-form-item
                label=" 组件选择 "
                :labelCol="{span: 7}"
```

```html
            :wrapperCol="{span: 10}"
        >
            <a-select
                show-search
                placeholder=" 请选择组件 "
                option-filter-prop="children"
                :filter-option="filterOption"
                v-decorator="['appId', { rules: [{ required: true, message: ' 请选择发布的组件！'}] }]"
            >
                <a-select-option v-for="d in options" :key="d.value">
                    {{ d.label }}
                </a-select-option>
            </a-select>
        </a-form-item>
        <a-form-item
            label=" 发布分支 "
            :labelCol="{span: 7}"
            :wrapperCol="{span: 10}"
        >
            <a-input
                placeholder=" 请输入发布单分支 "
                v-decorator="['branch', { rules: [{ required: true, message: ' 请输入发布单分支！'}] }]"
            />
        </a-form-item>
        <a-form-item
            label=" 发布描述 "
            :labelCol="{span: 7}"
            :wrapperCol="{span: 10}"
        >
            <a-textarea rows="4"
                placeholder=" 请输入发布单描述 "
                v-decorator="['description', { rules: [{ required: false, message: ' 请输入发布单描述！'}] }]"
            />
        </a-form-item>
        <a-form-item style="margin-top: 24px" :wrapperCol="{span: 10, offset: 7}">
            <a-button type="primary" html-type="submit"> 创建 </a-button>
        </a-form-item>
    </a-form>
    </a-card>
</template>
```

在 create.vue 文件中添加 script 标签，新建发布单组件中实现的事件如下：

- fetch：请求接口获取所有项目与应用信息，供用户选择新建哪个应用的发布单。
- handleChange：当过滤条件中项目列表被选择时，更新组件列表数据。
- filterOption：select 过滤函数。
- submitHandler：校验并提交用户选择的相关信息，生成发布单。

相关代码如下：

```
<script>
import API from '@/service'
export default {
    name: 'createRelease',
    data () {
        return {
            value: 1,
            fetching:false,
            projectOption:[],
            options:[],
            projects:{}
        }
    },
    beforeCreate() {
        this.form = this.$form.createForm(this, { name: 'createRelease' });
    },
    created(){
        this.fetch()
    },
    computed: {
        desc() {
        }
    },
    methods: {
        fetch(){
            API.AppList({}).then((res)=>{
                let result = res.data
                if(res.status == 200 && result.code == 0){
                    result.data.results.forEach(item=>{
                        this.projects[item.project_name]?
                            this.projects[item.project_name].push({
                                label:item.name,
                                value:item.id
                            }):this.projects[item.project_name] = [{
                                label:item.name,
```

```javascript
                            value:item.id
                        }]
                    })
                    this.projectOption = Object.keys(this.projects)
                }
                else{
                    this.$message,error(" 无法获取应用列表 ~")
                }
            })
        },
        handleChange(value) {
            this.form.setFieldsValue({
                appId:""
            })
            this.options = this.projects[value]
        },
        submitHandler(e){
            e.preventDefault()
            this.form.validateFields((err, fieldsValue) => {
                if (err) {
                    return;
                }
                let data = {
                    name:"",
                    app_id:fieldsValue["appId"],
                    git_branch:fieldsValue["branch"],
                    description:fieldsValue["description"]
                }
                API.CreateRelease(data).then((res)=>{
                    let result = res.data
                    if(res.status == 200 && result.code == 0){
                        this.$message.success(" 发布单创建成功 ~")
                        this.$router.push("releaseList")
                    } else if(res.status == 200 && result.code == 2000){
                        this.$message.error(" 你没有创建此应用发布单的权限！ ")
                    } else {
                        this.$message.error(" 无法获取应用列表 ~")
                    }
                })
            });
        },
        filterOption(input, option) {
            return (
                option.componentOptions.children[0].text.
```

```
                    toLowerCase().indexOf(input.toLowerCase()) >= 0
                );
            },
        }
    }
</script>
```

3. 构建发布单

在发布单列表操作列中使用模板渲染了构建按钮，单击按钮弹出模态框，展示通过 slot-scope 获取的当前发布单数据。在 template 中新增代码如下：

```
<a-modal
            :visible="visiable"
            title=" 软件构建 "
            okText=" 确定 "
            cancelText=" 关闭 "
            @cancel="onReset"
            @ok="onReset"
    >
            <a-card>
                <p> 发布单：{{this.modelData.name}}</p>
                <p>app: {{this.modelData.app_name}}</p>
                <p>git 地址：{{this.modelData.git_url}}/
                    {{this.modelData.project_name}}/
                    {{this.modelData.app_name}}
                </p>
                <p>git 项目 ID: {{this.modelData.git_app_id}}</p>
                <p> 代码分支：{{this.modelData.git_branch}}</p>
                <div v-if="buildStatus == 'notBegin'">
                    <p> 构建状态：尚未开始 </p>
                    <a-button type="danger" @click="onBuild"> 开始构建 </a-button>
                </div>
                <div v-else-if="buildStatus == 'building'">
                    <p> 构建状态：构建中，请等候
                        <a-icon type="sync" :style="{ fontSize: '24px', color: '#00f' }" spin /></p>
                </div>
                <div v-else-if="buildStatus == 'failed'">
                    <p> 构建状态：构建失败，请调整后重试。
                        <a-icon type="close" :style="{ fontSize: '24px', color: '#f00' }" /></p>
                    <a-button type="danger" @click="onBuild"> 开始构建 </a-button>
                </div>
                <div v-else>
                    <p> 构建状态：构建完成，如有权限，可操作流转。
```

```
                    <a-icon type="check" style="{ fontSize: '24px', color: '#000' }" /></p>
                <a-button type="danger" @click="onSwitch"> 快速环境流转 </a-button>
            </div>
        </a-card>
</a-modal>
```

在 script 标签中新增事件如下：

- onReset：控制模态框展示隐藏开关。
- buildShow：获取单个发布单信息并赋值，弹出模态框展示发布单信息。
- onBuild：请求服务端开始进入发布单构建，并请求定时器函数刷新进度。
- getBuildStatus：定时器函数用于刷新发布单构建进度。
- clearBuildTimer：清除定时器。
- onSwitch：获取当前发布单，跳转环境流转页面携带指定发布单。

新增的事件代码如下：

```
<script>
... // 其他忽略
    methods: {
    ...// 其他忽略
    buildShow(data){
        this.modelData = data
        this.visiable = true
    },
    onSwitch(){
        this.$router.push({
            path:'/environment/environmentList',
            query:{
                releaseNo: this.modelData.name
            }
        })
    },
    onBuild(){
        let params = {
            app_name: this.modelData.app_name,
            release_name: this.modelData.name,
            git_branch: this.modelData.git_branch
        }
        API.BuildRelease(params).then((res)=>{
            let result = res.data
            if(res.status == 200 && result.code == 0){
                this.$message.success(" 开始构建 ......")
```

```js
                    this.getBuildStatus()
                } else if(res.status == 200 && result.code == 2000){
                        this.$message.error(" 你没有构建此发布单的权限！ ")
                } else {
                        this.$message.error(" 构建请求失败 ~")
                }
            })
        },
        getBuildStatus() {
            this.buildTimer = setInterval(() => {                    // 创建定时器
                    let params = {
                        app_name: this.modelData.app_name,
                        release_name: this.modelData.name
                    }
                    API.BuildReleaseStatus(params).then((res)=>{
                        if(res.status != 200 ){
                            console.log(res, "no 200")
                            this.buildTimer && this.clearBuildTimer();    // 关闭定时器
                        } else {
                            let resData = res.data.data
                            if(resData == "ing" ){
                                this.buildStatus = 'building'
                            } else if (resData == "success" ) {
                                this.buildStatus = 'success'
                                console.log(resData)
                                this.buildTimer && this.clearBuildTimer();
                            } else {
                                this.buildStatus = 'failed'
                                console.log(resData)
                                this.buildTimer && this.clearBuildTimer();
                            }
                        }
                    })
            }, 3000);
        },
        clearBuildTimer() {                    // 清除定时器
            clearInterval(this.buildTimer);
            this.buildTimer = null;
        },
        onReset(){
            this.visiable = false
        },
    }
}
</script>
```

18.3 发布单部署历史页面搭建

发布单部署历史页面展示发布单号、归属项目/应用、构建分支、创建人、当前状态等信息，右侧使用时间轴展示当前发布单的状态流程。在 release 目录下创建 history.vue 文件，此文件为发布单部署历史组件。具体的 template 代码如下：

```
<template>
    <a-card>
        <div>
            <a-row>
                <a-col :span="11">
                    <a-list item-layout="horizontal">
                        <a-list-item >
                            <a-list-item-meta
                                :description=dataSource.id
                            >
                                <a slot="title">id</a>
                            </a-list-item-meta>
                        </a-list-item>
                        <a-list-item >
                            <a-list-item-meta
                                :description=dataSource.name
                            >
                                <a slot="title"> 发布单号 </a>
                            </a-list-item-meta>
                        </a-list-item>
                        <a-list-item >
                            <a-list-item-meta
                                :description=dataSource.description
                            >
                                <a slot="title"> 描述 </a>
                            </a-list-item-meta>
                        </a-list-item>
                        <a-list-item >
                            <a-list-item-meta
                                :description=dataSource.project_name
                            >
                                <a slot="title"> 项目 </a>
                            </a-list-item-meta>
```

```
</a-list-item>
<a-list-item >
    <a-list-item-meta
        :description=dataSource.app_name
    >
    <a slot="title"> 应用 </a>
    </a-list-item-meta>
</a-list-item>
<a-list-item >
    <a-list-item-meta
        :description=dataSource.create_user_name
    >
    <a slot="title"> 创建人 </a>
    </a-list-item-meta>
</a-list-item>
<a-list-item >
    <a-list-item-meta
        :description=dataSource.env_name
    >
    <a slot="title"> 环境 </a>
    </a-list-item-meta>
</a-list-item>
<a-list-item >
    <a-list-item-meta
        :description=dataSource.deploy_status_name
    >
    <a slot="title"> 状态 </a>
    </a-list-item-meta>
</a-list-item>
<a-list-item >
    <a-list-item-meta
        :description= dataSource.git_url
    >
    <a slot="title">git 服务器 </a>
    </a-list-item-meta>
</a-list-item>
<a-list-item >
    <a-list-item-meta
        :description= dataSource.git_app_id
    >
    <a slot="title">git id</a>
    </a-list-item-meta>
</a-list-item>
```

```html
            <a-list-item >
                <a-list-item-meta
                    :description=dataSource.git_branch
                >
                    <a slot="title">git 版本 </a>
                </a-list-item-meta>
            </a-list-item>
            <a-list-item >
                <a-list-item-meta
                    :description=dataSource.pipeline_url
                >
                    <a slot="title"> 构建流水线 url</a>
                </a-list-item-meta>
            </a-list-item>
            <a-list-item >
                <a-list-item-meta
                    :description=dataSource.deploy_script_url
                >
                    <a slot="title"> 部署脚本 </a>
                </a-list-item-meta>
            </a-list-item>
            <a-list-item >
                <a-list-item-meta
                    :description=dataSource.zip_package_url
                >
                    <a slot="title"> 软件包 </a>
                </a-list-item-meta>
            </a-list-item>
        </a-list>
    </a-col>
    <a-col :span="2">
    </a-col>
    <a-col :span="11">
        <a-timeline>
            <a-timeline-item color="blue" v-for="(item, index) in historyDataSource"
                :key="item.id">
                <a-icon slot="dot" type="clock-circle-o" style="font-size: 16px;" />
                <a-tag color='blue'>{{item.release_status_name}}</a-tag>
                <span v-show="item.release_status_name == 'Ready'">
                    {{item.log}}</span>
                <a-tag>{{item.create_user}}</a-tag>
                {{item.create_date}}
            </a-timeline-item>
```

```
            </a-timeline>
          </a-col>
        </a-row>
      </div>
    </a-card>
</template>
```

发布单部署历史组件使用通用的卡片容器 a-card 作为根节点,其中左侧使用 a-list 组件渲染组件信息,右侧使用 a-timeline 渲染该发布单历史操作,使用的 Antd 组件的详细属性已在技术要点中介绍,在此不再赘述。

在 history.vue 文件中添加 script 标签,部署历史组件中实现的事件包含:

- fetchReleaseData:根据发布单 ID 请求服务端获取发布单详情。
- fetchReleaseHistoryData:根据发布单 ID 请求服务端获取发布单部署历史记录。

相关代码如下:

```
<script>
import API from '@/service'
import moment from 'moment'

export default {
    name: 'releaseHistory',
    data () {
        return {
            releaseId: "",
            dataSource: {},
            historyDataSource: []
        }
    },
    created(){
        this.releaseId = this.$route.params.releaseId
        this.fetchReleaseData()
        this.fetchReleaseHistoryData()
    },
    methods: {
        fetchReleaseData(){
            API.ReleaseDetail(this.releaseId).then((res)=>{
                let result = res.data
                if(res.status == 200 && result.code == 0){
                    this.dataSource = result.data;
                }
```

```
                })
            },
            fetchReleaseHistoryData(){
                let params = {
                    release_id: this.releaseId
                }
                API.ReleaseHistory(params).then((res)=>{
                    let result = res.data
                    if(res.status == 200 && result.code == 0){
                        this.historyDataSource = result.data.results;
                    }
                })
            }
        }
    }
</script>
```

18.4 发布单生成基础服务配置

发布单模块列表查询、创建以及历史查询页面开发完成后，就需要在浏览器中测试开发的页面展示以及功能是否有问题，需要配置路由、service 服务以及 mock 服务。

18.4.1 路由配置

要在浏览器能够展示发布单模块三个页面，只需把三个页面的组件配置到 router 目录下的 config.js 文件中，代码如下：

```
const routeConfig = [
    {
        path: '/release',
        name: '发布单',
        meta: {
            icon: 'profile'
        },
        children:[
            {
                path: '/releaseList',
                name: '列表',
```

```
                    meta:{
                        title:"发布单列表"
                    },
                    component: () => import('@/views/release/list')
                },
                {
                    path: '/createRelease',
                    name: '新建',
                    component: () => import('@/views/release/create')
                },
                {
                    path: '/releaseHistory',
                    name: '发布单部署历史',
                    meta:{
                        isShowMenu:false
                    },
                    component: () => import('@/views/release/history')
                }
            ]
        },
    ]
```

其中发布单部署历史页面入口是在发布单列表中单击发布单单号,不需要在菜单导航栏中展示,通过在 meta 中添加 isShowMenu:false 参数"告诉"菜单栏忽略此菜单。

在浏览器地址中添加 hash 值为 release/list,即访问 http://localhost:8080/#/release/list,便可以在浏览器中看到开发完的发布单列表页面,同理添加 /release/create 和 /release/history 即可查看发布单新建和历史查询页面。

18.4.2 mock服务配置

项目中的本地 mock 服务用于模拟服务端请求,并返回自定义的模拟数据。发布单生成模块请求服务端针对发布单的增删查改接口,所以需要在 mock 目录下新增 release 目录,目录中新增 index.js 文件,文件内容如下:

```
import Mock from 'mockjs'
const releaseList = Mock.mock({
        "count": "@integer(1, 100)",
        "results|100": [
            {
                "id":"@id",
```

```
                    "project_name": "@name",
                    "git_name": "@name",
                    "name": "@name",
                    "description": "@sentence(5)",
                    "update_date": "@date",
                    "create_date": "@date",
                    "base_status": "@boolean",
                    "cn_name": "@name",
                    "app_id": "@id",
                    "git_app_id": "@id",
                    "git_trigger_token": "@guid",
                    "build_script": "./buid.sh",
                    "deploy_script": "./deploy.sh",
                    "zip_package_name": "test.zip",
                    "service_port": "@integer(1000,30000)",
                    "service_username": "@name",
                    "service_group": "@name",
                    "op_no": "@id",
                    "create_user": "@id",
                    "git": "@id",
                    "project": "@id"
                }
        ]
})

Mock.mock(`${process.env.VUE_APP_API_BASE_URL}/release/list/`, 'get', ({body}) => {
        return {
                "code": 0,
                "message": "success",
                "data": releaseList
        }
})

const releaseDetail = Mock.mock({
        "id": "@id",
        "project_name": "@name",
        "git_name": "@name",
        "name": "@name",
        "description": "@sentence(5)",
        "update_date": "@date",
        "create_date": "@date",
        "base_status": "@boolean",
        "cn_name": "@name",
```

```js
            "app_id": "@id",
            "git_app_id": "@id",
            "git_trigger_token": "@guid",
            "build_script": "./buid.sh",
            "deploy_script": "./deploy.sh",
            "zip_package_name": "test.zip",
            "service_port": "@integer(1000,30000)",
            "service_username": "@name",
            "service_group": "@name",
            "op_no": "@id",
            "create_user": "@id",
            "git": "@id",
            "project": "@id"
})

Mock.mock(`${process.env.VUE_APP_API_BASE_URL}/release/detail/`, 'get', ({body}) => {
        return {
                "code": 0,
                "message": "success",
                "data": releaseDetail
        }
})

const historyList = Mock.mock({
        "count": "5",
        "results|5": [
                {
                        "id": "@id",
                        "release": "@guid",
                        "create_user": "@name",
                        "deploy_status_name": "@name",
                        "name": "@guid",
                        "description": "@sentence(5)",
                        "update_date": "@date",
                        "create_date": "@date",
                        "base_status": true,
                        "log": "Create",
                        "deploy_status": 1
                }
        ]
})

Mock.mock(`${process.env.VUE_APP_API_BASE_URL}/history/release/`, 'get', ({body}) => {
```

```js
        return {
                "code": 0,
                "message": "success",
                "data": historyList
        }
})

const addRelease = Mock.mock({
        "name": "@guid",
        "description": "@sentence(5)",
        "git_branch": "@name",
        "app": "@id",
        "deploy_status": 1,
        "create_user": "@id"
})

Mock.mock(`${process.env.VUE_APP_API_BASE_URL}/release/create/`, 'post', ({body}) => {
        return {
                "code": 0,
                "message": "success",
                "data": addRelease
        }
})

Mock.mock(`${process.env.VUE_APP_API_BASE_URL}/release/build/`, 'post', ({body}) => {
        return {
                "code": 0,
                "message": "success",
                "data": {}
        }
})

Mock.mock(`${process.env.VUE_APP_API_BASE_URL}/release/build_status/`, 'post', ({body}) => {
        return {
                "code": 0,
                "message": "success",
                "data": {
                        data:"success"
                }
        }
})
```

在 mock 服务的入口文件 index.js 中引入 release 目录，即完成了发布单生成模块 mock 服务的注册，当发布单组件通过 GET/POST/PUT/DELETE 访问 /

release/* 和 /history/* 两个路径时，mock 服务层会返回配置的数据。发布单生成组件访问 mock 服务层的接口数据需要通过 service 层，所以还需要在 service 层配置发布单生成模块。

18.4.3　service服务配置

service 服务统一封装了 BiFang 部署平台项目所需的所有接口路径以及请求方式，并通过 axios 调用下游的 mock 数据，所以 service 服务需要新增发布单生成模块。

在 services 目录下的 api.js 路径集中新增发布单生成模块所需的访问路径，代码如下：

```
const API = {
        RELEASELIST: `${BASE_URL}/release/list/`,            // 查询发布单列表
        CREATERELEASE: `${BASE_URL}/release/create/`,        // 创建发布单
        RELEASEDETAIL: `${BASE_URL}/release/detail/`,        // 发布单详情
        RELEASEBUILD: `${BASE_URL}/release/build/`,          // 构建发布单
// 查询发布单构建状态
        RELEASEBUILDSTATUS: `${BASE_URL}/release/build_status/`,
        RELEASEHISTORY: `${BASE_URL}/history/release/`,      // 获取发布单历史
}
```

在 services 目录下新增 release.js 文件，用于配置发布单模块所需服务，代码如下：

```
import {request,METHOD} from '@/utils/request'
import {API} from './api'
/**
 * 获取发布单列表
 * @param {*} params
 */
async function ReleaseList(params){
    return request(
        API.RELEASELIST,
        METHOD.GET,
        params
    )
}
/**
 * 获取发布单详情
 * @param {*} params
```

```
     */
    async function ReleaseDetail(params){
            URL = `${API.RELEASEDETAIL}${params}/`
            console.log(URL)
            return request(
                    URL,
                    METHOD.GET
            )
    }
    /**
     * 新建发布单
     * @param {*} params
     */
    async function CreateRelease(params){
        return request(
                API.CREATERELEASE,
                METHOD.POST,
                params
        )
    }
    /**
     * 编译发布单
     * @param {*} params
     */
    async function BuildRelease(params){
        return request(
                API.RELEASEBUILD,
                METHOD.POST,
                params
        )
    }
    /**
     * 获取发布单编译进度和状态
     * @param {*} params
     */
    async function BuildReleaseStatus(params){
        return request(
                API.RELEASEBUILDSTATUS,
                METHOD.POST,
                params
        )
    }
    export default {
```

```
    ReleaseList,
    ReleaseDetail,
    CreateRelease,
    BuildRelease,
    BuildReleaseStatus
}
```

在 services 目录下新增 history.js 文件，用于配置查询历史所需服务，代码如下：

```
import {request,METHOD} from '@/utils/request'
import {API} from './api'
/**
 * 获取发布单历史操作
 * @param {*} params
 */
async function ReleaseHistory(params){
    return request(
            API.RELEASEHISTORY,
            METHOD.GET,
            params
        )
}
/**
 * 获取服务器历史操作
 * @param {*} params
 */
async function ServerHistory(params){
    return request(
            API.SERVERHISTORY,
            METHOD.GET,
            params
        )
}
export default {
    ReleaseHistory,
    ServerHistory
}
```

在 service 目录下主入口文件 index.js 中引入发布单生成模块和历史查询模块，代码如下：

```
import application from './release'
import permission from './history'
```

```
export default {
    ...release,
    ...history,
}
```

发布单生成模块中调用service层配置的接口名称,axios发起request请求并通过promise.then(response=>{})的回调方式把获取到的数据返回给处理函数。

至此,发布单生成模块列表查询页面、新建发布单页面以及发布单历史页面相对应的mock服务、service服务代码都已全部编写完毕,启动项目,单击左侧菜单导航中的发布单菜单,页面即可呈现在读者面前,发布单新增、查询列表、构建以及查询历史操作调用接口返回成功标识。

在service服务层中的api.js中改变BASE_UR和REAL_URL即可完成mock数据和真实环境的请求切换,方便项目的调试。

18.5 环境流转模块

发布单创建成功后需要应用管理员(一般为测试人员)扭转发布单到指定的环境,例如开发人员开发完成后新建发布单,测试人员扭转发布单到测试环境;待测试人员测试功能没问题后,再扭转发布单到正式环境,把代码发布到正式环境。

环境流转模块主要提供环境流转功能,以备发布单在此环境进行接下来的部署操作。

1. 环境流转模块功能清单

(1)环境流转模块功能清单如表18-2所列。

表18-2 环境流转模块功能表

模块名称	功能名称	所属页面	功能描述
环境流转	环境流转	环境流转页面	①显示发布单的环境列表 ②选择指定发布单的指定环境进行流转 ③流转信息二次确认 ④显示流转成功或失败消息

(2)环境流转模块功能说明。

①当发布单不在所属环境时,不可以在该环境部署。

② 一般应由测试人员拥有此权限，以免随便部署，中断正在进行的测试任务。

（3）环境流转模块截图如图 18-5 所示。

图 18-5　BiFang 发布单环境流转

2. 功能技术要点

在环境流转模块页面搭建中，主要使用 Antd 框架中的组件进行搭建，详细功能点使用的组件如下：

- 查询操作：使用 a-form 表单、a-table 表格、a-pagination 分页组件。
- 流转操作：自定义 envExchange 组件。

上述使用的 Antd 组件常用属性已介绍，不再赘述。

- Form 表单是具有数据收集、校验和提交功能的表单，包含复选框、单选框、输入框和下拉选择框等元素。
- Table 表格用于当需要对数据进行排序、搜索、分页、自定义操作等复杂行为时展示行列数据。
- Pagination 分页采用分页的形式分隔长列表，每次只加载一个页面。
- Select 选择器用于弹出一个下拉菜单给用户选择操作，用于代替原生的选择器，或者需要一个更优雅的多选器。

自定义 envExchange 组件用于用户选择发布单进行环境扭转时更新对应 table 中的行数据，隔离流转多行发布单更新导致的数据混淆。

envExchange 组件开放三个配置属性：

- title：String 类型，用于标识按钮文案。
- items：Array 类型，用于提供可扭转环境列表数据。

- record：Object 类型，table 中的单个发布单数据，用于发布单流转时做数据同步。

在 components 下的 tool 目录新增 envExchange.vue 文件，envExchange 组件实现代码如下：

```html
<template>
    <div>
        <a-select
            show-search
            placeholder=" 环境 "
            option-filter-prop="children"
            style="width: 80px"
            v-model="env"
            @change="onSelectChange"
        >
            <a-select-option v-for="d in items" :key="d.value">
            {{ d.label }}
            </a-select-option>
        </a-select>
        <a-popconfirm
            :title="`是否将 ${record.name} 发布单流转到 ${env} 环境 ?`"
            ok-text=" 是 "
            cancel-text=" 否 "
            :visible="visible"
            @visibleChange="handleVisibleChange"
            @confirm="handleChange">
            <a-button type="danger">{{title}}</a-button>
        </a-popconfirm>
    </div>
</template>
<script>
export default {
    name:"envExchange",
    data(){
        return {
            env:"",
            visible:false
        }
    },
    props:{
        title:String,
        items: Array,
```

```
                record: Object
        },
        methods:{
                handleVisibleChange(visible) {
                        if (!visible) {
                                this.visible = false;
                                return;
                        }

                        if (!this.env) {
                                this.$message.error(" 请选择流转环境 ");
                        } else {
                                this.visible = true;
                        }
                },
                onSelectChange(value){
                        this.env = value
                },
                handleChange(){
                        this.$emit("onChange",this.record,this.env)
                        this.env = ""
                }
        }
}
</script>
```

以上为搭建环境流转模块使用的 Antd 组件和封装的 envExchange 组件介绍，接下来进入页面搭建。

18.6 环境流转页面的搭建

环境流转页面包含查询发布单列表、扭转单个发布单环境，这里将以实现每个功能的步骤来呈现。

在 src 中的 views 目录下创建 environment 目录，并在该目录下创建 envlist.vue 文件，此文件为环境流转组件。

查询待扭转发布单列表

环境流转列表包含发布单编号列、归属项目列、归属组件列、用户名称列、更新时间列、对应环境列以及操作列，其中操作列包含选择环境下拉列表以及流转按钮。同时列表顶部包含过滤条件，根据用户输入的归属项目、归属组件和发布单编号进行过滤。template 的代码如下：

```
<template>
    <a-card>
        <div class="search">
            <a-form layout="inline" :form="form" @submit="submitHandler">
                <a-form-item
                    label=" 项目选择 "
                >
                    <a-select
                        show-search
                        placeholder=" 请选择项目 "
                        option-filter-prop="children"
                        :filter-option="filterOption"
                        @change="handleChange"
                        style="width:200px"
                        v-decorator="['projectName', { rules: [{ required: false, message: ' 请选择项目 !' }] }]"
                    >
                        <a-select-option v-for="d in projectOption" :key="d">
                            {{ d }}
                        </a-select-option>
                    </a-select>
                </a-form-item>
                <a-form-item
                    label=" 组件选择 "
                >
                    <a-select
                        show-search
                        placeholder=" 请选择组件 "
                        option-filter-prop="children"
                        :filter-option="filterOption"
                        style="width:200px"
                        v-decorator="['appName', { rules: [{ required: false,
                            message: ' 请选择发布的组件 !' }] }]"
                    >
                        <a-select-option v-for="d in options" :key="d.label">
                            {{ d.label }}
                        </a-select-option>
                    </a-select>
                </a-form-item>
                <a-form-item
                    label=" 发布单号 "
```

```html
            >
                <a-input
                    placeholder=" 请输入发布单 "
                    v-decorator="['releaseNo', { rules: [{ required: false, message: ' 请输入发布单 '}] }]" />
            </a-form-item>
            <a-form-item>
                <a-button type="primary" html-type="submit">
                    查询
                </a-button>
            </a-form-item>
        </a-form>
    </div>
    <div>
        <a-table
            :columns="columns"
            :dataSource="dataSource"
            rowKey="name"
            @change="onChange"
            :pagination="{
                current: params.currentPage,
                pageSize: params.pageSize,
                total: total,
                showSizeChanger: true,
                showLessItems: true,
                showQuickJumper: true,
                showTotal: (total, range) => `第 ${range[0]}-${range[1]} 条，总计 ${total} 条`
            }"
        >
            <template slot="env_name" slot-scope="text,record">
                <a-tooltip>
                    <template slot="title">
                        {{record.description}}
                    </template>
                    <a-tag color='blue'>{{text}}</a-tag>
                </a-tooltip>
            </template>
            <div slot="action" slot-scope="text, record">
                <env-exchange
                    title=" 流转 "
                    :items="envOptions"
                    :record="record"
                    @onChange="onEnvExchange"
                ></env-exchange>
```

```
                </div>
            </a-table>
        </div>
    </a-card>
</template>
```

环境流转组件以通用的卡片容器 a-card 作为根节点，其中内嵌 a-form 和 a-table 组件，在 a-table 组件中嵌套了 slot 名为 env_name 和 action 的 template 模板，分别用于渲染发布单归属环境列以及操作列，相关 Antd 组件详细属性已在技术要点中介绍，在此不再赘述。

在 envlist.vue 文件中添加 script 标签，环境流转组件中实现的事件包含：

- fetchData：请求接口获取待流转发布单列表信息。
- fetchEnv：请求接口获取可流转环境。
- fetchAppList：请求接口获取应用列表，供过滤条件中的组件列表使用。
- handleChange：当过滤条件中项目列表被选择时，更新组件列表数据。
- filterOption：a-select 组件过滤函数。
- onEnvExchange：监听 envExchange 组件事件变更，请求服务端更新发布单流转对应环境。
- onChange：用户单击分页或表格个数 pageSize 变更时重新请求数据。
- submitHandler：用户单击查询按钮根据 form 表单输入的条件查询可流转发布单列表信息。

相关代码如下：

```
<script>
import envExchange from '@/components/tool/envExchange'
import API from '@/service'
import moment from 'moment'
const columns = [
    {
        title: ' 发布单编号 ',
        dataIndex: 'name'
    },
    {
        title: ' 项目 ',
        dataIndex: 'project_name'
    },
    {
        title: ' 组件 ',
```

```
                dataIndex: 'app_name'
        },
        {
                title: ' 用户 ',
                dataIndex: 'create_user_name'
        },
        {
                title: ' 更新时间 ',
                dataIndex: 'update_date',
                sorter: true,
                customRender: (date) =>{ return moment(date).format("YYYY-MM-DD hh:mm")}
        },
        {
                title: ' 环境 ',
                dataIndex: 'env_name',
                scopedSlots: { customRender: 'env_name' }
        },
        {
                title: ' 操作 ',
                scopedSlots: { customRender: 'action' }
        }
]

export default {
        name: 'envList',
        components: {envExchange},
        data () {
                return {
                        total:0,
                        columns: columns,
                        dataSource: [],
                        envOptions:[],
                        projects:{},
                        projectOption: [],
                        options:[],
                        params:{
                                name:"",
                                currentPage:1,
                                pageSize:20,
                                release_status: 'Build,Ready,Success',
                                sort:""
                        }
```

```js
        },
        beforeCreate() {
            this.form = this.$form.createForm(this, { name: 'releaseList' });
        },
        created(){
            this.fetchAppList()
            this.fetchEnv()
        },
        mounted(){
            this.form.setFieldsValue({
                    releaseNo:this.$route.query.releaseNo||""
            })
            this.fetchData()
        },
        methods: {
            submitHandler(e){
                e.preventDefault()
                this.fetchData()
            },
            fetchData(){
                    this.form.validateFields((err, fieldsValue) => {
                            if (err) {
                            return;
                            }
                            const rangeValue = fieldsValue['timePicker']
                            this.params.name = fieldsValue["releaseNo"]
                            this.params.project_name = fieldsValue["projectName"]
                            this.params.app_name = fieldsValue["appName"]

                            API.ReleaseList(this.params).then((res)=>{
                                    let result = res.data
                                    if(res.status == 200 && result.code == 0){
                                    this.total = result.data.count
                                    this.dataSource = result.data.results;
                                    console.log(this.dataSource)
                                    } else {
                                    this.dataSource = []
                                    }
                            })
                    });
            },
            fetchEnv(){
```

```js
API.EnvList({}).then((res)=>{
    let result = res.data
    if(res.status == 200 && result.code == 0){
        result.data.results.forEach(item=>{
            this.envOptions.push({
                label:item.name,
                value:item.id
            })
        })
    }
    else{
        this.$message,error(" 无法获取环境列表 ~")
    }
})
},
fetchAppList(){
    API.AppList({}).then((res)=>{
        let result = res.data
        if(res.status == 200 && result.code == 0){
            result.data.results.forEach(item=>{
                this.projects[item.project_name]?
                    this.projects[item.project_name].push({
                        label:item.name,
                        value:item.id
                    }):this.projects[item.project_name] = [{
                        label:item.name,
                        value:item.id
                    }]
            })
            this.projectOption = Object.keys(this.projects)
        }
        else{
            this.$message,error(" 无法获取应用列表 ~")
        }
    })
},
handleChange(value) {
    this.form.setFieldsValue({
        appName:""
    })
    this.options = this.projects[value]
},
filterOption(input, option) {
    return (
```

```
                    option.componentOptions.children[0].text
                        .toLowerCase().indexOf(input.toLowerCase()) >= 0
                );
        },
        onChange(pagination, filters, sorter) {
            console.log('Various parameters', pagination, filters, sorter);
            let {
                current,
                pageSize
            } = pagination
            let {
                order,
                field
            } = sorter
            this.params.currentPage = current
            this.params.pageSize = pageSize
            this.params.sorter = (field?field:"")
            this.fetchData()
        },
        onEnvExchange(record,env) {
            let params = {
                env_id: env,
                release_name: record.name,
                app_id: record.app
            }
            console.log(params)
            API.EnvExchange(params).then((res)=>{
                let result = res.data
                if(res.status == 200 && result.code == 0){
                    this.$message.success(" 环境流转完成，此发布单可以在指定环境进行部署。")
                    this.fetchData()
                } else if(res.status == 200 && result.code == 2000){
            this.$message.error(" 你没有此发布单的流转环境权限！ ")
        } else {
                    this.$message.error(" 环境流转错误，请联系系统管理员 ~")
                }
            })
        },
    }
}
</script>
```

通过 import 引入 service API，调用的方法为其中的服务器列表请求，在 service 服务配置中介绍具体的实现。

18.7 环境流转基础服务配置

环境流转页面开发完成后，若要在浏览器中测试开发的页面展示以及功能是否有问题，需要配置路由、service 以及 mock 服务。

18.7.1 路由配置

若要在浏览器能够展示环境流转页面，只需把组件配置到 router 目录下的 config.js 文件中，代码如下：

```
const routeConfig = [
    {
        path: '/environment',
        name: ' 环境 ',
        meta: {
            icon: 'apartment'
        },
        children:[
            {
                path: '/envlist',
                name: ' 流转环境 ',
                component: () => import('@/views/environment/envlist')
            }
        ]
    },
]
```

在浏览器地址中添加 hash 值为 environment/envlist，即访问 http://localhost:8080/#/environment/envlist，便可以在浏览器中看到开发完的环境流转页面。

18.7.2 mock服务配置

项目中的本地 mock 服务用于模拟服务端请求，并返回自定义的模拟数据。环境流转页面请求服务端针对发布单环境的查询和流转接口，所以需要在 mock 目录下新增 env 目录，目录中新增 index.js 文件，文件内容如下：

```
import Mock from 'mockjs'
const envList = Mock.mock({
    "count": 3,
```

```
            "results": [
                {
                    "id": 1,
                    "name": "dev",
                    "description": "@sentence(5)",
                    "update_date": "@date",
                    "create_date": "@date"
                },
                {
                    "id": 2,
                    "name": "prd",
                    "description": "@sentence(5)",
                    "update_date": "@date",
                    "create_date": "@date"
                },
                {
                    "id": 3,
                    "name": "uat",
                    "description": "@sentence(5)",
                    "update_date": "@date",
                    "create_date": "@date"
                }
            ]
        })

Mock.mock(`${process.env.VUE_APP_API_BASE_URL}/env/list/`, 'get', ({body}) => {
        return {
            "code": 0,
            "message": "success",
            "data": envList
        }
})

Mock.mock(`${process.env.VUE_APP_API_BASE_URL}/env/exchange/`, 'post', ({body}) => {
        return {
            "code": 0,
            "message": "success",
            "data": {}
        }
})
```

在 mock 服务的入口文件 index.js 中引入 env 目录,即完成了环境流转模块 mock 服务的注册,当环境流转页面通过 GET/POST 访问 /env/list 和 /env/

exchange 两个路径时，mock 服务层会返回配置的数据。环境管理模块访问 mock 服务层的接口数据需要通过 service 层，所以还需要在 service 层配置环境流转模块。

18.7.3　service服务配置

service 服务统一封装了 BiFang 部署平台项目所需的所有接口路径以及请求方式，并通过 axios 调用下游的 mock 数据，所以 service 服务需要新增环境流转模块。

在 services 目录下的 api.js 路径集中新增环境流转所需的访问路径，代码如下：

```
const API = {
        ENVLIST: `${REAL_URL}/env/list/`,              // 获取流转环境列表
        ENVEXCHANGE: `${REAL_URL}/env/exchange/`,      // 单发布单流转
}
```

在 services 目录下新增 environment.js 文件，用于配置环境流转页面所需服务，代码如下：

```
import {request,METHOD} from '@/utils/request'
import {API} from './api'
/**
 * 获取环境列表
 * @param {*} params
 */
async function EnvList(params){
        return request(
                API.ENVLIST,
                METHOD.GET,
                params
        )
}
/**
 * 环境流转
 * @param {*} params
 */
async function EnvExchange(params){
        return request(
                API.ENVEXCHANGE,
                METHOD.POST,
```

```
                    params
            )
    }
export default {
            EnvList,
            EnvExchange
    }
```

在 service 目录下主入口文件 index.js 中引入环境流转模块，代码如下：

```
import environment from './environment'

export default {
            ...environment,
    }
```

环境流转页面中调用 service 层配置的接口，axios 发起 request 请求并通过 promise.then(response=>{}) 的回调方式把获取到的数据返回给处理函数。

至此，环境流转页面相对应的 mock 服务、service 服务代码都已全部编写完毕，启动项目，单击左侧菜单导航中的环境流转菜单，页面即可呈现在读者面前，查询发布单、扭转环境操作调用接口返回成功标识。

在 service 服务层中的 api.js 中改变 BASE_UR 和 BASE_URL 即可完成 mock 数据和真实环境的请求切换，方便项目的调试。

18.8 小 结

本章带领读者熟悉了 BiFang 部署平台中发布单生成模块以及环境流转模块的页面设计、页面搭建到功能实现的流程，掌握了如何在公共服务路由、mock 服务以及 service 服务中添加发布单生成模块以及环境流转，中间也穿插着用到的 ant-design-vue 框架相关的组件知识点以及自定义封装 envExchange 组件用于单个发布单状态的转换。

第 19 章将带领大家进入 BiFang 部署平台发布单部署和 Dashboard 数据面板的开发。发布单创建并流转到指定环境后如何部署至指定的服务器？为什么需要 Dashboard 数据面板？让我们带着这些问题，前进！

生活总是让我们遍体鳞伤，但之后这些受伤的地方将变成我们最强壮的盔甲！

第 19 章 发布单部署、Dashboard 模块设计与搭建

业与有肝胆人共事，从无字句处读书。

——周恩来

本章 GitHub 代码地址：https://github.com/aguncn/bifang-book/tree/main/bifang-ch19。

第 19 章主要介绍了 BiFang 部署平台管理员模块的设计与搭建，开发主界面布局组件和相关组件，在 mock 服务层、service 服务层以及路由层添加用户组和用户页面专属的模块，熟悉整个流程配置方法。接下来将带领大家进入项目与应用模块的搭建，再次实践开发流程配置。

本章将在前端项目中设计搭建项目应用管理模块以及服务器管理模块，使用 ant-design-vue 提供的组件构建页面组件；新增项目管理和应用管理页面，在 mock 服务、service 服务和路由层添加页面对应的项目与应用模块；新增服务器管理页面，在 mock 服务、service 服务和路由层添加页面对应的服务器管理模块。

19.1 发布单部署

开发人员通过 BiFang 部署平台成功构建应用发布单并扭转发布单状态为 ready 状态，运维部署人员可单击发布单进入部署环节，部署平台后台服务使用 salt 连接应用部署服务器执行相关部署命令。

部署模块主要提供发布单的部署回滚，日志查看等功能。

1. 功能说明与清单

（1）部署模块功能清单如表 19-1 所列。

表 19-1　部署模块功能表

模块名称	功能名称	所属页面	功能描述
部署	部署列表	部署列表页面	显示可部署发布单列表 显示发布单部署状态 进入发布单部署页面 搜索具体发布单
	发布单部署	发布单部署页面	显示发布单信息 显示部署批次 显示及选择部署服务器 显示服务器上已部署的主备发布单 进入查看服务器日志页面 进入部署页面 进入回滚页面
	服务器日志	日志查看页面	显示部署批次 显示操作细节 显示出错细节（如有）
	部署	实时部署进度页面	显示部署服务器 显示部署状态 显示出错细节（如有） 更新部署批次
	回滚	实时回滚进度页面	显示回滚服务器 显示回滚状态 显示出错细节（如有） 更新部署批次

（2）部署模块功能说明。

① 发布单的状态，是以其在指定环境里所有的服务器上的部署进度而言。

② 服务器上的备用发布单，以 tips 的方式显示。

③ 只支持回滚到上一次的部署，不支持无限回滚上上次。

④ 每选择好服务器，单击一次部署按钮，就算一个部署批次。所以，同一个发布单可能产生多个部署批次。

⑤ 服务器日志会显示当前发布单的所有批次内容，并且给出批次的标示。

⑥ 部署和回滚，在后端实现上大体相同，只是传的参数稍有差别。

（3）部署模块截图（回滚或部署等截图，在书后配合代码讲解时给出）如图 19-1～图 19-3 所示。

第 19 章 发布单部署、Dashboard 模块设计与搭建

图 19-1 BiFang 部署列表

图 19-2 BiFang 发布单部署

图 19-3 BiFang 服务器部署日志

2. 功能技术要点

在项目与应用模块项目管理页面以及应用管理页面搭建中，主要使用 Antd 框架中的组件进行搭建，详细功能点使用的组件如下：

- 查询操作：使用 a-form 表单、a-table 表格、a-pagination 分页组件。
- 新建操作：使用 a-form 表单、a-select 选择框、a-input 输入框组件。
- 构建操作：使用 a-modal 模态框、a-button 按钮组件。
- 查看详情操作：使用 a-list 列表、a-timeline 时间轴组件。

上述使用的 Antd 组件常用属性如下，方便读者清晰地知道各属性的作用。

Form 表单是具有数据收集、校验和提交功能的表单，包含复选框、单选框、输入框、下拉选择框等元素，常用的 API 如下：

- layout：表单元素排列方式，水平排列（vertical）、垂直排列（horizontal，默认）、行内排列（inline）。
- labelAlign：label 标签的文本对齐方式，左对齐（left）、右对齐（right）。
- labelCol：label 标签布局，设置 span offset 值，如 {span: 3, offset: 12}。
- wrapperCol：需要为输入控件设置布局样式时，使用该属性，用法同 labelCol。
- getFieldsValue：获取一组输入控件的值，如不传入参数，则获取全部组件的值。
- setFieldsValue：设置一组输入控件的值。
- validateFields：校验并获取一组输入域的值与 error，若 fieldNames 参数为空，则校验全部组件。

Table 表格用于当需要对数据进行排序、搜索、分页、自定义操作等复杂行为时展示行列数据，常用的 API 如下：

- dataSource：表格数据源，对应服务端返回的数据。
- columns：表格列的配置，Object 类型数组。
- title：列头显示文字。
- align：内容的对齐方式，'left' | 'right' | 'center'。
- dataIndex：列数据在数据项中对应的 key。
- customRender：生成复杂数据的渲染函数，参数分别为当前行的值、当前行数据、行索引，@return 里面可以设置表格行 / 列合并。
- scopedSlots：使用 columns 时，可以通过该属性配置支持 slot-scope 的属

性，如 scopedSlots: {customRender: 'XXX'}。
- loading：页面是否加载中。
- pagination：分页器，参考配置项或 pagination 文档，设为 false 时不展示和进行分页。
- rowKey：表格行 key 的取值，可以是字符串或一个函数。
- change：分页、排序、筛选变化时触发。

Pagination 分页采用分页的形式分隔长列表，每次只加载一个页面，常用的 API 如下：
- current：当前页数。
- pageSize：每页条数。
- total：数据总数。
- showSizeChanger：是否可以改变 pageSize。
- showQuickJumper：是否可以快速跳转至某页。
- showSizeChange：pageSize 变化的回调。

Button 按钮标记了一个（或封装一组）操作命令，响应用户单击行为，触发相应的业务逻辑，常用的 API 如下：
- disabled：按钮失效状态。
- icon：设置按钮的图标类型。
- loading：设置按钮载入状态。
- type：设置按钮类型，可选值为 primary dashed danger link。
- click：单击按钮时的回调。

Modal 模态对话框在当前页面正中打开一个浮层，承载用户处理事务，又不希望跳转页面以致打断工作流程，常用 API 如下：
- cancelText：取消按钮文字。
- closable：是否显示右上角的关闭按钮。
- destroyOnClose：关闭时销毁 Modal 里的子元素 footer 底部内容，当不需要默认底部按钮时，可以设为 ":footer="null""。
- mask：是否展示遮罩 maskClosable，单击蒙层是否允许关闭。
- okText：确认按钮文字。
- okType：确认按钮类型。
- title：标题。

- width：宽度。

Popconfirm 气泡框在目标元素附近弹出浮层提示，询问用户进一步确认操作，常用 API 如下：

- cancelText：取消按钮文字。
- okText：确认按钮文字。
- title：确认框的描述。
- confirm：单击确认的回调。
- cancel：单击取消的回调。

List 通用列表可承载文字、列表、图片、段落，常用于后台数据展示页面，常用 API 如下：

- footer：列表底部。
- grid：列表栅格配置。
- itemLayout：设置 List.Item 布局，设置成 vertical 则竖直样式显示，默认横排。
- loadMore：加载更多。
- split：是否展示分割线。
- dataSource：列表数据源。

Timeline 垂直展示的时间流信息，当有一系列信息需按时间排列时，可正序和倒序，常用 API 如下：

- pending：指定最后一个幽灵节点是否存在内容。
- reverse：节点排序。
- pendingDot：当最后一个幽灵节点存在时，指定其时间图点。
- mode：通过设置 mode 可以改变时间轴和内容的相对位置。

以上为搭建发布单部署模块使用的 Antd 组件介绍，接下来进入部署模块页面搭建。

19.2 待部署列表页面搭建

待部署发布单列表页面包含查询待部署发布单列表以及单个发布单部署按钮，单击部署发布单将跳转发布单部署页面部署发布单，这里将以实现每个功能的步骤来呈现。

在 src 中的 views 目录下创建 deployment 目录，并在该目录下创建 deployList.vue 文件，此文件为待部署发布单列表组件。

查询待部署列表

待部署发布单列表包含发布单编号列、归属项目列、归属应用列、创建用户列、部署环境列、发布单状态列以及部署操作按钮列，单击部署按钮进入发布单部署页面。同时列表顶部包含过滤条件，根据用户选择的项目名称、应用名称和发布单编号进行过滤。template 的代码如下：

```html
<template>
    <a-card>
        <div class="search">
            <a-form layout="inline" :form="form" @submit="submitHandler">
                <a-form-item
                    label=" 项目选择 "
                >
                    <a-select
                        show-search
                        placeholder=" 请选择项目 "
                        option-filter-prop="children"
                        :filter-option="filterOption"
                        @change="handleChange"
                        style="width:200px"
                        v-decorator="['projectName', { rules: [{ required: false,
                            message: ' 请选择项目 !' }] }]"
                    >
                        <a-select-option v-for="d in projectOption" :key="d">
                            {{ d }}
                        </a-select-option>
                    </a-select>
                </a-form-item>
                <a-form-item
                    label=" 组件选择 "
                >
                    <a-select
                        show-search
                        placeholder=" 请选择组件 "
                        option-filter-prop="children"
                        :filter-option="filterOption"
                        style="width:200px"
                        v-decorator="['appName', { rules: [{ required: false,
                            message: ' 请选择发布的组件 !' }] }]"
```

```html
                    >
                        <a-select-option v-for="d in options" :key="d.label">
                            {{ d.label }}
                        </a-select-option>
                    </a-select>
                </a-form-item>
                <a-form-item
                    label=" 发布单号 "
                >
                    <a-input
                        placeholder=" 请输入发布单 "
                        v-decorator="['releaseNo', { rules: [{ required: false,
                            message: ' 请输入发布单 !' }] }]" />
                </a-form-item>
                <a-form-item>
                    <a-button type="primary" html-type="submit">
                        查询
                    </a-button>
                </a-form-item>
            </a-form>
        </div>
        <div>
            <a-table
                :columns="columns"
                :dataSource="dataSource"
                rowKey="name"
                @change="onChange"
                :pagination="{
                    current: params.currentPage,
                    pageSize: params.pageSize,
                    total: total,
                    showSizeChanger: true,
                    showLessItems: true,
                    showQuickJumper: true,
                    showTotal: (total, range) => `第 ${range[0]}-${range[1]} 条，总计 ${total} 条`
                }"
            >
                <template slot="deploy_status_name" slot-scope="text,record">
                    <a-tooltip>
                        <template slot="title">
                            {{record.description}}
                        </template>
                        <a-tag color='blue' v-if="text==='Ready'"> 准备就绪 </a-tag>
```

```
                            <a-tag color='green' v-if="text==='Success'"> 部署完成 </a-tag>
                            <a-tag color='orange' v-if="text==='Ongoing'"> 部署中 ...</a-tag>
                            <a-tag color='red' v-if="text==='Failed'"> 部署异常 </a-tag>
                        </a-tooltip>
                    </template>
                    <div slot="action" slot-scope="text, record">
                        <a-button type="primary" @click="goDeploy(record)"> 部署 </a-button>
                    </div>
                </a-table>
            </div>
        </a-card>
</template>
```

待部署发布单列表组件以通用的卡片容器 a-card 作为根节点，其中内嵌 a-form 和 a-table 组件，在 a-table 组件中嵌套了 slot 名为 deploy_status_name 和 action 的 template 模板。

- deploy_status_name 模板用于渲染发布单状态，供用户对当前发布单状态一目了然。
- action 模板用于渲染部署操作列，发布单可以被多次部署，若服务器应用程序出现问题时，可使用相同的发布单再次部署。

其他详细的属性已在技术要点中介绍，在此不再赘述。

在 deployList.vue 文件中添加 script 标签，部署列表组件中实现的事件包含：

- goDeploy：单击部署按钮获取当前行发布单相关信息，通过 router 跳转发布单部署页面。
- fetchData：请求接口获取待部署的发布单列表信息。
- fetchAppList：请求接口获取应用列表，供过滤条件中的组件列表使用。
- handleChange：当过滤条件中项目列表被选择时，更新组件列表数据。
- filterOption：a-select 组件过滤函数。
- onChange：用户单击分页或表格个数 pageSize 变更时触发重新请求数据
- submitHandler：用户单击查询按钮根据 form 表单输入的条件查询待部署发布单列表信息。

相关代码如下：

```
export default {
    name: 'deployList',
```

```
        components: {BfTable},
        data () {
            return {
                total:0,                    // 待部署发布单总数量
                columns: columns,           // 列配置项
                dataSource: [],             // 数据源
                projects:{},                // 项目列表
                projectOption: [],          // 项目选择数据源
                options:[],                 // 应用选择数据源
                params:{                    // 分页配置
                    name:"",
                    currentPage:1,
                    pageSize:20,
                    deploy_status: 'Ready,Ongoing,Failed,Success',
                    sort:""
                }
            }
        },
        beforeCreate() {
            this.form = this.$form.createForm(this, { name: 'releaseList' });
        },
        created(){
            this.fetchAppList()
            this.fetchData()
        },
        methods: {
            goDeploy(data){
                this.$router.push({ path: '/deployment/deploy', query:
{ releaseId: data.id,    appId: data.app, envId: data.env}});
            },
            submitHandler(e){
                e.preventDefault()
                this.form.validateFields((err, fieldsValue) => {
                    if (err) {
                        return;
                    }
                    const rangeValue = fieldsValue['timePicker']
                    this.params.name = fieldsValue["releaseNo"]
                    this.params.project_name = fieldsValue["projectName"]
                    this.params.app_name = fieldsValue["appName"]
                    this.fetchData()
                });
            },
```

```
fetchData(){
    API.ReleaseList(this.params).then((res)=>{
        let result = res.data
        if(res.status == 200 && result.code == 0){
            this.total = result.data.count
            this.dataSource = result.data.results;
        } else {
            this.dataSource = []
        }
    })
},
fetchAppList(){
    API.AppList({}).then((res)=>{
        let result = res.data
        if(res.status == 200 && result.code == 0){
            result.data.results.forEach(item=>{
                this.projects[item.project_name]?
                    this.projects[item.project_name].push({
                        label:item.name,
                        value:item.id
                    }):this.projects[item.project_name] = [{
                        label:item.name,
                        value:item.id
                    }]
            })
            this.projectOption = Object.keys(this.projects)
        }
        else{
            this.$message,error(" 无法获取应用列表 ~")
        }
    })
},
handleChange(value) {
    this.form.setFieldsValue({
        appName:""
    })
    this.options = this.projects[value]
},
filterOption(input, option) {
    return (
        option.componentOptions.children[0].text
.toLowerCase().indexOf(input.toLowerCase()) >= 0
    );
```

```
            },
            onChange(pagination, filters, sorter) {
                let {
                    current,
                    pageSize
                } = pagination
                let {
                    order,
                    field
                } = sorter
                this.params.currentPage = current
                this.params.pageSize = pageSize
                this.params.sorter = (field?field:"")
                this.fetchData()
            }
        }
    }
</script>
```

通过 import 引入 service API，调用的方法为 service 服务配置的请求发布单列表的 request 请求，在 service 服务配置中介绍具体的实现。

19.3 部署发布单

在待部署发布单列表操作列中使用模板渲染了部署按钮，单击按钮跳转到发布单部署页面。在发布单部署页面顶部使用 DetailList 组件展示发布单详情信息，下方展示归属应用的服务器 IP 列表，应用服务器 IP 是在服务器部署模块中添加关联的。

发布单部署页面包含以下功能：

- 发布单部署，选择需要部署的服务器 IP，单击部署按钮通知服务器执行部署操作。
- 查看服务器日志，显示部署操作细节，若部署出错展示错误信息。
- 实时部署进度，显示各部署的服务器实时进度。
- 实时回滚进度，当发布后的应用服务异常，单击回滚至上一个发布版本，实时更新回滚进度。

在 release 目录下创建 deploy.vue 文件，此文件为发布单部署组件。

1. 发布单详情与部署

发布单部署页面顶部展示发布单归属项目、归属应用、对应环境、发布单描述、发布单号、部署批次信息，下方表格中展示服务器 IP+端口、服务器所属系统、归属环境、当前部署应用发布单号、服务器日志以及更新事件，单击服务器日志按钮查看服务器日志。

在部署组件 deploy.vue 中新增 template 代码如下：

```
<template>
    <a-card>
        <div>
            <a-card type="inner" title=" 发布单信息 ">
                <detail-list size="small">
                    <detail-list-item term=" 项目 ">{{ title.project_name }}</detail-list-item>
                    <detail-list-item term=" 应用 ">
                        {{ title.app_name }}
                        <a :href="title.deploy_script_url" target="_blank">
                            <a-tag color="green">Deploy</a-tag>
                        </a>
                    </detail-list-item>
                    <detail-list-item term=" 环境 ">{{ title.env_name }}</detail-list-item>
                    <detail-list-item term=" 发布单 ">{{ title.name }}</detail-list-item>
                    <detail-list-item term=" 发布描述 ">{{ title.description }}</detail-list-item>
                    <detail-list-item term=" 部署批次 ">{{ title.deploy_no }}</detail-list-item>
                </detail-list>
            </a-card>
            <a-table
                :columns="columns"
                :dataSource="dataSource"
                rowKey="ip"
                :row-selection="{ selectedRowKeys: selectedRow, onChange: onSelectChange }"
                @change="onChange"
                :pagination="{
                    current: params.currentPage,
                    pageSize: params.pageSize,
                    total: total,
                    showSizeChanger: true,
                    showLessItems: true,
                    showQuickJumper: true,
                    showTotal: (total, range) => `第 ${range[0]}-${range[1]} 条，总计 ${total} 条`
                }"
            >
                <template slot="release_name" slot-scope="record">
```

```html
                        <a-tooltip>
                            <template slot="title">
                                {{ record.back_release_name }}
                            </template>
                            <a-tag color='green'
                                v-if="record.main_release_name===title.name">
                                {{ record.main_release_name }}</a-tag>
                            <a-tag color='red'
                                v-if="record.main_release_name!==title.name">
                                {{ record.main_release_name }}</a-tag>
                        </a-tooltip>
                    </template>
                    <template slot="server_log" slot-scope="text, record">
                        <a-button type="primary" @click="logShow(record)">
                            日志
                        </a-button>
                    </template>
                </a-table>
            </div>
        </a-card>
</template>
```

在 deploy.vue 文件中添加 script 标签，发布单信息以及服务器列表展示实现的事件包含：

- fetchReleaseData：请求接口获取发布单详情信息。
- onChange：用户单击分页或表格个数 pageSize 变更时触发重新请求数据。
- fetchServerData：请求接口获取当前发布单对应的服务器 IP 列表。
- fetchAllData：合并 fetchReleaseData 和 fetchServerData。
- onSelectChange：表格中服务器列表行被选中时触发选择事件。

相关代码如下：

```
<script>
import DetailList from '@/components/tool/DetailList'
import API from '@/service'
import moment from 'moment'

const DetailListItem = DetailList.Item
const columns = [
    {
        title: 'IP_Port',
        dataIndex: 'name'
```

```
    },
    {
        title: ' 系统 ',
        dataIndex: 'system_type'
    },
    {
        title: ' 环境 ',
        dataIndex: 'env_name'
    },
    {
        title: ' 发布单 ',
        scopedSlots: { customRender: 'release_name' }
    },
    {
        title: ' 部署日志 ',
        scopedSlots: { customRender: 'server_log' }
    },
    {
        title: ' 更新时间 ',
        dataIndex: 'update_date',
        sorter: true,
        customRender: (date) =>{ return
moment(date).format("YYYY-MM-DD hh:mm")}
    }
]
export default {
    name: 'serviceDeploy',
    components: {DetailList,DetailListItem},
    data () {
        return {
            total: 0,                      // 发布单服务器总数量
            visiableLog:false,             // 服务器日志模态框开关
            visiableDeploy:false,          // 部署、回滚按钮开关
            releaseId: "",                 // 当前发布单号
            appId: "",                     // 发布单归属应用 ID
            envId: "",                     // 发布单归属环境 ID
            title: "",                     // 发布单详情
            columns: columns,              // 服务器表格列配置
            dataSource: [],                // 服务器列表数据源
            serverHistorySource: [],       // 服务器日志数据源
            selectServer: "",              // 选择的服务器日志归属的 IP
            selectedRow:[],                // 服务器列表选择的行数据
            deployTitle: "",               // 部署标题
```

```
                    deployType: "",          // 部署类型
                    deployResult: "",        // 部署结果
                    deployLog: "",           // 部署日志
                    params:{                 // 分页数据
                        currentPage:1,
                        pageSize:20,
                        sort:""
                    }
                }
        },
        created(){
            this.releaseId = this.$route.query.releaseId
            this.appId = this.$route.query.appId
            this.envId = this.$route.query.envId
            this.fetchAllData()
        },
        methods: {
            fetchAllData(){
                this.fetchReleaseData()
                this.fetchServerData()

            },
            fetchReleaseData(){
                API.ReleaseDetail(this.releaseId).then((res)=>{
                    let result = res.data
                    if(res.status == 200 && result.code == 0){
                        this.title =    result.data;
                        this.fetchServerHistoryData()
                    }
                })
            },
            fetchServerData(){
                let params = {
                    app_id: this.appId,
                    env_id: this.envId
                }
                API.ServerList(params).then((res)=>{
                    if(res.status == 200 ){
                        let result = res.data
                        this.total = result.data.count
                        this.dataSource = result.data.results;
                    } else {
                        this.dataSource = []
```

```
            }
        })
    },
    fetchServerHistoryData(){
        let params = {
            release_id: this.releaseId,
            env_id: this.envId,
            pageSize: 200
        }
        API.ServerHistory(params).then((res)=>{
            let result = res.data
            if(res.status == 200 && result.code == 0){
                this.serverHistorySource = result.data.results
            }
        })
    },
    onSelectChange(selectedRowKeys,selectedRows) {
        this.selectedRow = selectedRowKeys;
    },
    onChange(pagination, filters, sorter) {
        let {
            current,
            pageSize
        } = pagination
        let {
            order,
            field
        } = sorter
        this.params.currentPage = current
        this.params.pageSize = pageSize
        this.params.sorter = (field?field:"")
        this.fetchData()
    }
  }
}
</script>
```

2. 发布单部署 / 回滚

选择需要部署的服务器 IP, 部署 / 回滚按钮变更为可单击状态, 单击按钮二次确认后请求部署 / 回滚服务接口。在 template 中新增代码如下:

```
<template>
    <div class="alert">
```

```html
<a-alert type="info" style="line-height:2.4" :show-icon="true" v-if="selectedRow">
    <div class="message" slot="message">
        <a-row>
            <a-col :span="2">
                已选择  <a>{{selectedRow.length}}</a>  项
            </a-col>
            <a-col :span="22">
                <a-popconfirm
                    title=" 确定执行回滚操作么？"
                    ok-text=" 是 "
                    cancel-text=" 否 "
                    @confirm="confirmRollback"
                >
                    <a-button type="primary" :disabled="selectedRow.length == 0"
                        style="float:left"> 回滚 </a-button>
                </a-popconfirm>
                <a-popconfirm
                    title=" 确定执行部署操作么？"
                    ok-text=" 是 "
                    cancel-text=" 否 "
                    @confirm="confirmDeploy"
                >
                    <a-button type="danger" :disabled="selectedRow.length == 0"
                        style="float:right"> 部署 </a-button>
                </a-popconfirm>
            </a-col>
        </a-row>
    </div>
</a-alert>
</div>
<a-modal
    :visible="visiableDeploy"
    :title="deployTitle"
    okText=" 确定 "
    cancelText=" 关闭 "
    @cancel="onDeployReset"
    @ok="onDeployReset"
>
<a-card>
    服务器列表：{{this.selectedRow}}<br/>
    <span v-if="deployResult === 'wait'"> 状态：部署中
    <a-icon type="sync" :style="{ fontSize: '24px', color: '#00f' }" spin /><br/></span>
    <span v-else-if="deployResult === 'success'"> 状态：部署完成
```

```
                <a-icon type="check" :style="{ fontSize: '24px', color: '#000' }" /> <br/></span>
                <span v-else-if="deployResult === 'noPermission'"> 状态：无部署权限
                <a-icon type="close"  style="{ fontSize: '24px', color: '#f00' }" /> <br/></span>
                <span v-else> 状态：部署出错，请查看服务器日志
                <a-icon type="close"  :style="{ fontSize: '24px', color: '#f00' }" /> <br/></span>
                输出记录：<span> {{deployLog}} <br/></span>
        </a-card>
    </a-modal>
</template>
```

在 script 标签中新增事件包含：

- confirmDeploy：确认执行部署操作，调用 onDeploy 执行部署。
- confirmRollback：确认执行回滚操作，调用 onDeploy 执行回滚。
- onDeploy：根据发布单信息请求服务端执行部署/回滚操作，更新发布单状态。
- onDeployReset：重新请求发布单相关信息，重置标志位数据。

新增的事件代码如下：

```
<script>
... // 其他忽略
        methods: {
        ...// 其他忽略
        onDeployReset(){
            this.fetchAllData()
            this.deployResult = ""
            this.deployLog = ""
            this.visiableDeploy = false
        },
        confirmDeploy(e) {
            this.deployTitle = " 部署 "
            this.deployType = "deploy"
            this.onDeploy()
        },
        confirmRollback(e) {
            this.deployTitle = " 回滚 "
            this.deployType = "rollback"
            this.onDeploy()
        },
        onDeploy(){
            let params = {
                target_list: this.selectedRow.toString(),
```

```
                user_id: this.title.create_user,
                app_name: this.title.app_name,
                service_port: this.title.service_port,
                env_name: this.title.env_name,
                release_name: this.title.name,
                deploy_no: this.title.deploy_no,
                deploy_type: this.deployType,
                op_type: 'deploy',
            }
            this.deployResult = "wait"
            this.visiableDeploy = true
            API.Deploy(params).then((res)=>{
                let result = res.data
                if(res.status == 200 && result.code == 0){
                    this.deployResult = "success"
                } else if(res.status == 200 && result.code == 2000){
                    this.deployResult = "noPermission"
                } else {
                    this.deployResult = "failed"
                }
                this.deployLog = res.data.data
            })
        },
    }
</script>
```

3. 服务器日志

在服务器表格中单击某一行中的"日志"按钮可查看当前服务器部署的日志详情，使用时间轴方式展示部署每一步的详情操作。在 template 中新增代码如下：

```
<a-modal
            :visible="visiableLog"
            title=" 服务器日志 "
            okText=" 确定 "
            cancelText=" 关闭 "
            @cancel="onLogReset"
            @ok="onLogReset"
        >
            <a-card>
                <a-timeline>
                    <div v-for="(item, index) in serverHistorySource" :key="item.id">
                        <a-timeline-item color="blue" v-if="item.server==selectServer">
```

```
                    <a-icon slot="dot" type="clock-circle-o" style="font-size: 16px;" />
                    <a-tag color='blue'>{{item.server}}</a-tag>
                    <a-tag>{{item.env}} {{item.release}} {{item.op_type}} {{item.action_type}}
                    </a-tag><br/>
                    {{item.log}}<br/>
                    <a-tag color='green'>{{item.create_user}}</a-tag>
                    {{item.create_date}}
                    <a-badge :count="item.log_no" />
                </a-timeline-item>
            </div>
        </a-timeline>
    </a-card>
</a-modal>
```

在 script 标签中新增事件包含：

- logShow：设置当前选择的服务器 IP，展示服务器日志模态框。
- fetchServerHistoryData：请求服务端接口获取当前 IP 下的操作日志。
- onLogReset：关闭服务器日志模态框。

新增的事件代码如下：

```
<script>
... // 其他忽略
    methods: {
        ...// 其他忽略
        fetchServerHistoryData(){
            let params = {
                release_id: this.releaseId,
                env_id: this.envId,
                pageSize: 200
            }
            API.ServerHistory(params).then((res)=>{
                let result = res.data
                if(res.status == 200 && result.code == 0){
                    this.serverHistorySource = result.data.results
                }
            })
        },
        logShow(data){
            this.selectServer = data.name
            this.visiableLog = true
        },
        onLogReset(){
```

```
                    this.visiableLog = false
                },
        }
</script>
```

通过 import 引入 service API,调用的 Deploy、ServerHistory 方法为 service 服务配置的请求发布单信息的 request 请求,在 service 服务配置中介绍具体的实现方式。

19.4　部署模块基础服务配置

发布单部署模块待部署列表页面以及部署页面开发完成后,就需要在浏览器中测试开发的页面展示以及功能是否有问题,即进入开发本地测试阶段,需要用到第 14 章介绍的路由、service 服务以及 mock 服务配置。

19.4.1　路由配置

若要在浏览器能够展示发布单部署模块两个页面,需要把两个页面的组件配置到 router 目录下的 config.js 文件中,具体代码如下:

```
const routeConfig = [
    {
        path: '/deployment',
        name: '部署',
        meta: {
            icon: 'cloud-upload'
        },
        children:[
            {
                path: '/deployList',
                name: '服务部署',
                component: () => import('@/views/deployment/deployList')
            },
            {
                path: '/deploy',
                name: '发布单部署',
                meta:{
                    isShowMenu:false
                },
```

```
                    component: () => import('@/views/deployment/deploy')
                }
            ]
        },
    ]
```

其中发布单部署页面入口是在待部署发布单列表中,不需要在菜单导航栏中展示,通过在 meta 中添加 isShowMenu:false 参数"告诉"菜单栏忽略此菜单。

通过在浏览器地址中添加 hash 值为 deployment/deployList,即访问 http://localhost:8080/#/deployment/deployList,便可以在浏览器中看到开发完的发布单列表页面,同理添加 deployment/deploy,可查看发布单部署页面。

19.4.2　mock服务配置

项目中的本地 mock 服务用于模拟服务端请求,并返回自定义的模拟数据。发布单部署页面需要请求后端的部署和历史查询接口,所以需要在 mock 目录下新增 deployment 目录,目录中新增 index.js 文件,内容如下:

```
import Mock from 'mockjs'
Mock.mock(`${process.env.VUE_APP_API_BASE_URL}/deploy/deploy/`, 'post', ({body}) => {
        return {
                "code": 0,
                "message": "success",
                "data": {}
        }
})

const serverHistory = Mock.mock({
        "count": 20,
        "results|20": [
                {
                        "id": "@id",
                        "server": "@ip",
                        "env|1": [
                                "dev",
                                "prd",
                                "uat"
                        ],
                        "release": "@guid",
                        "create_user": "@name",
```

```
                    "name": "@guid",
                    "description": "@sentence(5)",
                    "update_date": "@date",
                    "create_date": "@date",
                    "base_status": true,
                    "op_type": "deploy",
                    "log": "@sentence(5)"
            }
        ]
    })

Mock.mock(`${process.env.VUE_APP_API_BASE_URL}/history/server/`, 'get', ({body}) => {
        return {
                "code": 0,
                "message": "success",
                "data": serverHistory
        }
    })
```

在 mock 服务的入口文件 index.js 中引入 deployment 目录，即完成了部署模块 mock 服务的注册，当部署页面访问 /deploy/* 路径时，mock 服务层会返回配置的数据。部署页面访问 mock 服务层的接口数据需要通过 service 层，所以还需要在 service 层配置发布单部署模块。

19.4.3　service服务配置

service 服务统一封装了 BiFang 部署平台项目所需要的所有接口路径以及请求方式，并通过 axios 调用下游的 mock 数据，所以 service 服务需要新增发布单部署模块。

在 services 目录下的 api.js 路径集中新增部署模块所需的访问路径，代码如下：

```
const API = {
    DEPLOY: `${REAL_URL}/deploy/deploy/`,           // 部署、回滚接口
    SERVERHISTORY: `${REAL_URL}/history/server/`,   // 服务器日志
}
```

在 services 目录下新增 deploy.js 文件，用于配置发布单部署模块所需服务，代码如下：

```
import {request,METHOD} from '@/utils/request'
```

```
import {API} from './api'
/**
 * 部署软件
 * @param {*} params
 */
async function Deploy(params){
    return request(
        API.DEPLOY,
        METHOD.POST,
        params
    )
}
export default {
    Deploy
}
```

在 services 目录下的 history.js 文件中新增用于查询服务器日志所需服务，代码如下：

```
import {request,METHOD} from '@/utils/request'
import {API} from './api'
/**
 * 获取发布单历史操作
 * @param {*} params
 */
async function ReleaseHistory(params){
    return request(
        API.RELEASEHISTORY,
        METHOD.GET,
        params
    )
}
/**
 * 获取服务器历史操作
 * @param {*} params
 */
async function ServerHistory(params){
    return request(
        API.SERVERHISTORY,
        METHOD.GET,
        params
    )
}
```

```
export default {
        ReleaseHistory,
        ServerHistory
}
```

service 目录下主入口文件 index.js 中引入发布单部署模块和历史查询模块，代码如下：

```
import application from './release'
import permission from './history'
export default {
        ...release,
        ...history,
}
```

在发布单部署模块中调用 service 层配置的接口名称，并通过 promise.then(response=>{}) 的回调方式把获取的数据返回给处理函数。

至此，发布单部署模块待部署列表页面以及发布单部署页面相对应的 mock 服务、service 服务代码都已全部编写完毕，启动项目，单击左侧菜单导航中的服务部署菜单，页面即可呈现在读者面前，待部署发布单查询列表以及部署/回滚操作调用接口返回成功标识。

在 service 服务层中的 api.js 中改变 BASE_UR 和 REAL_URL 即可完成 mock 数据和真实环境的请求切换，方便项目的调试。

19.5 Dashboard 数据面板模块

Dashboard 是商业智能仪表盘（business intelligence dashboard，BI dashboard）的简称，它是一般商业智能都拥有的实现数据可视化的模块，是向企业展示度量信息和关键绩效指标（key performance index，KPI）现状的数据虚拟化工具。在网页中，它主要是由多个卡片组合而成的一个数据信息展示页面。

BiFang 部署平台中的 Dashboard 模块主要提供几个统计数据，如项目、应用、发布单总数以及发布最频繁的项目、出错率最高的项目、最近的操作等，同时页面中包含快捷入口模块，可供用户更快地进入对应的功能页面。

1. Dashboard 模块功能清单

（1）Dashboard 模块功能清单如表 19-2 所列。

第 19 章　发布单部署、Dashboard 模块设计与搭建

表 19-2　Dashboard 模块功能表

模块名称	功能名称	所属页面	功能描述
Dashboard	Dashboard	Dashboard 页面	显示项目、应用、发布单总数 显示发布单数量最多的 TOP5 应用 显示发布出错数量最多的 TOP5 应用 显示用户最近的操作记录

（2）Dashboard 模块功能说明。

① 服务器和端口组合，形成一条唯一记录。

② 同一个服务器的不同端口，可以部署不同的应用。

③ 服务器和端口组合必须属于一个环境 (若测试环境、线上环境等)。

④ 服务器和端口组合必须属于一个应用。

（3）Dashboard 模块截图如图 19-4 所示。

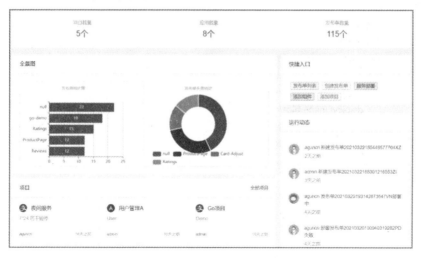

图 19-4　BiFang Dashboard

2. 功能技术要点

在 Dashboard 模块相关页面搭建中，主要使用 Antd 框架中的组件进行搭建，详细功能点使用的组件如下：

● 项目应用概览：自定义 headInfo 组件。

● 全景图：echart 画图。

在 components 下的 tool 目录新增 headInfo.vue 文件，开放两个属性：title 和 content，代码如下：

```
<template>
    <div class="head-info">
        <span>{{title}}</span>
        <p>{{content}}</p>
    </div>
</template>

<script>
export default {
    name: 'HeadInfo',
    props: ['title', 'content']
}
</script>
<style lang="less" scoped>
    .head-info{
        text-align: center;
        padding: 0 24px;
        flex-grow: 1;
        flex-shrink: 0;
        align-self: center;
        span{
            color: @bf-text-color-second;
            display: inline-block;
            font-size: 14px;
            margin-bottom: 4px;
        }
        p{
            color: @bf-text-color;
            font-size: 24px;
            margin: 0;
        }
    }
</style>
```

ECharts 是一个使用 JavaScript 实现的开源可视化库，涵盖各行业图表，满足各种需求，它遵循 Apache-2.0 开源协议，免费商用并兼容当前绝大部分浏览器（IE8/9/10/11、Chrome、Firefox、Safari 等）及兼容多种设备，可随时随地任意展示。

Echart 常用的属性如下：

```
option = {
    tooltip:{              // 提示框浮层
```

```
        // position:[10,10],             // 默认是在同一个地方的
        formatter:'{c}',                 //{a} 系列名 series[0].name {b} 横坐标 {c} 纵坐标
        // backgroundColor:'blue',       // 背景颜色
},
legend:{
        data:[' 良率 '] // 图形顶部的标签，单击可显示隐藏整条线，必须与 series[index]
                        内的 name 同名
},
xAxis: {
        type: 'category',                //xAxis 轴类型
        data: ['08:00', '10:00', '12:00', '14:00', '16:00', '18:00', '20:00', '22:00', '00:00', '02:00',
            '04:00', '06:00'],
        splitLine: { show: false },      // 坐标轴中间的分割线，隐藏
        axisLabel: {
            color: '#79a5a5'
        },                               // 坐标轴刻度标签
        axisLine: { show: false },       // 坐标轴轴线，隐藏
        axisTick: { show: false }        // 坐标轴刻度，隐藏
},
yAxis: {
        type: 'value',
        name: ' 良率 (%)',                //y 轴，名字
        nameTextStyle: {
            color: '#79a5a5'
        },                               //y 轴，名字、颜色
        splitLine: { show: false },
        axisLabel: {
            color: '#79a5a5'
        },
        axisLine: { show: false },
        axisTick: { show: false },
        min: 0,
        max: 100,
//      splitNumber:5                    // 分割段数
},
series: [{
        name:' 良率 ',
        data: [90, 95, 90, 91, 96, 81, 91, 92, 78, 96, 91, 91],         // 数据
        type: 'line',
        symbol: 'pin',                   // 数值点的形状
        symbolSize:8,                    // 数值点的大小
        symbolRotate:180,                // 数值点的旋转角度
        showSymbol:true,                 // 数值点的显示隐藏
```

```
            hoverAnimation:true,                  // 数值点 hover 是否有动画
//          stack:'',                             // 可堆叠前面的数值到最后一项
            label: {
                normal: {
                    show: true                    // 数值点的数据
                }
            },
            lineStyle: {
                color: '#c7e821'                  // 折线的样式、颜色
            },
            itemStyle: {
                color: ({ data }) => {
                    if (data <= 80) { return '#ff0000 ' }   // 不同点的颜色
                    return '#c7e821'
                }
            },
            markLine: {    // 划线
                symbol: 'none',                   // 取消最后的箭头
                data: [
                    [{
                        value: 80,                // 线的数值
                        lineStyle: {              // 线样式、颜色
                            color: '#36dcdc'
                        },
                        label: {                  // 线标签
                            color: '#000',        // 颜色
                            position: 'middle',   // 位置
                            formatter: '{c} %',   // 格式
                            backgroundColor: '#36dcdc',    // 背景色
                            padding: 5,
                            borderRadius: 5
                        },
                        coord: ['08:00', 80]      // 起点 [ 横坐标，数值 ]
                    }, {
                        coord: ['06:00', 80]      // 终点 [ 横坐标，数值 ]
                    }]
                ]
            },
            areaStyle: {                          // 区域填充样式
                color: {                          // 颜色
                    type: 'linear',
                    x: 0,
                    y: 0,
                    x2: 0,
```

```
                    y2: 1,
                    colorStops: [{
                        offset: 0, color: '#505E0D'          // 0% 处的颜色
                    }, {
                        offset: 1, color: '#171A14'          // 100% 处的颜色
                    }],
                    global: false // 缺省为 false
                }
            }
        }]
}
```

Echart 配置图标有几百个属性，这里不再一一介绍，有兴趣的读者可以前往 echart 官网查看详细教程和 API 配置。

19.6 Dashboard 页面搭建

Dashboard 页面包含项目概览模块、全景图模块、项目列表模块、运行动态模块以及快捷入口模块发布单构建，这里将以实现每个模块功能的步骤来呈现。

在 src 中的 views 目录下创建 dashboard 目录，并在该目录下创建 workspace.vue 文件，此文件为 Dashboard 工作台组件。

1. 项目概览

项目概览展示 BiFang 平台配置的项目数量、应用数量以及构建发布单的数量。对于此类常用的样式，我们封装为单独的组件供 Dashboard 使用。

在工作台组件 workspace.vue 中引入 19.5 节封装的 headInfo 组件，配置项目数量、应用数量和发布单数量，代码如下：

```
<template>
    <div class="dashboard">
        <a-card :bordered="false">
            <div style="display: flex; flex-wrap: wrap">
                <head-info title=" 项目数量 " :content="allCount.projectCount+' 个 '" />
                <head-info title=" 应用数量 " :content="allCount.appCount+' 个 '" />
                <head-info title=" 发布单数量 " :content="allCount.releaseCount+' 个 '"/>
            </div>
        </a-card>
    </div>
</template>
```

工作台组件内嵌通用的卡片容器 a-card，a-card 中内嵌 headInfo 组件，设置标题和内容。

在 workspace.vue 文件中添加 script 标签，工作台组件中实现的事件包含：

● fetchAllCount：请求接口获取项目数、应用数以及发布单数量。

相关代码如下：

```
<script>
export default {
        name:"dashboard",
        components: {HeadInfo},
        created(){
                this.fetchAllCount()
        },
        data(){
                return {
                        allCount:{},           // 项目应用与发布单数据统计
                        releaseData:[],        // 发布单详情统计
                        projectData:[]         // 项目统计
                }
        },
        methods:{
            fetchAllCount(){
                API.AllCount().then((res)=>{
                        let result = res.data
                        if(res.status == 200 && result.code == 0){
                            // 后端传过来的是一个 Json 对象
                            this.allCount.projectCount = result.data.project
                            this.allCount.appCount = result.data.app
                            this.allCount.releaseCount = result.data.release
                        } else {
                                this.allCount = {}
                        }
                })
            },
        }
}
</script>
```

2. 全景图

全景图使用 echart 绘制两个图形，一个是使用柱状图展示项目构建的发布单数量，选取前五名并由大到小展示；另一个是使用饼状图展示构建的发布单

失败比例，用户可以一目了然地观察到哪些项目经常出现发布单构建失败，这需要重点关注。

在 components 下的 tool 目录新增 overallView.vue 文件，该文件为全景图组件，内部使用 echart 渲染柱状图和饼状图。在全景图组件中添加 template 标签，代码如下：

```
<template>
    <div class="Echarts">
        <a-card-grid  style="width:50%;height:250px;padding:0">
                <div ref="main" style="width:330px;height:250px"></div>
        </a-card-grid>
        <a-card-grid style="width:50%;height:250px;padding:0">
                <div ref="main1" style="width:330px;height:250px"></div>
        </a-card-grid>
    </div>
</template>
```

在 script 标签中新增事件包含：

- fetchReleaseTop5：请求服务端接口获取发布单发布数量 Top5 的项目。
- fetchReleaseFailedTop5：请求服务端接口获取发布单发布失败数量 Top5 的项目。
- releaseTop5Chart：渲染发布单 Top5 项目柱状图。
- failedTop5Chart：渲染发布失败数量 Top5 的项目饼状图。

全景图组件新增 script 标签，代码如下：

```
<script>
import API from '@/service'
// 指定图表的配置项和数据
const releaseTop5Option = {
        title: {
                //top:"bottom",
                subtext: ' 发布单统计图 ',
                left: 'center',
        },
        tooltip: {
                trigger: 'axis',
                axisPointer: {
                        type: 'shadow'
                }
        },
```

```javascript
        grid: {
            left: '3%',
            right: '4%',
            bottom: '3%',
            containLabel: true
        },
        xAxis: {
            type: 'value'
        }
};
const faiedTop5Option = {
        title: {
            //top:"bottom",
            left: 'center',
            subtext: ' 发布单失败统计 '
        },
        tooltip: {
            trigger: 'item'
        },
        legend: {
            top: '5%',
            left: 'center',
            top:"bottom"
        },
        series: [
            {
                name: ' 发布单失败数量 ',
                type: 'pie',
                radius: ['40%', '70%'],
                avoidLabelOverlap: false,
                itemStyle: {
                    borderRadius: 10,
                    borderColor: '#fff',
                    borderWidth: 2
                },
                label: {
                    show: false,
                },
                labelLine: {
                    show: false
                },
                data: []
            }
```

```js
    ]
};
export default {
    name: 'OverallView',
    created(){
            this.fetchReleaseTop5()
            this.fetchReleaseFailedTop5()
    },
    methods:{
        releaseTop5Chart(option){
            // 基于准备好的 dom，初始化 echarts 实例
            var top5Chart = this.$echarts.init(this.$refs['main']);
                Object.assign(releaseTop5Option,option)
            // 使用刚指定的配置项和数据显示图表
            top5Chart.setOption(releaseTop5Option);
        },
        failedTop5Chart(){
            // 基于准备好的 dom，初始化 echarts 实例
            var failedChart = this.$echarts.init(this.$refs['main1']);
            // 使用刚指定的配置项和数据显示图表
                failedChart.setOption(faiedTop5Option);
        },
        fetchReleaseTop5(){
                API.ReleaseTop5().then((res)=>{
                    let result = res.data
                    if(res.status == 200 && result.code == 0){
                        console.log(result)
                        this.releaseTop5Chart({
                            yAxis: {
                                    type: 'category',
                                    data: result.data.app_name.reverse()
                            },
                            series: [
                                {
                                        name: ' 发布单量 ',
                                        type: 'bar',
                                        label: {
                                                show: true
                                        },
                                        emphasis: {
                                                focus: 'series'
                                        },
                                        data: result.data.release_count.reverse()
```

```
                                        }
                                    ]
                                });
                            }
                        })
                },
                fetchReleaseFailedTop5(){
                    API.ReleaseFailedTop5().then((res)=>{
                        let result = res.data
                        if(res.status == 200 && result.code == 0){
                            let series_data = []
                            result.data["app_name"].forEach((item,index)=>{
                                series_data.push({
                                    value:result.data["release_count"][index],
                                    name:String(item)
                                })
                            })
                            faiedTop5Option.series[0].data = series_data
                            this.failedTop5Chart()
                        }
                    })
                }
            }
        }
</script>
```

在工作台组件 workspace.vue 中引入 overallView 全景图组件，放置在项目概览下方左侧，使用 a-row 和 a-col 布局组件按照 2(左侧):1(右侧) 分割页面，左侧展示全景图与项目列表，右侧展示快捷入口和运行动态模块。在 template 标签中添加如下代码：

```
<template>
<div>
    <a-card :bordered="false">
        <div style="display: flex; flex-wrap: wrap">
            <head-info title=" 项目数量 " :content="allCount.projectCount+' 个 '" :bordered="true"/>
            <head-info title=" 应用数量 " :content="allCount.appCount+' 个 '" :bordered="true"/>
            <head-info title=" 发布单数量 " :content="allCount.releaseCount+' 个 '"/>
        </div>
    </a-card>
        <a-row style="margin-top:20px">
            <a-col style="padding-right:2px" :xl="16" :lg="24" :md="24" :sm="24" :xs="24">
                <a-card title=" 全景图 " :bordered="false">
```

```
                    <overall-view />
                </a-card>
                <a-card class="project-list" style="margin: 12px 0;" :bordered="false"
                    title=" 项目 "
                    :body-style="{padding: 0}">
                        <a href="#/application/project" slot="extra"> 全部项目 </a>
                </a-card>
            </a-col>
            <a-col style="padding-left: 12px" :xl="8" :lg="24" :md="24" :sm="24" :xs="24">
                <a-card title=" 快捷入口 " style="margin-bottom: 24px"
                    :bordered="false" :body-style="{padding: 0}">
                </a-card>
                <a-card title=" 运行动态 " style="margin-bottom: 24px" :bordered="false">
                </a-card>
            </a-col>
        </a-row>
    </div>
</template>
```

3. 项目列表模块

项目列表模块展示最近更新的六个项目，每个项目卡片展示项目名称、项目描述和创建人等信息。在 template 标签中项目列表通用卡片中加入如下代码：

```
<a-card class="project-list" style="margin: 12px 0;"
:bordered="false" title=" 项目 " :body-style="{padding: 0}">
    <a href="#/application/project" slot="extra"> 全部项目 </a>
    <div>
    <a-card-grid :key="i" v-for="(item, i) in projectData">
        <a-card :bordered="false" :body-style="{padding: 0}">
            <a-card-meta :description="item.description">
                <div slot="title" class="card-title">
                    <a-avatar size="small" :src="item.logo" />
                    <span>{{item.cn_name}}</span>
                </div>
            </a-card-meta>
            <div class="project-item">
                <a class="group" href="/#/">{{item.create_user_name}}</a>
                <span class="datetime">{{item.time}}</span>
            </div>
        </a-card>
    </a-card-grid>
    </div>
</a-card>
```

在 script 标签中新增事件包含：
- fetchProjectData：请求接口获取项目列表。

新增的事件代码如下：

```
<script>
... // 其他忽略
        methods: {
        ...// 其他忽略
        fetchProjectData(){
            let param = {
                currentPage: 1,
                pageSize: 6,
                sort: '-update_date'
            }
            API.ProjectList(param).then((res)=>{
                let result = res.data
                if(res.status == 200 && result.code == 0){
                    this.projectTotal = result.data.count
                    result.data.results.forEach(item=>{
                        this.projectData.push({
                            ...item,
                            logo: projectIcons[Math.floor(Math.random()*10)%3],
                            time:this.formatTime(item.update_date)
                        })
                    })
                }
                else{
                    this.projectData = []
                }
            })
        },
    }
</script>
```

4. 快捷入口模块

快捷入口模块展示用户常用的发布单列表、创建发布单、服务部署、添加组件和添加项目等功能，单击可直接进入相关功能。在 template 标签中快捷入口通用卡片中加入如下代码：

```
<a-card title=" 快捷入口 " style="margin-bottom: 24px"
:bordered="false" :body-style="{padding: 0}">
    <div class="item-group">
```

第 19 章　发布单部署、Dashboard 模块设计与搭建

```
                <a-tag color="red">
                        <a href="#/release/releaseList"> 发布单列表 </a>
                </a-tag>
                <a-tag color="green">
                        <a href="#/release/createRelease"> 创建发布单 </a>
                </a-tag>
                <a-tag color="blue">
                        <a href="#/deployment/deploy"> 服务部署 </a>
                </a-tag><br/>
                <a-tag color="blue">
                        <a href="#/application/app"> 添加组件 </a>
                </a-tag>
                <a-tag color="green">
                        <a href="#/application/project"> 添加项目 </a>
                </a-tag>
        </div>
</a-card>
```

5. 运行动态模块

运行动态模块展示近期操作的发布单信息，展示创建发布单用户名称、发布单编号以及对应的操作，如新建、流转或部署等。在 template 标签中运行动态通用卡片中加入如下代码：

```
<a-card title=" 运行动态 " style="margin-bottom: 24px" :bordered="false">
        <div style="min-height: 400px;">
        <a-list>
        <a-list-item :key="index" v-for="(item, index) in releaseData">
                <a-list-item-meta>
                        <a-avatar slot="avatar" :src="item.avatar" />
                        <div slot="title" v-html="item.action" />
                        <div slot="description">{{item.time}}</div>
                </a-list-item-meta>
        </a-list-item>
        </a-list>
        </div>
</a-card>
```

在 script 标签中新增事件包含：

- fetchReleaseData：请求接口获取最近操作信息。
- formatTime：根据时间匹配几分钟前、几小时前、几天前。
- actionTranformer：根据用户和发布单信息拼接运行操作信息。

新增的事件代码如下：

```
<script>
...// 其他忽略
    methods: {
    ...// 其他忽略
        fetchReleaseData(){
            let params = {
                currentPage:1,
                pageSize:5,
                sort:"-update_date"
            }
            API.ReleaseList(params).then((res)=>{
                let result = res.data
                if(res.status == 200 && result.code == 0){
                    this.releaseTotal = result.data.count
                    result.data.results.forEach(item=>{
                        this.releaseData.push({
                            avatar: avatars[Math.floor(Math.random()*10)%4],
                            user:item.create_user_name,
                            name:item.name,
                            action:this.actionTranformer(item.create_user_name
                                ,item.release_status_name,item.name),
                            time: this.formatTime(item.update_date)
                        })
                    })
                } else {
                    this.releaseData = []
                }
            })
        },
        formatTime(time){
            let str = "";
            let current = moment(),
                date = moment(time),
                du = moment.duration(current-date,'ms'),
                days = du.get('days'),
                hours = du.get('hours'),
                mins = du.get('minutes');
            if(days > 0){
                str = days+" 天之前 "
            }
            else if(hours > 0){
                str = hours+" 小时之前 "
```

```
            }
            else{
                str = mins+" 分钟之前 "
            }
            return str
        },
        actionTranformer(username,status,releaseNo){
            let action=""
            let release = `<a href="#/release/releaseList">${releaseNo}</a>`
            switch(status.toLowerCase()){
                case "create":
                    action=`${username} 新建发布单 ${release}`
                    break;
                case "build":
                    action=`${username} 构建发布单 ${release}`
                    break;
                case "ready":
                    action=`${username} 扭转发布单 ${release}`
                    break;
                case "success":
                    action=`${username} 部署发布单 ${release} 成功`
                    break;
                case "failed":
                    action=`${username} 部署发布单 ${release} 失败`
                    break;
                case "ongoing":
                    action=`${username} 发布单 ${release} 部署中`
                    break;
                default:
                    action=`${username} 操作发布单 ${release}`
                    break;
            }
            return action
        }
    }
}
</script>
```

19.7 Dashboard 基础服务配置

Dashboard 数据面板页面开发完成后,就需要在浏览器中测试开发的页面展示以及功能是否有问题,需要配置路由、service 以及 mock 服务。

19.7.1 路由配置

若要在浏览器能够展示 Dashboard 数据面板页面,只需要将 Dashboard 组件配置到 router 目录下的 config.js 文件中,代码如下:

```
const routeConfig = [
    {
        path: '/',
        name: 'Home',
        component: MainLayout,
        redirect: '/dashboard/workspace',
        children:[
            {
                path: '/dashboard',
                name: 'Dashboard',
                meta:{
                    icon: 'dashboard'
                },
                children:[
                    {
                        path: '/workspace',
                        name: ' 工作台 ',
                        meta:{
                            title: " 工作台 "
                        },
                        component: () => import('@/views/dashboard/workspace')
                    }
                ]
            }
        ]
    }
]
```

在浏览器地址中添加 hash 值为 server/list,访问 http://localhost:8080/#/server/list,便可以在浏览器中看到开发完的服务器管理页面。

19.7.2 mock服务配置

项目中的本地 mock 服务用于模拟服务端请求，并返回自定义的模拟数据。Dashboard 页面请求服务端针对项目的统计接口，所以需要在 mock 目录下新增 dashboard 目录，目录中新增 index.js 文件，内容如下：

```javascript
import Mock from 'mockjs'
Mock.mock(`${process.env.VUE_APP_API_BASE_URL}/stats/all_count/`, 'get', ({body}) => {
    return {
        "code": 0,
        "message": " 操作成功 ",
        "data": {
            "project": "@integer(1,100)",
            "app": "@integer(1,100)",
            "release": "@integer(1,100)"
        }
    }
})

const top5Release = Mock.mock({
    "app_name|1-10": [
        "project1",
        "project2"
    ],
    "release_count": @range(1,10,2)
})

Mock.mock(`${process.env.VUE_APP_API_BASE_URL}/stats/release_top5/`, 'get', ({body}) => {
    return {
        "code": 0,
        "message": "success",
        "data": top5Release
    }
})

const top5FailedRelease = Mock.mock({
    "app_name|1-10": [
        "project5",
        "project6"
    ],
    "release_count": @range(1,10,2)
})
```

```
Mock.mock(`${process.env.VUE_APP_API_BASE_URL}/stats/release_failed_top5/`, 'get', ({body}) => {
    return {
        "code": 0,
        "message": "success",
        "data": top5FailedRelease
    }
})
```

在 mock 服务的入口文件 index.js 中引入 Dashboard 目录，即完成了 Dashboard 模块 mock 服务的注册，当工作台 Dashboar 页面访问 /stat/* 路径时，mock 服务层会返回配置的数据。Dashboard 模块访问 mock 服务层的接口数据需要通过 service 层，所以还需要在 service 层配置 Dashboard 模块。

19.7.3　service服务配置

service 服务统一封装了 BiFang 部署平台项目所需要的所有接口路径以及请求方式，并通过 axios 调用下游的 mock 数据，所以 service 服务需要新增 Dashboard 模块。在 services 目录下的 api.js 路径集中新增 Dashboard 所需的访问路径，代码如下：

```
const API = {
    // 项目应用与发布单总数
    ALLCOUNT: `${REAL_URL}/stats/all_count/`,
    // 发布单 Top5 项目
    RELEASETOP5: `${REAL_URL}/stats/release_top5/`,
    // 失败 Top5 项目
    RELEASEFAILEDTOP5: `${REAL_URL}/stats/release_failed_top5/`
}
```

在 services 目录下新增 state.js 文件，用于配置 Dashboard 组件所需服务，代码如下：

```
import {request,METHOD} from '@/utils/request'
import {urlFormat} from '@/utils/util'
import {API} from './api'
/**
 * 获取项目，应用，发布单总数
 */
async function AllCount(){
    return request(
        API.ALLCOUNT,
```

```
                    METHOD.GET,
        )
    }
    async function ReleaseTop5(){
        return request(
                    API.RELEASETOP5,
                    METHOD.GET,
        )
    }
    async function ReleaseFailedTop5(){
        return request(
                    API.RELEASEFAILEDTOP5,
                    METHOD.GET,
        )
    }
    export default {
        AllCount,
        ReleaseTop5,
        ReleaseFailedTop5
    }
```

在 service 目录下主入口文件 index.js 中引入 Dashboard 模块，代码如下：

```
    import stats from './stats'
    export default {
            ...stats,
    }
```

Dashboard 组件中调用 service 层配置的接口，axios 发起 request 请求并通过 promise.then(response=>{}) 的回调方式把获取到的数据返回给处理函数。

至此，Dashboard 页面相对应的 mock 服务、service 服务代码都已全部编写完毕，启动项目，单击左侧菜单导航中的 Dashboard→工作台菜单，页面即可呈现在读者面前，获取项目、应用、发布单以及运行动态信息调用接口返回成功标识。

在 service 服务层的 api.js 中改变 BASE_UR 和 REAL_URL 即可完成 mock 数据和真实环境的请求切换，方便项目的调试。

19.8 小　结

本章带领读者熟悉了 BiFang 部署平台中项目与应用模块以及服务器模块的页面设计、页面搭建到功能实现的流程，掌握了如何在公共服务路由、mock 服务以及 service 服务中添加项目应用模块和服务器模块，中间也穿插了用到的 ant-design-vue 框架相关的组件知识点，相信没有了解过相关知识点的读者能对使用框架组件布局页面有全面的认识。

第 20 章将带领大家进入 BiFang 部署平台发布单的管理和环境流转，发布单是什么？如何流转发布单状态以及它跟项目应用、服务器关系是怎样的？让我们带着这些问题，前进！

再长的路，一步步也能走完；再短的路，不迈开双脚也无法完成！

第 20 章 前后端服务联调

> 耐心是一切聪明才智的基础。
>
> ——柏拉图

前述章节实现了 BiFang 部署平台后台服务和前端页面展示的搭建,接下来将进入开发流程的下一阶段:前后端联调。联调阶段是验证后台服务和前端请求接口是否一致的关键阶段,理论上来说以 JSON 结构定义为契约,前后端服务严格按照接口定义进行前后端分离开发,联调应该是一遍就能通过,但在实际开发中该阶段是出现问题最多的阶段,出现的问题需要进行大量的沟通,甚至更改 JSON 契约结构,导致前后端重新开发也可能出现,因此,JSON 契约的合理性和可扩展性是个人的开发经验和技术能力的体现。

本章将介绍如何在 Vue 项目配置转发到后端接口服务,启动部署平台后台 Python 服务后,切换 mock 服务与真实服务地址实现前后端的串联。

20.1 前端接口服务转发

前后端分离项目发布到生成环境,前端服务和后端服务使用不是相同的 IP 和端口,如前端服务使用 8080 端口,后台服务使用 8000 端口,对于浏览器同源策略的限制,即不同的域名、端口,不同的协议不允许共享资源,以保障浏览器安全。如何解决跨域,常用以下三种方案:

① JSONP 请求。
② CORS 跨域。
③ 反向代理。

接下来将详细介绍它们的实现方式和不同之处。

20.1.1 JSONP

JSONP 方式主要通过动态插入一个 script 标签,浏览器对 script 的资源引用没有同源限制，同时资源加载到页面后会立即执行。实际项目中 JSONP 通常用来获取 JSON 格式数据，这时前后端通常会约定一个参数 callback，该参数的值就是处理返回数据的函数名称。JSONP 方式模拟代码如下：

```
<script>
    var jsonp_script = document.createElement('script');
    jsonp_script.type = "text/javascript";
    jsonp_script.src = "http://localhost:8000/jsonp?callback=cb";
    document.head.appendChild(jsonp_script);
    var cb = function(data){
        alert(data.name);
    }
</script>
```

缺点：

- 这种方式无法发送 post 请求。
- 另外要确定 JSONP 的请求是否失败并不容易，大多数框架的实现都是结合超时时间来判定的。

20.1.2 CORS

CORS 是现代浏览器支持跨域资源请求的一种方式，当使用 XMLHttpRequest 发送请求时，浏览器发现该请求不符合同源策略，会给该请求加一个请求头：Origin，后台进行一系列处理，如果确定接受请求则在返回结果中加入一个响应头：Access-Control-Allow-Origin。浏览器判断该响应头中是否包含 Origin 的值，如果有，则浏览器处理响应，我们就可以拿到响应数据；如果不包含，则浏览器直接驳回，这时无法拿到响应数据。浏览器兼容性如图 20-1 所示。

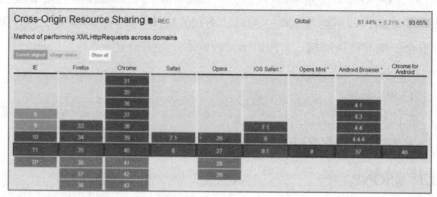

图 20-1 CORS 浏览器兼容性

使用 CORS 跨域请求，浏览器会发起两次请求，第一次为浏览器发起的 OPTIONS 请求，OPTIONS 请求头部中包含以下头部：Origin、Access-Control-Request-Method、Access-Control-Request-Headers，发送这个请求后，服务器可以设置如下头部与浏览器沟通来判断是否允许这个请求。

- Access-Control-Allow-Origin：指定域名（如 a.com）或 *（不推荐）。
- Access-Control-Allow-Method：'GET,POST,OPTIONS'。
- Access-Control-Allow-Headers：'X-Requested-With, Content-Type'。

第二次才是真正的请求，携带请求参数请求服务端，服务端处理后返回对应的数据。

缺点：
- 需要服务端额外设置参数，若正式环境使用同域名，造成不必要的资源浪费。
- 需要浏览器额外发起请求。

20.1.3 反向代理

反向代理首先将请求发送给后台服务器，通过服务器来发送请求，然后将请求的结果传递给前端，而且访问数据时就好像在访问本地服务器一样。如此，就可以直接获得 cookie 等数据。

反向代理可以使用后台服务单独实现，也可以使用三方高性能高并发代理服务 Nginx，若使用 Nginx 作为代理，需要在 nginx.conf 中添加路径代理，代码如下：

```
location / {
        root /code/bifang/; # 前端代码路径
        index    index.html index.htm;
}
location ^~ /api
{
        proxy_pass http://localhost:8000;
        proxy_cookie_path    /;
        proxy_set_header    Host            $host;
        proxy_set_header    Referer         $http_referer;
        proxy_set_header    Cookie          $http_cookie;
        proxy_set_header    X-Real-IP       $remote_addr;
        proxy_set_header    X-Forwarded-For $proxy_add_x_forwarded_for;
}
```

缺点：

需要使用三方模块或自己开发。

20.1.4 Vue代理

使用 Vue CLI 脚手架搭建项目时，Vue 框架本身内置了反向代理模块，方便进行接口转发，在项目根目录下的 vue.config.js 中添加如下代码实现反向代理配置：

```
module.exports = {
    devServer: {
        proxy: {
            // 此处要与 /services/api.js 中的 API_PROXY_PREFIX 值保持一致
            '/api': {
                target: 'http://localhost:8000',
                changeOrigin: true,
                pathRewrite: {          // 路径替换
                    '^/api': '/'
                }
            }
        }
    }
}
```

通过添加 devServer 配置，Vue 项目可以方便地进行反向代理配置，轻松解决跨域问题。

20.2 后端服务本地启动

启动 BiFang 部署平台后端服务，用于本地联调测试。

20.2.1 下载或clone代码

代码地址：https://github.com/aguncn/bifang。

假定保存的本地地址为 D:\Code\bifang，此目录下的 bifangback 就是接下来操作的主目录。

使用 git 命令 clone 代码，命令如下：

```
git clone https://github.com/aguncn/bifang
```

20.2.2　安装Python虚拟机环境

（1）在 bifangback 的目录下运行如下命令，建立一个 Python 虚拟机环境：

```
python -m venv venv  //windows
python3 -m venv venv  //macos
```

（2）运行命令，启动此虚拟环境：

```
venv\bin\activate  //windows
source venv/bin/activate  //macos
```

（3）运行如下命令，安装依赖第三方库：

```
pip install -r requirements.txt
```

如图 20-2 所示。

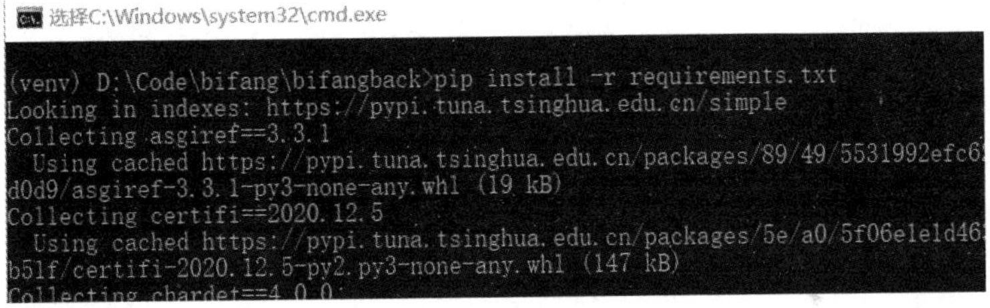

图 20-2　安装项目依赖

（4）更新依赖

当安排好依赖包后，整个 Python 基本环境就建好了。

如果以后 requirements.txt 文件有更新，可继续运行 pip install-r requirements.txt。每次运行 bifangback 项目时，都需要选择运行 venv\bin\activate 进入此虚拟环境。

20.2.3　运行程序

运行程序有两种方式，这两种方式用于日常测试 API，任选一种即可（在查看 swagger API 时选第一种。在测试异步效果时选第二种）。

（1）第一种：python manage.py runserver，如图 20-3 所示。

图 20-3 运行程序方式 1

（2）第二种：uvicorn bifangback.asgi:application-reload，如图 20-4 所示。

图 20-4 运行程序方式 2

20.2.4 运行测试

由于 bifangback 是一个前后端分离的项目，所以只提供了以下浏览器网址供访问。

（1）访问首页：

```
http://127.0.0.1:8000/            #首页
```

这只是一个 demo，输出"Hello, BiFang!"

（2）访谈后台管理页面：

```
http://127.0.0.1:8000/admin/      #后台管理
```

用户名/密码：admin/password，如图 20-5 所示。

图 20-5 后台管理

（3）访问 Swagger API：

http://127.0.0.1:8000/swagger/　　　# swagger 地址

如图 20-6 所示。

图 20-6　swagger 文档

（4）访问说明文档：

http://127.0.0.1:8000/redoc　# 说明文档

如图 20-7 所示。

图 20-7　接口说明文档

（5）使用 postman 测试工具访问登录接口，测试服务连通性。访问地址：http://localhost:8000/jwt_auth/。post 请求数据：

```
{
    "username": "admin",
    "password": "password"
}
```

返回数据如图 20-8 所示。

图 20-8　登录接口测试

20.3　前后端联调

通过前面两节的操作，本地启动了 BiFang 部署平台后端服务（端口 8000）和前端服务（端口 8080），并在前端项目配置了反向代理，访问 /api 路径指向后台服务端口。

20.3.1　mock 与真实接口切换

在之前章节开发前端各模块页面时也提到了可以通过切换 BASE_URL 和 REAL_URL 实现 mock 与真实接口的切换，代码如下：

```
const BASE_URL = process.env.VUE_APP_API_BASE_URL;
const REAL_URL = process.env.VUE_APP_API_REAL_URL;
const API = {
    LOGIN: `${REAL_URL}/jwt_auth/`,
    REGISTER: `${BASE_URL}/account/register/`,
}
```

调用登录接口使用的是真实环境后台服务的接口，而注册接口则调用 mock 服务。细心的读者会发现，BASE_URL 和 REAL_URL 配置的路径不同，这是在项目中的环境配置文件 .env.development 中实现的：

```
//.env.development
```

```
VUE_APP_API_BASE_URL=http://127.0.0.1:8080/mock
VUE_APP_API_REAL_URL=http://127.0.0.1:8080/api
```

访问 BASE_URL 的路径是 /mock，对应的是 mock 服务中的路径配置，而访问 /api 会被反向代理到本地后端服务端口 8000，即 http://127.0.0.1:8000。

通常在正式环境不需要 /mock 路径，访问的二级路径也与开发环境不同，因此在 .env.prd 中增加 REAL_URL 的配置，改为正式环境访问路径，如 BiFang 部署平台后端服务注册接口正式环境访问路径为 http://www.bifang.com/restapi/account/register，则 REAL_URL 配置路径为：

```
//.env.prd
VUE_APP_API_REAL_URL=http://www.bifang.com/restapi
```

使用不同路径访问 mock 和真实服务，可以方便地切换，使后端服务可以做到实现一个接口就可以实时测试，而不影响前端项目的运行。

20.3.2 登录接口测试

前端代理配置好真实后台服务路径后，使用登录模块测试前后端的连通性和正确性，其他模块就不再一一调试，这一步通用。

打开浏览器输入 http://localhost:8080，进入登录页面，输入登录用户名 admin 与密码 password，单击登录按钮，在代码中设置断点，即可看到如图 20-9 和图 20-10 所示的请求。

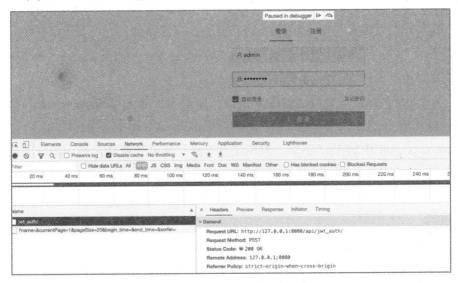

图 20-9　登录页面测试

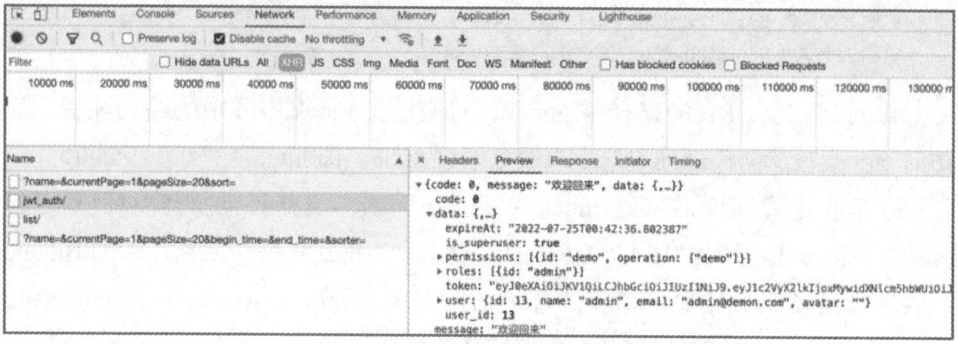

图 20-10　登录接口返回值

前端调用 /api/jwt_auth/ 路径，后台服务收到请求，校验处理完毕后返回 status：200 以及对应的数据，至此前后端服务已联通且功能正常，接下来需联调测试发布单、项目与应用等模块各接口，确保服务的连通性和正确性。

20.4　前端项目打包部署

前后端联调完成后，各模块功能已可正常运行，接下来进入项目开发的最后一个阶段——打包部署。

20.4.1　前端项目打包配置介绍

前端项目开发环境用 node 环境进行开发，但发布到正式环境还是打包后的静态资源文件：HTML、JS、CSS 等，这就需要在 Vue 项目配置文件 vue.config.js 进行打包配置。

vue.config.js 是一个可选的配置文件，如果项目的（和 package.json 同级）根目录中存在这个文件，那么它会被 @vue/cli-service 自动加载。也可以使用 package.json 中的 Vue 字段，但是注意这种写法需要严格遵照 JSON 的格式来写。

这个文件应该导出一个包含选项的对象：

```
// vue.config.js
/**
 * @type {import('@vue/cli-service').ProjectOptions}
 */
```

```
module.exports = {
    // 选项 ...
    publicPath: isProd ? '/bifang/' : '/',
    outputDir: 'build',
    assetsDir: 'static',
    productionSourceMap: false
}
```

常用的配置项如下：

- publicPath：部署应用包时的基本 URL。
- outputDir：当运行 vue-cli-service build 时生成的生产环境构建文件的目录。
- assetsDir：放置生成的静态资源（js、css、img、fonts）的（相对于 outputDir 的）目录。
- productionSourceMap：如果你不需要生产环境的 source map，可以将其设置为 false 以加速生产环境构建。
- indexPath：指定生成的 index.html 的输出路径（相对于 outputDir）。
- chainWebpack：是一个函数，会接收一个基于 webpack-chain 的 ChainableConfig 实例。允许对内部的 webpack 配置进行更细粒度的修改。
- pluginOptions：用来传递任何第三方插件选项。

20.4.2 添加 chainWebpack

在 webpack 中配置文件引用路径和 CDN 配置，在 vue.config.js 文件中的 chainWebpack 函数中添加如下代码：

```
chainWebpack: (config)=>{
            // 修改文件引入自定义路径
            config.resolve.alias
                    .set('@', path.resolve(__dirname, "src"))
            // 生产环境下使用 CDN
            if (isProd) {
                    config.plugin('html')
                            .tap(args => {
                                    args[0].cdn = assetsCDN
                                    return args
                            })
            }
}
```

配置 cdn 路径，添加 assetsCDN 变量配置如下：

```
const assetsCDN = {
    // webpack build externals
    externals: {
        vue: 'Vue',
        'vue-router': 'VueRouter',
        vuex: 'Vuex',
        axios: 'axios',
        '@antv/data-set': 'DataSet',
        'js-cookie': 'Cookies'
    },
    js: [
        '//cdn.jsdelivr.net/npm/vue@2.6.11/dist/vue.min.js',
        '//cdn.jsdelivr.net/npm/vue-router@3.3.4/dist/vue-router.min.js',
        '//cdn.jsdelivr.net/npm/vuex@3.4.0/dist/vuex.min.js',
        '//cdn.jsdelivr.net/npm/axios@0.19.2/dist/axios.min.js',
        '//cdn.jsdelivr.net/npm/nprogress@0.2.0/nprogress.min.js',
        '//cdn.jsdelivr.net/npm/clipboard@2.0.6/dist/clipboard.min.js',
        '//cdn.jsdelivr.net/npm/@antv/data-set@0.11.4/build/data-set.min.js',
        '//cdn.jsdelivr.net/npm/js-cookie@2.2.1/src/js.cookie.min.js'
    ]
}
```

20.4.3 添加LESS预处理

在使用 Vue CLI 脚手架配置时，选择使用 LESS 作为样式开发语言，这时需要添加 CSS 全局预处理器 style-resources-loader，它可以导入 CSS 预处理器的一些公共的样式文件变量，比如 variables、mixins、functions，避免在每个样式文件中手动地 @import 导入，然后在各个 CSS 文件中直接使用变量。安装插件：

```
npm install style-resources-loader
```

在 vue.config.js 中添加配置如下：

```
pluginOptions: {
            'style-resources-loader': {
                    preProcessor: 'less',
                    patterns: [path.resolve(__dirname, "./src/theme/index.less")],
            }
}
```

20.4.4 项目打包

使用如下命令打包前端项目：

```
npm run build
```

打包时间较长，经过 webpack 打包后在项目中生成 build 目录，资源文件在 build 目录下的 static 目录中，入口文件为 index.html，build 目录结构如下：

```
favicon.ico  index.html  static

./static:
css  img  js

./static/css:
app.945209e7.css
chunk-27667bfb.401e1840.css
chunk-3447e4a8.821fe226.css
chunk-40f44018.73fb0ccb.css
chunk-4c50c064.2202ae93.css
chunk-63643a08.bc3756d5.css
chunk-701ebbab.cbd75255.css
chunk-787ae1c4.53cbf201.css

chunk-8e459656.7d38555a.css
chunk-96aa4714.87f7f667.css
chunk-9f8b12ec.a3fee2d3.css
chunk-c68b585c.6470ec2c.css
chunk-e6229912.6aa03a7c.css
chunk-f8c69cc2.2cd6f581.css
chunk-vendors.a0f42029.css

./static/img:
logo.f9b9adf5.png

./static/js:
about.4920804f.js
app.24ebffe7.js
chunk-27667bfb.1ef4ef9d.js
chunk-2d0a349c.b312938a.js
chunk-2d0c4a73.e90a4b25.js
chunk-3447e4a8.62d0dd99.js
chunk-40f44018.5c03525c.js
chunk-4c50c064.b4f2d858.js
chunk-63643a08.caa176cc.js
chunk-701ebbab.72be00fc.js

chunk-787ae1c4.3a3b4144.js
chunk-7f69ce18.7fedb991.js
chunk-8e459656.7f6d3682.js
chunk-96aa4714.8b9faa9c.js
chunk-9f8b12ec.d0ff42d7.js
chunk-c68b585c.8ca4b083.js
chunk-e6229912.c27bd0ba.js
chunk-f8c69cc2.75586c65.js
chunk-vendors.5fa8fe7d.js
```

使用命令把前端项目资源文件打包至 build 目录后，可以把 build 目录中的所有文件和子目录压缩为 build.zip，上传到部署服务器并解压至指定目录，如此整个项目的发布部署已完成。BiFang 部署平台第一次手动发布后，可以把 BiFang 部署平台项目录入平台中，前端和后端分为两个不同的应用，后续可以通过部署平台自动打包发布。

20.5 小 结

本章介绍了前后端服务联调的基础配置，首先了解了解决跨域的三种方式以及如何在 Vue 项目中方便地配置代理；接下来通过本地启动 BiFang 平台后端服务，实现前端和后端服务的本地联调。联调完成后，需要把前端项目本地构建打包发布至服务器，介绍如何在 vue.config.js 中添加基础配置，构建打包完成后，把前端项目发布至服务器即可。

征服自己，就能征服一切！